Proceedings in
Information and Communications Technology   2

F. Peper H. Umeo
N. Matsui T. Isokawa (Eds.)

Natural Computing

4th International Workshop on Natural Computing
Himeji, Japan, September 2009
Proceedings

 Springer

Editor

Ferdinand Peper
National Institute of Information and Communications Technology, Japan
E-mail: peper@nict.go.jp

Hiroshi Umeo
Osaka Electro-Communications University, Japan
E-mail: umeo@cyt.osakac.ac.jp

Nobuyuki Matsui
University of Hyogo, Japan
E-mail: matsui@eng.u-hyogo.ac.jp

Teijiro Isokawa
University of Hyogo, Japan
E-mail: isokawa@eng.u-hyogo.ac.jp

Library of Congress Control Number: 2009942769
CR Subject Classification(1998): F.1.1, J.2, J.3, C.1.3

ISSN   1867-2914
ISBN   978-4-431-53867-7 Springer Tokyo Berlin Heidelberg New York

Springer is a part of Springer Science+Business Media
springer.com

©Springer 2010

Typesetting: by authors and data conversion by Scientific Publishing Services, Chennai

Printed on acid-free paper   SPIN: 12720200   5 4 3 2 1 0

Preface

The complex behavior of systems in nature is rooted in intricate mechanisms of interactions that, although having factors in common with human-made systems of computation, often supersede them in terms of reliability, power efficiency, and computational capacity. It is thus no surprise that natural systems have become the inspiration of novel algorithms that are based on nonstandard computational mechanisms.

The International Workshop on Natural Computing (IWNC) is focused on theoretical and experimental studies of nature-inspired principles of information processing, novel and emerging paradigms of computation and computing architectures, and case studies of simulated or real-world computing devices implemented in biological, social, chemical, engineering and physical systems. Topics include cellular automata, DNA computation, the physics of computation, computation in living cells, nanocomputing, artificial chemistry, swarm systems, evolutionary computing, reaction-diffusion processors, plasmodium computers, neural networks, chaotic systems, noise-driven computation, and others.

The goal of the IWNC is to offer computer scientists, biologists, mathematicians, electronic engineers, physicists, and social scientists a platform to critically assess present findings in the field, and to outline future developments in nature-inspired computing.

This year's workshop hosted, for the first time, a special session on Unconventional Models of Communication. The reason for this is that the characteristics making natural systems so interesting for computation also offer important hints for the design of communication systems. A fundamental research challenge in this context is the design of robust decentralized communication systems capable of operating in changing environments with noisy input.

IWNC is organized annually and the details of the past three meetings are as follows:

- December 14-15, 2006: University of the West of England, Bristol, UK
- December 10-13, 2007: Nagoya University, Nagoya, Japan
- September 23, 2008: Yokohama National University, Yokohama, Japan

The latest workshop, held in Himeji, Japan, September 23-25, 2009, brought together more than 70 distinguished researchers (from more than 15 countries) with a wide variety of backgrounds working in the field of natural computing. The workshop involved the cooperation between the institutions of the members of the organizing committee, i.e., the National Institute of Information and Communications Technology (NICT), Osaka Electro-Communication University, and the University of Hyogo, Himeji, all located in Japan.

This volume contains 19 refereed papers for oral presentation, and 14 refereed papers for poster presentation, as well as four invited papers and six papers from invited speakers of worldwide reputation. The special session on Unconventional Models of Communication included one invited talk (by Masayuki Murata) and two presentations (by Sasitharan Balasubramaniam et al. and by Yuki Moritani et al.) as part of the invited papers. The special session was organized by Naoki Wakamiya and Kenji Leibnitz, both from Osaka University in Japan, and we would like to express our gratitude for their efforts in making it a great success.

Each paper was reviewed by three members of the program committee. We are extremely grateful to the reviewers, whose expertise and efficiency guaranteed the high quality of the workshop.

We would like to take this opportunity to express our sincere thanks to the invited speakers and the presenters of invited papers for having accepted our invitations to present their research. The invited speakers were:

- Anirban Bandyopadhyay of the National Institute of Materials Science (NIMS; Japan)
- Laszlo B. Kish of Texas A&M University (USA)
- Bruce J. MacLennan of the University of Tennessee (USA)
- Masayuki Murata of Osaka University (Japan)
- Toshio Nakagaki of Hokkaido University (Japan)
- Milan N. Stojanovic of Columbia University (USA)

The presenters of invited papers were:

- Sasitharan Balasubramaniam of the Waterford Institute of Technology (Ireland)
- Andrew Kilinga Kikombo of Hokkaido University (Japan); paper not included in this proceedings
- Jia Lee of NICT (Japan) and Chong-Qing University (China)
- Yuki Moritani of NTT DoCoMo (Japan)
- Manish Dev Shrimali of the LNM Institute of Information Technology (India)

We would like to thank those authors who showed interest in IWNC 2009 by submitting their papers for review. It is a pleasure to express our sincere thanks to our colleagues in the Program Committee and the International Steering Committee. This workshop would not have been possible without their help, advice, and continuous encouragement. We would also like to express our gratitude to Naotake Kamiura of the University of Hyogo in Japan, who provided us with valuable advice and support concerning the organization of the workshop.

Finally, the organization of IWNC 2009 was made possible thanks to financial and technical support from the National Institute of Information and Communications Technology (NICT), Osaka Electro-Communication University, the University of Hyogo in Himeji, the city of Himeji, the Himeji Convention & Visitors Bureau, the Support Center for Advanced Telecommunications Technology Research (SCAT), and the Society of Instrument and Control Engineers (SICE), all located in Japan.

November 2009

Ferdinand Peper
Hiroshi Umeo
Nobuyuki Matsui
Teijiro Isokawa

Organization

4th IWNC (IWNC2009) Organization

Organizing Committee

Chair	Ferdinand Peper	NICT (Japan)
Co-chair	Hiroshi Umeo	Osaka Electro-Communication University (Japan)
Co-chair	Nobuyuki Matsui	University of Hyogo (Japan)
Conference Secretary	Teijiro Isokawa	University of Hyogo (Japan)

International Streeing Committee

Andrew Adamatzky	University of West-England, UK
Cristian Calude	The University of Auckland, New Zealand
Masami Hagiya	Tokyo University, Japan
Nobuyuki Matsui	University of Hyogo, Japan
Kenichi Morita	Hiroshima University, Japan
Ferdinand Peper	NICT, Japan
Grzegorz Rozenberg	Leiden University, The Netherlands
Yasuhiro Suzuki	Nagoya University, Japan
Hiroshi Umeo	Osaka Electro-Communication University, Japan

Program Committee

Susumu Adachi (Japan)
Hiroyasu Andoh (Japan)
Masanori Arita (Japan)
Takaya Arita (Japan)
Tetsuya Asai (Japan)
Daniela Besozzi (Italy)
Ed Blakey (UK)
Peter Dittrich (Germany)
Nazim Fatès (France)
Giuditta Franco (Italy)
Katsunobu Imai (Japan)
Teijiro Isokawa (Japan)
Osamu Katai (Japan)
Satoshi Kobayashi (Japan)
Eisuke Kita (Japan)

Jia Lee (NICT, Japan)
Kenji Leibnitz (Japan)
Jian-Qin Liu (Japan)
Vincenzo Manca (Italy)
Nobuyuki Matsui (Japan)
Giancarlo Mauri (Italy)
Makoto Naruse (Japan)
Haruhiko Nishimura (Japan)
Katsuhiro Nishinari (Japan)
Yasumasa Nishiura (Japan)
Marion Oswald (Austria)
Ferdinand Peper (Japan)
Hiroki Sayama (USA)
Tatsuya Suda (USA)
Yuki Sugiyama (Japan)

Hideaki Suzuki (Japan)   Hiroshi Umeo (Japan)
Yasuhiro Suzuki (Japan)   Naoki Wakamiya (Japan)
Junji Takabayashi (Japan)   Masayuki Yamamura (Japan)
Christof Teuscher (USA)   Masafumi Yamashita (Japan)
Keiichiro Tokita (Japan)   Kenichi Yoshikawa (Japan)
Kazuto Tominaga (Japan)   Claudio Zandron (Italy)
Kazunori Ueda (Japan)   Klaus Peter Zauner (UK)

Sponsoring Institutions

National Institute of Information and Communications Technology (NICT),
    Japan
Osaka Electro-Communication University, Japan
University of Hyogo, Japan
Himeji City, Japan
Support Center for Advanced Telecommunications Technology Research (SCAT),
    Japan
The Society of Instrument and Control Engineers (SICE), Japan

Table of Contents

Contributed Papers

Investigating Universal Computability of Conventional Cellular Automata Problems on an Organic Molecular Matrix

Anirban Bandyopadhyay*, Rishi Bhartiya, Satyajit Sahu, and Daisuke Fujita

Advanced Nano Characterization Center, National Institute for Materials Science,
1-2-1 Sengen, Tsukuba 305-0047 Japan
Tel.: +81-29-859-1370
anirban.bandyo@gmail.com

Abstract. We have self-assembled organic molecular multi-level switch *2, 3-Dichloro-5,6-dicyano-1,4-benzoquinone* (DDQ) as a functional cellular automaton circuit on an atomic flat gold 111 substrate where each molecule processes two bits of information. Since the logic-state transport rules could explain continuous transition of molecular states at every single molecular site in the molecular assembly, we can represent these transport rules as the cellular automata rules. Therefore, by analyzing the relative contrast of the molecules in the scanning tunneling current image it has been possible to reveal the spontaneous transport rules of molecular conductance states in terms of bits or digital signs. A dedicated program has been constructed to analyze the local changes (of single molecule) in the Scanning Tunneling Microscopic (STM) image of a molecular layer autonomously and the program is capable of extracting the elementary information transport rules from a series of STM images that depicts the evolution of a solution/phenomenon. Using this program, a rigorous statistical analysis of the surface transport of bits has been carried out. We have further programmed some of these rules to analyze the conventional CA computability on the organic molecular layer. Since we can tune the composition of molecular circuits, we can make some rules active and some rules passive. This feature has enabled us to carry out versatile and robust CA computation in the simulator. We have shown here with 10 examples how effectively conventional CA problems could be addressed exploring a few sets of rules of this 2 bit molecular cellular automata.

1 Introduction

Currently, great efforts are being made to create supercomputers able to perform high performance spatio-temporal calculations. Cellular Automaton (CA) is an important tool for such computation [1]. CA is a model for truly parallel computing. Truly parallel computing does not mean dividing a particular job into many parts and executing them in different machines at a time. Rather the entire information is processed in a single hardware and all different elements of information influence each other at a time to reach a solution collectively. This particular feature of computing essentially separates all existing biological processors from the manmade computing devices, where the solution of a problem is logically defined even before the computing begins.

---

* Corresponding author.

F. Peper et al. (Eds.): IWNC 2009, PICT 2, pp. 1–12, 2010.

In the last half a century, a large number of mathematical models of cellular automaton have been proposed, universal simulators have been created and universal computability has been studied extensively [2,3]. However, a physical system does not exist that exhibits features of cellular automaton. To build a realistic cellular automaton we have to assemble single CA cells into a matrix that will have a defined set of rules for the spatio-temporal information evolution. Here we have used single molecule conductance switch as a cellular automata cell to perform computing. The reason for choosing a molecular switch is the fact that switch can update its conducting state simply by exchanging electron from the neighboring molecules. By self-assembling the molecules on gold 111 atomic flat substrate, we have created the CA circuit. The experimental details of the realization of an effectively operating CA circuit has been discussed elsewhere [4].

In the last one decade, there have been a number of reports for the building of a single cellular automaton cell using a single molecule [5]. However, it has not been possible to building a cellular automaton grid or circuit that evolves a pattern following a set of abstract rules. Rather, the direction of research has also been shifted towards building a logic gate to replace particular components of the existing hardware of an integrated chip [6]. Such an attempt would not add to the development of a truly parallel computing machine because the sequential processing principles would remain the same in its computational mechanism. Thus, unfortunately, the entire approach of practical CA cell realization has been confined to advancing the existing computers to the next generation; no conceptual advancement is seeded in the approach, which is essential if the current civilization progresses to match the biological computability.

In our 2 D molecular bi-layer, one molecule has six neighbors. Since each molecule is a 2 bit conductance switch, they can have any one of four conducting states, assigned as 0, 1, 2, 3 and represented in the CA grid as blue, green, yellow and red balls respectively (Figure 1 a). By pointing an atomically sharp scanning tunneling microscope tip on the molecules one by one we have applied an electric bias and written 0, 1, 2, 3 states of our choice. When we scanned the surface to read the states at a regular time interval, we found that the input pattern evolves with time. From the dynamics of pattern evolution we have grown the formulation how to control the patterns for computing. Our computational principle is that we will draw an input pattern and the output pattern will naturally change with time as if the solution of a problem. We have solved ten distinct computational problems using the molecular cellular automata simulator wherein the CA rules for the Quinone based molecular CA have been encoded.

2   Universality and Computability

When we solve a problem using an existing computer, the logical algorithm to derive solution is strictly defined. By flowing current in a particular fashion through the circuit, we perform series of logical operations one after another and reach solution spending much less time than it would take if we do it manually. Speeding up this process has been the key to the development of CMOS based architectures in the last half a century. During computation, only a small fraction of large number of switches of a processing chip functions at a time.

Surprisingly, no such strict logical circuit is seen in the structure of a biological processor. Therein circuits are dynamic, continuously evolving and hardly retain identical structural parameters for a long time. Then how do the bio-processors compute? There are many existing theories. Several researchers have made distinct attempts to model global behavior of bio-processors. However, the mechanism of logical reasoning has not revealed concretely. Information is written physically in the biological processors. The pattern of written information evolves, but a replica remains there that retains memory. The sub-patterns overlap physically and generate a unique pattern characteristic of the input pattern. When current flows through a circuit of a computer, all components are abandoned except the one under operation.

The universality is achieved both in the conventional computer and in the advanced bio-processors. Since large number of bits can operate at a time in CA, it resembles brain better than conventional computers. Here we study a number of conventional computing paradigms that requires distinct dynamics for practical realization on the molecular CA. In the conventional CA, as CA rules are predefined, physical realization of only one kind of dynamics is possible [7]. However, since in the molecular cellular automaton, one can switch over from one set of CA rules to another, it is possible to extend computability towards universality (Figure 1 b).

a
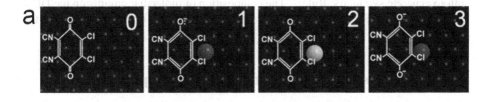

b
1. Write an input pattern using 0, 1, 2, 3 (STM tip)
2. Trigger the pattern (scan at 0.9 V)
3. Read the evolving pattern periodically (scan at 30 mV)
Identify the circuit and measure density of electronic charge on the surface
4. Identify the rules (abstract transport of molecular state)
5. Encode the rule in a program or build simulator
6. Run simulator understand dynamics
7. Write an input pattern physically again and read the dynamics
8. Tune input pattern so that output is the solution of the problem

Fig. 1. a. Four logic states 0, 1, 2 and 3 denoted as blue, green, yellow and red colored balls respectively. b. The computing scheme and computational procedure.

3 Molecular Cellular Automaton Rules

Molecular Cellular Automaton follows a large number of CA abstract rules categorized into eight classes of rules. Complete abstraction is not possible for molecular CA, since ionic charges have long ranges of interaction and modulation transport of cellular states long distance apart. This factor effectively increases the effective neighbors, thus generating significantly large number of rules. We have used particular mathematical formulations to simulate proper updating of cells in the molecular cellular automaton.

1. Rule 1 in the DDQ CA originates from the fact that in a distributed negative charged states (state 1 and state 3), a pseudo positive charge is created which attracts other charges from outside and destabilize the system. 2. Molecular circuits tune minimum spatial and temporal limit. When two logic states collide, they move apart in a constant motion continuously. This rule adds a restriction into the motion, functions as if a logical resistor on the surface, slowing down the motion of logic states. 3. The third rule is billiard ball collision [8]. Two kinds of negatively charged states, state 1 and state 3 collide and move apart like billiard balls. Groups or clusters of logic states also move in the same manner. Rule 4. When negatively charged states move as a single particle, they leave a trail of state 2. On the surface, state 2 density increases as all particles move. State 2 is a meta-stable state of State 0. Therefore, state 2 has a tend to return to the state 0. This often happens on the surface, cluster of state 2 symmetrically vanishes to state 0. Rule 5. If during motion, a number of particles come into contact, then they might form a group and move as a standalone system or a single unit. If they collide with maximum contact points, they disintegrate. Rule 6. In a cluster of states 3, 1 one state 3 might break into two state 1s and two state 1s might fuse into one state 3, keeping the total number of extra electrons conserved on the surface. They fuse or split to generate symmetry. Rule 7. The composition of molecular circuits on the DDQ CA determines the active CA rules and controls the dynamics of charge motion on the surface.

4   Dirichlet Tessellation (DT) or Voronoi Decomposition

DT is an extremely important tool used for computation in almost every field of science and economics [9]. Using this tool a space is sub-divided into multiple polygon spaces in such a way that there exists one point in every sub-division and any two neighboring points are equally separated from the nearest neighbor (Figure 2 a). The advantage of using this tool in computation is that an extremely complex or dynamic information space could be represented in terms of a few points. Thus, a highly interactive system where several local system points interact and influence each other at a time to evolve a particular phenomenon could be addressed in a realistic manner. For an example, if a robot wants to decide obstacle free routes towards destination in an obstacle dense zone, it can decompose entire space following Voronoi algorithm. Similarly, it could be used to explain phenomena where spatially distributed interactive system points evolve the property globally, like protein folding, rainfall forecasting, capacity of a wireless network etc. Therefore, this is a tool enables us to take into account massive parallelism that is existing in every natural system around us.

An organic monolayer when STM scanned at a low bias reveals the circuit assembly. If we draw the border of each circuit region, they mostly generate linear boundaries as shown in the figure 2 b. "As we write a matrix on the DDQ CA, depending on the excess electron (state 1s and/or 3s) concentration, DDQs in a particular matrix region re-orient to the nearest energetically favorable circuit. Hence, for every new input matrix, a unique arrangement of several circuit-domains is created automatically that change

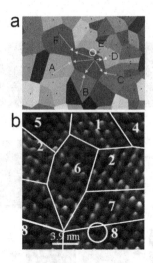

Fig. 2. a. Basic concepts of Dirichlet tessellation. b. An STM image and how this image could be seen as a voronoi-decomposed diagram is demonstrated.

under electron flow." With an example, we show that the circuit boundaries are straight line creating polygons and it is possible to put points in these regions satisfying the condition of Voronoi diagram. Since Voronoi decomposition changes continuously we can possibly use this technique as an alternative representation of solving a problem.

5 Directed Propagation of Information

To perform directed propagation of Information, first we need to change the molecular surface into circuit type 7/8 by scanning the entire surface at a particular tip bias. The reason for such changing is that we want to make sure that collision and repulsion rules dominate over the pseudo charge rule, which inhibits directed propagation in many cases. On the particular surface with specific circuit features, we need to create a mirror image of the information packet in contact with the sending packet of information on the surface. Whenever, two such packets are in contact then they repulse each other, this becomes the initial triggering force for activating the propagation. To send complicated information the information containing packet has to be designed so that it forms a group following Rule 5 and behave like a single unit on the CA surface. Then along the path we need to create reflectors as shown in the figure 3a which will guide the information packet by deflecting it towards a particular direction.

Even in presence of pseudo charge rules or Rule 1, it is possible to send a packet of information with complex and detailed information to a particular direction for an infinite motion in principle (figure 3 a below).

6 Information Writing, Storage and Retrieval

If a matrix is written on the molecular surface, by bringing the STM tip on the top of each molecule and applying a suitable threshold pulse, states will neither move nor

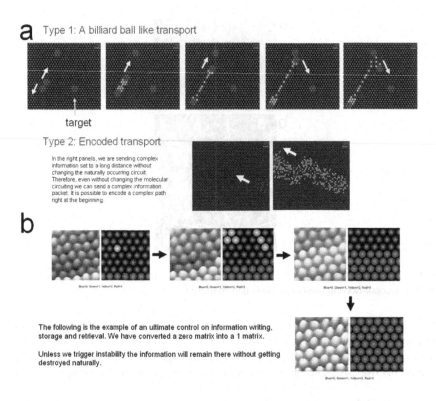

Fig. 3. a. Billiard ball like motions of small groups of states of state 1 and state 3 (above). A set of ungrouped state 1s are directed by increasing density of states towards the direction of motion (below). b. Starting from the lowest row we have written state 1s row by row to complete the state 1 matrix. STM image (left) and the digital conversion (right) is shown side by side.

die out (storage) till the surface is scanned with more than a particular bias and the spontaneous pattern evolution is triggered. Simply by scanning the surface at −1.68 V one can remove all the states from the surface or erase all the information (reset). To retrieve information one requires to simply scanning the surface at a very low bias (few mV) so that the written information is not damaged and at the same time, encoded information is revealed. There could be several other means to get the information out. For an example, one could easily direct a small information packet through the surface as a probe and after colliding with the unknown information packet one could easily come to know about the information content by reading the output by analyzing the evolved pattern.

Information could also be written in the form of a packet of logic bits. If we could write the information so that all the bits together form a group and move through the surface together as an unit. Therefore, we can write, store, and retrieve information in the surface in several ways, some of them are direct, and some of them are indirect detection of information. In the Figure 3 b we demonstrate creation of a matrix of state 1 from state 0.

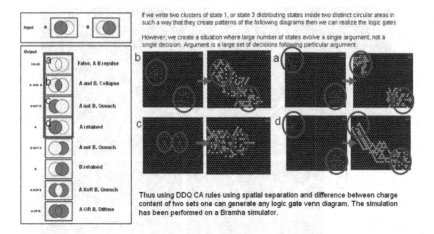

Fig. 4. Venn diagram of logic operations (left). Here a, b, c, d are four major operations of four logic gates. Red rings denote the location of input signal.

7   Logic Gate

We have established the mechanism to guide an information packet towards a particular direction. To explore this mechanism we need to create a circuit that enables passivating rule 1 to and activate rule 3. Now to construct logic gate or any functionality on the CA circuit, we need to put suitable triggering and deflection points, which would enable conditional deflection [10]. This particular feature enables us to create logic gate. This is all about guiding balls to a particular point say D under certain conditions, and diverting them under certain other conditions from D. So, if we have a detector at D, we can get the output. If billiard ball rules are working on a surface, we can create any logic gate of our choice. In the literature, billiard ball based logic gates have been shown explicitly.

However, there is a simpler and more versatile way to realize any logic gate without changing the naturally obtained circuit. If we realize the Venn diagram (Figure 4 left) of logic gates on the molecular layer, we can mimic entire logic operation with all solutions of the input set. A sensor detecting ionic concentration can easily detect output of the logic gates.

Here on the simulator, we have written two clusters of state 1, or state 3 distributing states inside two distinct circular areas in such a way that they create patterns of the following diagrams then we can realize the Venn diagram of logic gates.

However, we create a situation where large number of states, evolve a single argument, not a single decision. Logic output is a large set of decisions following particular argument as demonstrated in the Figure 4.

8   Coalescence and Giant Explosion

This is a phenomenon where large number of elements collapse into the smallest available space [11]. Coalescence derives directly from Rule 3 and an example is shown in

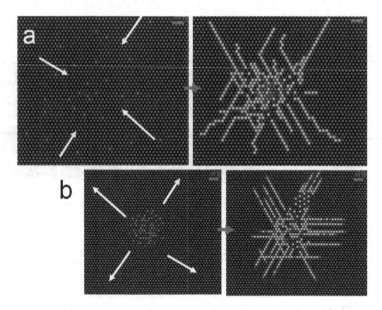

Fig. 5. a. All green balls depicting state 1 collapses into a small region. b. A mixed collection of state 1 and state 3 bursts out all around.

the Figure 5 a. The Rule describes that if clusters of states 1, 3 by any chance collapses together into a particular symmetry then it automatically forms a group and then behaves as a single unit or particular particle. Coalescence is used to mimic clouds, fog, atmospheric polution, polymerisation, school of fish, flock of birds etc.

We can create an opposite phenomenon or explosion using the same simulator simply by minimizing the effect of rule 1 on the surface as demonstrated in the Figure 5 b.

9   Density Classification Task (DCT)

Density classification is one of the most commonly used tools required for classifying particular element in the mixture, which could be in any form in Nature [12]. The task though appears simple becomes extremely challenging when we consider that elements of different densities interact and influence each other at a time. Obviously, a collective approach in the massively parallel computing becomes essential to accurately execute density classification. In the density classification task, one needs to find the CA rule that would isolate a majority state from a randomized initial configuration. Our Cellular Automata is synchronous because the interaction of negatively charged molecules does not stop for a certain time and start again. Once the motion of the logic state is triggered by scanning the surface at a particular bias, updating of cell states becomes a continuous process. Therefore, cells would continue to update infinitely or infinitely inhibiting the probability of asynchronous CA operation. DCT should not be confined into selecting a particular majority state in a CA, the concept of classification could be universalized. For an example, in our CA, we can isolate particular local region in terms of density

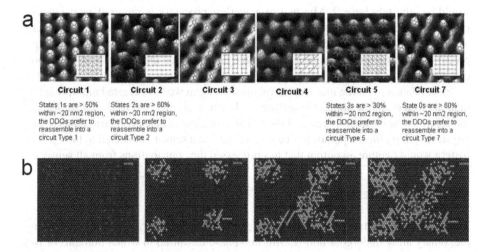

Fig. 6. a. STM image of molecular circuits, quantitative conditions for creating the circuits are given adjacent below the STM image. **b.** Four mixed collection of information creates a new composition in four updated events of the CA matrix.

of charge and at the same time, we can categorize the region, based on the nature of distribution of charge. Thus in the present molecular CA, density classification means states on a surface would evolve in such a way that the characteristics of the states would represent the density of states distinctly.

The molecular layer can switch between eight different circuits. If we define them as 1, 2, 3, 4, 5 ⋯ 8 then we can explain the density classification in technical terms "In a matrix if states 0 or 2 are > 60% within ~20 nm2 region, the DDQs prefer to reassemble into a circuit Type 7 or 2 respectively. If state 1s are > 50% or state 3s are > 30%, then a circuit Type 1 or 5 is created respectively. As we write a matrix, depending on the excess electron (state 1s and/or 3s) concentration, DDQs in a particular matrix region re-orient to the nearest energetically favorable circuit. Hence, for every new input matrix, a unique arrangement of several circuit-domains is created automatically that change under electron flow."

Therefore, density classification is automatically done in the current CA by the states emerging within a particular circuit region. The circuit type distinctly provides the class within which the electronic charge density of the region belongs to as demonstrated in the Figure 6 a.

10   Multi-agent Robotics

To send robots to the Mars every single milligram is important. We cannot send a large supercomputer of the size of a big room connected at the head of a robot. There are three fundamental problems for practical applications with miniature robots. 1. It needs to solve many problems at a time, even for a single problem it needs to consider several parameters at a time, and we can not write a program to do this. Because the program would be infinitely long. 2. Minimal or powerless processor is required, as we cannot put large batteries due to the weight limit of a robot. In addition, the processor must survive under extreme noise to the level of processing signals. 3. Information storage, processing, instruction generation for the final decision-making should be a single processing unit.

To simulate a real time multi-agent robotics problem we have created multiple information clusters on the simulator surface as shown in the figure 6 b. In course of time, the information clusters merge and they evolve in a particular geometric pattern. In a physical system, these operations occur at a time and a sensor connected to the evolved pattern would capture an output that has processed the contributions from all sensors.

11   Phase Ordering

There are several instances in nature where phase ordering takes place [13]. For an example, binary alloy undergoes non-equilibrium phase transition in which domains of two stable phases grow. A cellular automaton demonstrating phase ordering should have three fundamental criteria. First, the logic states will have local tendency to segregate. Second, the segregated bulk phase will be stable after accumulation. Third total number of states depicting a phase would remain conserved.

There are two kinds of phase ordering, conserved and non-conserved. In case of conserved phase ordering, elements will change order parameter with their neighbors, as a result, entire surface appears as homogeneously spaced phases. However, if order parameter remains constant throughout the surface and there is no instance of order parameter exchange, then entire surface becomes inhomogeneous.

To create non-conserved phase ordering on a surface, we need to ensure that first, on the CA surface the logic states must not die out. There should either be state 1 or state 3 but not both together. If they are together, then due to Rule 6, inter state conversion might take place and total number of logic states would not be conserved. Then, we cannot create the phase ordering. Therefore, phase ordering is created between the normal state blue and combination of yellow and red. This is necessary to include the state 2 or yellow balls as it tunes the resistance of the path for the state 1 or state 3 transport. Here, generation of state 2 effectively plays the role of depicting the order parameter and its exchange mechanism. A combination of Yellow and red balls or Yellow and green balls represent a single phase. The other phase is always blue. Here in the Figure 7 a, we have demonstrated an example of such phase ordering.

12   Ant Colony

To form an Ant colony we have to include two important features of an Ant society. First, they can suddenly start moving if they feel the necessity to move somewhere and their future motions are not dominated by collision [14]. With DDQ CA we can not generate spontaneous motion of state 1, 3 randomly to any other direction unless it collides or is activated by a pseudo charge (Rule 1). DDQ CA only decides next position of a state by collision rule and after that pseudo charge rule takes over therefore we can realize ant colony even when all rules are active. DDQ CA basically governs motion of

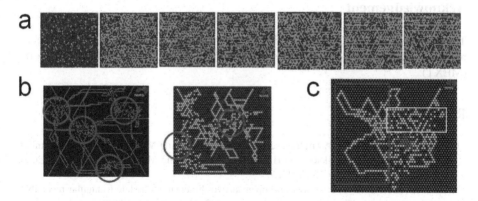

Fig. 7. a. The phase ordering event that generates from an absolute random matrix. b. Examples of two ant colonies and how ants create those colonies could be understood following the yellow state paths. c. The event of chemotaxis is demonstrated.

an ant (red or green balls, state 1, 3) following extremely complex interactions. Some of the factors it include is that, it responds to any local activity among colony members (global convergence), and if concentration becomes more then it can decide to leave that region. During the course of motion if by chance ants leave chemicals in a certain way, it responds (see chemotaxis for specific studies on chemical effect).

The essential condition for the formation of a true ant colony is the fact that an ant coming from a colony far away would be able to analyse the activities in different sub-locals and intelligently decide its own path. We demonstrate this feature in the two examples in the Figure 7 b.

13   Chemotaxis

When an ant moves it leaves some chemicals behind. This chemical triggers other ants to move [15]. In open air the chemicals can not survive for long. They slowly disappear. DDQ CA rules show that whenever red or green balls move, they leave a trace of yellow balls. So the green and red balls are ants and we do not differentiate between red and green colored ants for this computation, they represent similar species. The yellow balls are the chemicals released by the ants. The clusters of state 2 or chemicals generated from the trails of red and green balls/ants slowly disappear following DDQ CA rules. However as mentioned in describing the DDQ CA rules, yellow balls are the metastable states of DDQ. Therefore, they are prone to accept extra electron/s from the red and green balls and convert to red or green. In this way, yellow trails serve as a less resistance path to the red and green balls. This would appear to an observer as if yellow trails are exciting the red and green balls/ants similar to chemicals for ants.

In the figure 7c, inside the red region the path is intact, the environment has not diluted away the chemicals as happened inside the yellow region. Non-dilution is necessary for green ants to locate/retrace other's path.

Statistically we have found that at single update, the ant analyses the complex interaction going on inside the neighboring assembly states and finally decides the route.

Acknowledgement

The work is funded by Grants in Aid for Scientific Research (A) for 2009-2011 Grant number 21681015 awarded by Ministry of Science and education, culture and sports (MEXT).

References

1. Spezzano, G., Talia, D.: A High-Level Cellular Programming Model for Massively Parallel Processing. In: 1997 Workshop on High-Level Programming Models and Supportive Environments (HIPS 1997), p. 55 (1997)
2. Imai, K., Morita, K.: A computation-universal two-dimensional 8-state triangular reversible cellular automaton. Theoretical Computer Science 231(2), 181–191 (2000)
3. Wolfram, S.: Universality and complexity in cellular automata. Physica D 10, 1–35 (1984)
4. Bandyopadhyay, A., Fujita, D., Pati, R.: Architecture of a massive parallel processing nano brain operating 100 billion molecular neurons simultaneously. Int. J. Nanotech. and Mol. Comp. 1, 50–80 (2009)
5. Orlov, A.O., Amlani, I., Bernstein, G.H., Lent, C.S., Snider, G.L.: Realization of a functional cell for Quantum-dot Cellular Automata. Science 277, 928–930 (1997)
6. Imre, A., Csaba, G., Ji, L., Orlov, A., Bernstein, G.H., Porod, W.: Majority Logic gate for magnetic Quantum-dot Cellular Automata. Science 311, 205–208 (2006)
7. Gaylord, R.J., Nishidate, K.: Modeling Nature. Springer, Santa Clara (1996)
8. Durand-Lose, J.: Computing inside the billiard ball model, pp. 135–160. Springer, London (2001)
9. Korobov, A.: Discrete Versus Continual Description of Solid State Reaction Dynamics from the Angle of Meaningful Simulation. Discrete Dynamics in Nature and Society 4, 165–179 (2000)
10. Stone, C., Toth, R., Costello, B.D.L., Adamatzky, A., Bull, L.: Coevolving Cellular Automata with Memory for Chemical Computing: Boolean Logic Gates in the B-Z Reaction. In: Rudolph, G., Jansen, T., Lucas, S., Poloni, C., Beume, N. (eds.) PPSN 2008. LNCS, vol. 5199, pp. 579–588. Springer, Heidelberg (2008)
11. Cieplak, M.: Rupture and Coalescence in two-dimensional cellular automata fluids. Phys. Rev. E. 51, 4353–4361 (1995)
12. Morales, F.J., Crutchfield, J.P., Mitchell, M.: Evolving two-dimensional cellular automata to perform density classification: A report on work in progress. Parallel Computing 27(5), 571–585 (2001)
13. Oono, Y., Puri, S.: Computationally efficient modeling of ordering of quenched phases. Phys. Rev. Lett. 58, 836–839 (1987)
14. Deborah, M.G.: The development of organization of an ant colony. American Scientist 83, 50–57 (1995)
15. Pennisi, E.: The secret language of bacteria. The new scientist, 30–33 (1995)

Noise-Based Logic and Computing: From Boolean Logic Gates to Brain Circuitry and Its Possible Hardware Realization

Laszlo B. Kish[1], S.M. Bezrukov[2], S.P. Khatri[1], Z. Gingl[3], and S. Sethuraman[1]

[1] Department of Electrical and Computer Engineering
Texas A&M University
College Station, TX 77845, USA
[2] Laboratory of Physical and Structural Biology
Program in Physical Biology, NICHD
National Institutes of Health
Bethesda, MD 20892, USA
[3] Department of Experimental Physics
University of Szeged
Dom ter 9, Szeged, H-6720, Hungary

Abstract. When noise dominates an information system, like in nano-electronic systems of the foreseeable future, a natural question occurs: Can we perhaps utilize the noise as information carrier? Another question is: Can a deterministic logic scheme be constructed that may explain how the brain efficiently processes information, with random neural spike trains of less than 100 Hz frequency, and with a similar number of human brain neurons as the number of transistors in a 16 GB Flash dive? The answers to these questions are yes. Related developments indicate reduced power consumption with noise-based deterministic Boolean logic gates and the more powerful multivalued logic versions with an arbitrary number of logic values. Similar schemes as the Hilbert space of quantum informatics can also be constructed with noise-based logic by utilizing the noise-bits and their multidimensional hyperspace without the limitations of quantum-collapse of wavefunctions. A noise-based string search algorithm faster than Grover's quantum search algorithm can be obtained, with the same hardware complexity class as the quantum engine. This logic hyperspace scheme has also been utilized to construct the noise-based neuro-bits and a deterministic multivalued logic scheme for the brain. Some of the corresponding circuitry of neurons is shown. Some questions and answers about a chip realization of such a random spike based deterministic multivalued logic scheme are presented.

In this paper, we present a short summary of our ongoing efforts in the introduction, development and design of noise-based deterministic multivalued logic schemes and elements.

The numerous problems faced by current microprocessors to follow Moore's law [1,2,3,4] intensify the academic search for non-conventional computing alternatives. Today's computer logic circuitry is a system of coupled DC amplifier

F. Peper et al. (Eds.): IWNC 2009, PICT 2, pp. 13–22, 2010.

stages and this situation represents enhanced vulnerability against *variability* [1] of fabrication parameters such as threshold voltage inaccuracies in CMOS. Thin oxides imply greater power dissipation due to leakage currents[1]. Thermal noise and its induced errors (see Figure 1) and the related power dissipation limits are problems of a fundamental nature [2,3,4]. If we want to increase the logic values in a traditional CMOS gate while the error rate and clock frequency is kept constant, idealistically, with zero MOS threshold voltage, the number of required supply voltages scales with the number of logic values. The power dissipation scales with the square of the number of logic values, see Figure 1. Such a high price is unacceptable.

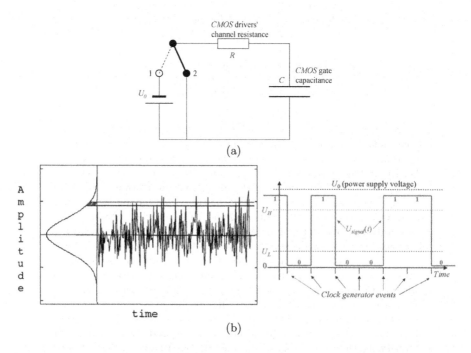

Fig. 1. (a) Model for the MOSFET gate and its driving. (b) Left: Gaussian noise (thermal, 1/f, shot) and its amplitude distribution and its level-crossing events. Right: Critical levels in binary logic: when the signal is beyond U_H it is interpreted as H and when it is below U_L , it is interpreted as L.

Because the speed (bandwidth), energy dissipation and error probability are interrelated issues, it is important to emphasize that:

- Claims about *high performance* without considering *error rate* and *energy efficiency* aspects are interesting but meaningless for practical developments.
- Claims about *high energy efficiency* without considering *error rate* and *performance drop* aspects are interesting but meaningless for practical developments.

Table 1. Comparison of digital computers and the brain

Typical Laptop	Human Brain
Processor dissipation: 40 W	Brain dissipation: 12-20 W
Deterministic digital signal	Stochastic signal: analog/digital?
Very high bandwidth (GHz range)	Low bandwidth (<100 Hz)
Sensitive for errors (freezing)	Error robust
Deterministic binary logic	Unknown logic
Potential-well based memory	Unknown memory mechanism
Addressed memory access	Associative memory access (?)

- Claims about *efficient error correction* without considering *energy require-ment* and *performance drop* aspects are interesting but meaningless for prac-tical developments.
- Finally, all these *performance-error-energy* implications *must be addressed at the system level* otherwise they are meaningless for practical developments. Maybe we can *win at the single gate level* which is interesting, but unimpor-tant if we *lose at the system level*.

Exploring *Natural Computing*, we can ask: How does biology do it? Let us make a comparison between a digital computer and the brain; see Table 1. Most im-portantly, the brain utilizes noise as information carrier. The result is a logic engine with a relatively low number of neurons (100 billions), extraordinary per-formance, acceptable error probability and extremely low power dissipation [5].

Motivated by these observations and the fact that *stochastic logic is very slow and produces too many errors*, our goal has been to study the possibility of constructing *deterministic high-performance logic schemes* utilizing noise as the information carrier [6,7,8,9,10].

The first open question is about what modeling to use: a *continuum noise based logic* [7,8], or a *logic with random spike trains* [9], like the brain? Some of the other concerns are listed here:

- Deterministic logic from noise: Averaging (statistical) slowdown?
- Speed in general?
- Number of logic values?
- Energy need; power dissipation versus performance?
- Error probability?
- Devices and logic gates?

In Figure 2, the schematics of continuum-noise-based logic gates is shown. The reference noises should be small to reduce power dissipation originating from their distribution. The scheme is very robust against the accumulation and prop-agation of fast switching errors. Thus the DC switches can be of extremely poor quality with enormously high error probability, such as 0.1. This situation helps to reduce switching power dissipation [7].

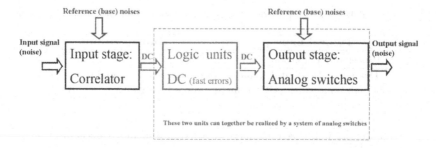

Fig. 2. Schematics of continuum based logic gates [7]

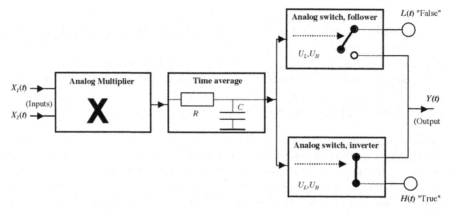

Fig. 3. Circuitry of the continuum noise based XOR logic gate

In Figure 3, the circuitry of the continuum-noise-based XOR logic gate is shown [7]. Similarly, gates can be constructed in either the binary or multivalued logic systems [7].

Another advantage of the continuum-noise-based logic is that a multidimensional logic hyperspace [7,8] can be built from the products of the original reference noises, which we now call *noise-bits* (similarly to the qubits of quantum computing) without the limitations [8] of quantum-collapse of wavefunctions, see Figure 4. Noise-based string search algorithms [8] faster than Grover's quantum search algorithm are obtained with the same hardware complexity class as the quantum engine.

The continuum-noise-based logic has the following potential advantages [7,8]

1. Arbitrary number of logic values in a single wire. Utilization of quantum-like hyperspace is possible.
2. Due to the zero mean of the stochastic processes, the logic values are AC signals, and AC coupling can ensure that the variability-related vulnerabilities are strongly reduced.

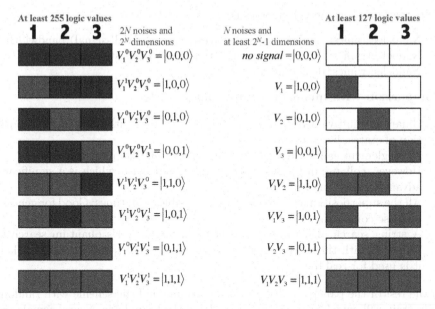

Fig. 4. Multidimensional logic hyperspace can be built from the products of the original reference noises [7,8]. Left: complete quantum Hilbert space emulation including vacuum states [8] with three noise-bits. The zero elements have their own noise. Right: hyperspace emulation with three simplified noise-bits where the zero element is represented by nonexistent corresponding noise term in the product [8].

3. Robust against noise and interferenc. The different basic logic values are orthogonal not only to each other but also to any transients/spikes or any background noise including thermal noise or circuit noise, such as 1/f, shot, generation-recombination, etc., processes. Moreover, the usual binary switching errors do not propagate and accumulate.
4. Due to the orthogonality and AC aspects (points 2 and 3 above), the logic signal on the data bus can have a much lower effective value than the power supply voltage of the chip. This property, along with the robustness against switching errors, has a potential to reduce the energy consumption.

Disadvantages

1. The continuum noise based logic is slower due to the need to average at the output of the correlator. In binary gates, its digital bandwidth is about 1/500 of the bandwidth of the small-signal-bandwidth of the noise [7]. That results in about 0.1-0.5 GHz clock frequency. This deficiency may be compensated by the multivalued logic abilities of such gates.
2. The need for more complex hardware. This deficiency may also be compensated by the multivalued logic abilities.

Its relevance for nano-electronics is obvious:

- Nanoelectronics: smaller device size, thus higher (small-signal) bandwidth.
- Nanoelectronics: smaller device size, thus more transistors on the chip.
- Nanoelectronics: deafening noise.

Comparison with quantum computing [8]

- Binary or multivalued systems can be constructed, with optional superposition of states, like quantum.
- Entanglement can be made in the superposition, like quantum.
- However, collapse of the wavefunction does not exist, which is a significant advantage over quantum.
- All the superposition components are accessible at all times: Good for general purpose computing.
- A string search algorithm which outperforms Grover's quantum search is proposed with the same hardware complexity as the quantum engine when it is used for real data.

In the rest of the paper, we discuss our deterministic logic scheme with random spike trains [9], which is also a simple model of the brain logic. Neural signals are stochastic unipolar, overlapping spikes trains: their product has non-zero mean value. This situation rules out multiplication as a solution. Multiplying with neurons seems to be difficult, anyway. Thus the basic question is: Can we build an orthogonal noise-bit type multidimensional Hilbert space here? The answer is yes, by using set-theoretical operations between partially overlapping neural spike trains which we call *neuro-bits*. For two and three neuro-bit examples, see Figure 5.

For neural circuit realization, the key building element (out of the neurons) is the *orthon* [9] which consists of two neurons, see Figure 6. The *orthognator* unit which prepares the multidimensional hyperspace from the neuro-bits as inputs, see Figure 5 as example, can be built of orthons and neurons, see Figure 7 as example. However, there are two families of orthognators:

1. *Nonsequential orthogonator* (for neural systems): Such a circuitry [9] has N parallel inputs run by parallel, partially overlapping random spikes trains and it generates all variations of the available set-theoretical intersections of the input spikes. There are $M = 2^N - 1$ output wires with non-overlapping spike trains [9]. The advantage of the nonsequential orthogonator is that it may be faster and it can transform a set of partially overlapping spike trains into a set of orthogonal trains.
2. *Sequential orthogonator* for chip applications [10]: It has a single input to be fed by a single (infinite) spike sequence. A *controlled rotary switch* will distribute the subsequent spikes to M output wires in a cyclic way (demultiplexing). This circuitry maybe more advantageous in certain chip designs [10].

The great potential of this logic scheme comes from the fact that, when analyzing multidimensional superpositions, the very first spike of a searched hyperspace

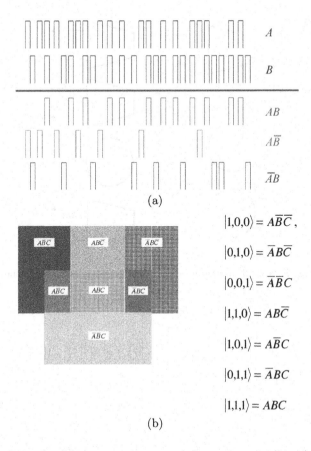

$$|1,0,0\rangle = A\overline{B}\overline{C},$$

$$|0,1,0\rangle = \overline{A}B\overline{C}$$

$$|0,0,1\rangle = \overline{A}\overline{B}C$$

$$|1,1,0\rangle = AB\overline{C}$$

$$|1,0,1\rangle = A\overline{B}C$$

$$|0,1,1\rangle = \overline{A}BC$$

$$|1,1,1\rangle = ABC$$

(b)

Fig. 5. (a) 3-dimensional hyperspace generated from two neuro-bits (partially overlapping neural spike trains) utilizing set-theoretical operations [9]. (b) the same with three neuro-bits and seven-dimensional hyperspace, and its illustration by overlapping squares as neuro-bits.

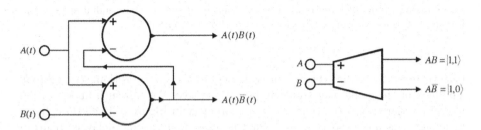

Fig. 6. The orthon [9] consists of two neurons and has two inputs and two outputs. Left: its neural circuitry where the plus stands for the excitatory input and the minus for the inhibitory input of the neuron. Right: Its symbol.

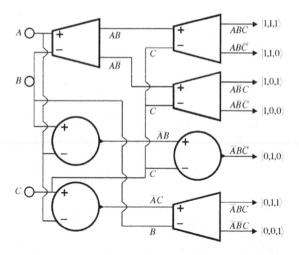

Fig. 7. Neural orthogonator [9]. Third order, non-sequential.

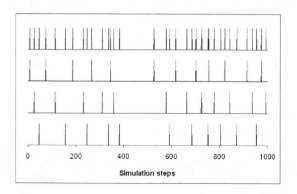

Fig. 8. Input and output signals of second-order sequential orthogonator [10] driven by zero-crossing events of band-limited white noise. Upper line: the original spike train. Lower lines: the orthogonal sub-trains at the three outputs [10].

base vector will tell that this vector is element of the superposition, see Figure 9. The lack of this spike will similarly be informative. We call such an operation neural Fourier transformation [9], see its circuitry in Figure 9. Even though the logic signal is noise, no time averaging is needed, just some delay is incurred until the first relevant spike appears (or is found to be missing).

The essential open question is about this idealistic scheme is: Can these circuit elements be found in the brain?

Very encouraging facts are observations of the role of the first spike in sensory neural signals (called sometimes first-spike-coding) [11,12].

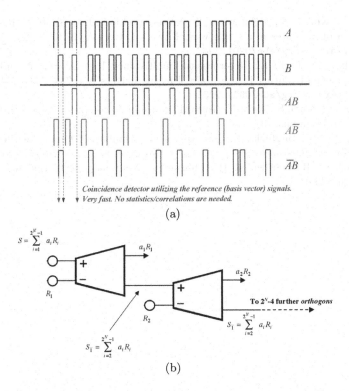

Fig. 9. (a) The principle of neural Fourier transformation. The first arriving spike of that vector will indicate the presence of that vector in the superposition. This subfigure illustrates the detecting of the orthogonal elements of a two neuro-bits system. (b) The circuitry of neural Fourier transformation [9].

References

1. Khatri, S., Shenoy, N.: Logic Synthesis. In: EDA for IC Implementation, Circuit Design and Process Technology. CRC Press, Boca Raton (2006)
2. Kish, L.B.: End of Moore's law; Thermal (noise) death of integration in micro and nano electronics. Physics Letters A (305), 144–149 (2002)
3. Porod, W., Grondin, R., Ferry, D.: Dissipation in computation. Physical Review Letters (52), 232–235 (1984)
4. Cavin, R., Zhirnov, V., Hutcby, J.: Energy barriers, demons and minimum energy operation of electronic devices. Fluctuation and Noise Letters (5), C29–C38 (2005)
5. Bezrukov, S., Kish, L.: How much power does neural signal propagation need? Smart Mater. Struct. (11), 800–803 (2002)
6. Kish, L.: Thermal noise driven computing. Applied Physics Letters (89) (2006), http://arxiv.org/abs/physics/0607007
7. Kish, L.: Noise-based logic: binary, multi-valued, or fuzzy, with optional superposition of logic states. Physics Letters A (373), 911–918 (2009), http://arxiv.org/abs/0808.3162

8. Kish, L., Khatri, S., Sethuraman, S.: Noise-based logic hyperspace with the super-position of 2^n states in a single wire. Physics Letters A (373), 1928–1934 (2009), http://arxiv.org/abs/0901.3947
9. Bezrukov, S., Kish, L.: Deterministic multivalued logic scheme for information processing and routing in the brain. Physics Letters A (373), 2338–2342 (2009), http://arxiv.org/abs/0902.2033
10. Gingl, Z., Khatri, S., Kish, L.: Toward brain-mimic computer chips (August 2009) (draft, to be published)
11. Johansson, R., Birznieks, I.: First spikes in ensembles of human tactile afferents code complex spatial fingertip events. Nature Neurosci. (7), 170–177 (2004)
12. Guyonneau, R., VanRullen, R., Thorpe, S.: Neurons tune to the earliest spikes through STDP

Models and Mechanisms for Artificial Morphogenesis

Bruce J. MacLennan

Department of Electrical Engineering and Computer Science
University of Tennessee, Knoxville
http://www.cs.utk.edu/~mclennan

Abstract. Embryological development provides an inspiring example of the creation of complex hierarchical structures by self-organization. Likewise, biological metamorphosis shows how these complex systems can radically restructure themselves. Our research investigates these principles and their application to artificial systems in order to create intricately structured systems that are ordered from the nanoscale up to the macroscale. However these processes depend on mutually interdependent unfoldings of an information process and of the "body" in which it is occurring. Such *embodied computation* provides challenges as well as opportunities, and in order to fulfill its promise, we need both formal and informal models for conceptualizing, designing, and reasoning about embodied computation. This paper presents a preliminary design for one such model especially oriented toward artificial morphogenesis.

Keywords: Algorithmic assembly, embodied computation, embodiment, embryological development, metamorphosis, morphogenesis, nanotechnology, post-Moore's Law computing, reconfigurable systems, self-assembly, self-organization.

1 Background

1.1 Embodied Computation

Pfeifer, Lungarella, and Iida [29, p. 1088] provide a concise definition of *embodiment*: "the interplay of information and physical processes." Hence, *embodied computation* may be defined as computation whose physical realization is directly involved in the computational process or its goals. It includes computational processes that directly exploit physical processes for computational ends and those processes in which information representation and computation are implicit in the physics of the system and its environment, which effectively represent themselves. It also includes computational processes in which the intended effects of the computation include the growth, assembly, development, transformation, reconfiguration, or disassembly of the physical system embodying the computation. Embodied computation is based on some of the insights from *embodied cognition* and *embodied artificial intelligence* [6,7,13,28], but extends them to all computation [21,25].

F. Peper et al. (Eds.): IWNC 2009, PICT 2, pp. 23–33, 2010.

The most common model of computation (binary digital logic) is far removed from the physical processes by which it is implemented, and this has facilitated a beneficial independence of computer design from device technology. Thus our technological investment in computer design has been preserved through several generations of device technology (from relays to ICs). However, as in the coming decades we enter the era of post-Moore's Law computing, increasing the density and speed of computation will require a greater assimilation between computational and physical processes [21,22,24,25]. In part this will be accomplished by developing new physical systems and processes, but the other half of the equation is to develop new models of computation that are closer to the laws of physics. The challenge is to identify models that are sufficiently low level to be readily implementable and sufficiently high level to be relatively independent of particular implementation technologies.

One of the advantages of embodied computing is that many computational processes are performed by physical systems "for free." For example, neural networks often make use of universal approximation theorems based on linear combinations and simple sigmoidal functions [12, pp. 208–94]. However, sigmoidal behavior is a common consequence of many physical processes, in which some resource is exhausted, and therefore sigmoids are available "for free" without additional computation. Similarly, negative feedback often arises from the natural degradation or dissipation of physical substances, and so these processes are directly available for embodied computation. Similarly, stochastic effects, which are unavoidable in many physical systems, especially at the nanoscale, can be exploited profitably in algorithms such as stochastic resonance [4] and simulated annealing [14]. Finally, in embodied computation the concurrency typical of physical, chemical, and biological processes can be directly exploited to achieve parallelism "for free." There are many other examples.

Typically, the structure of a computational system governs its function, that is, the computation it performs. Conversely, in *algorithmic assembly* a computational process governs the assembly of some physical structure [3,15,16,30,31]. However, some embodied systems integrate these two processes. For example, in embryological morphogenesis, the physical structure of the embryo governs the very computational processes that create the physical structure of the embryo. Structure governs function, and function creates structure.

Further, since embodied computation systems are potentially capable of modifying their own structure, they can be naturally adaptive. Beyond this, they may be "radically reconfigurable," that is, able to reorganize their physical structure to adapt to changing circumstances and objectives [18,19,20,23]. A related, but very important, property is self-repair, since acceptable configurations often can be defined as stationary states to which the system reconfigures after damage. Finally, embodied computation systems can be designed for self-destruction, which is especially important for nanoscale systems.

We are used to programming in an idealized world of perfect logic, independent of physical realization, and therefore embodied computing presents unfamiliar challenges, for we have to pay more attention to the physical realization of an embodied computation and its environment. However, embodied computation is not simply applied physics, for although embodied computation makes more direct use of physical processes than does conventional computation, its focus is on processes that are relatively independent of specific physical systems, so that they can be applied in a variety of physical substrates. For example, reaction-diffusion systems can be instantiated in a wide variety of media [1]. Further, since natural computation is embodied, we often can look to natural systems to get ideas for implementing artificial embodied computing systems.

1.2   Artificial Morphogenesis

Our research is currently focused on a particular application of embodied computation that is of central importance in nanotechnology: artificial morphogenesis [22]. While a variety of self-assembly processes have been applied to nanomaterials, these have not solved the problem of assembling systems that are hierarchically structured from the nanoscale up to the macroscale. Some of the issues are being addressed in the context of research in programmable matter and reconfigurable robotics, but we believe that these efforts would be improved by a more morphogenetic approach, for embryological morphogenesis is the best example that we have of successful self-assembly of physical systems structured from the nanoscale up to the macroscale.

Morphogenesis has a number of special characteristics that distinguish it from most other self-organizing processes, and we believe that these characteristics will be important in embodied computation. For example, we commonly think of computation as taking place in a fixed substrate, and many self-assembly processes also assume a fixed substrate or matrix in which agents move. In morphogenesis, in contrast, the computational medium is assembled by the computational process, as a zygote develops into a hollow blastula and then into a more complex structure of tissues, which govern the information and control processes in the medium. Generally speaking, in nature self-organization proceeds without the benefit of fixed, predefined reference frames and coordinate systems, which is one source of the robustness of these processes.

In morphogenesis, *tissues* (groups of cells with a common function) form and reform under control of their inherent self-organizing processes. We think of the embryo as solid, but most of the tissues are elastic, at least during development, and elastic properties influence the forms that develop [35, ch. 6]. In other cases, tissues behave more like viscous fluids, perhaps percolating through a more solid matrix, and this fluid motion is essential in cell migration [5][9, pp. 92–4][34]. Non-cellular substances, such as morphogens and other signaling chemicals diffuse like gases through anisotropic media, but cells also exhibit facilitated diffusion [9, pp. 13–15, 156, 252]. In many cases, tissues occupy a middle ground, with viscoelastic properties [9, pp. 21–2, 133]. In general terms, morphogenesis takes place in the relatively unexplored realm of *soft matter* [8][9, p. 2], and our theories of embodied computation, at least as applied to morphogenesis, need to take account of its characteristics.

Developmental biologists have identified about twenty fundamental processes involved in morphogenesis [9, pp. 158–9][32]. If indeed, these processes are sufficient to produce complex, hierarchically-structured systems, such as vertebrate organisms, then they define an agenda for embodied computation applied to artificial morphogenesis.

2   Requirements for a Formalism

The goal of our research is to develop and evaluate formal methods for embodied computation oriented toward artificial morphogenesis. This implies several requirements.

First, the goals of embodied computation are best served by a continuous perspective, since the laws of physics are primarily continuous. This is especially true in our intended application area, morphogenesis, for tissues and their environments are naturally treated as continua (e.g., epithelia, mesenchyme, blood).

Nevertheless, the two perspectives — the discrete and the continuous — are complementary. In many applications of embodied computation, especially when the computational process is implemented by very large numbers of computational elements, we will want to be able to move fluently between the two perspectives. This is especially the case in morphogenesis, where in the early stages of development we are faced with discrete phenomena — 1, 2, 4, 8, etc. cells with specific shapes and arrangements — whereas in later stages (when there are more than $\sim 10\,000$ cells) it is more convenient to treat the cell masses as viscoelastic tissues and apply continuum mechanics [9]. Therefore a formalism for embodied computation should support a systematic ambiguity between discrete and continuous models. Thus, the formalism should support complementarity by treating bodies, tissues, and other macroscopic masses as comprising an indefinitely large number of *elements*, which we interpret ambiguously as infinitesimal points in a continuum or as members of a finite, discrete set of units.

To maintain these complementary interpretations, the formalism should treat the elements as having a small, indeterminate size, and therefore the formalism should describe the properties of elements in terms of *intensive quantities*, such as number density and mass density, which do not depend on size, in preference to *extensive quantities*, such as volume and mass, which do.

The formalism should be suitable for describing processes in all kinds of media, including solids, liquids, gases, and physical fields. Further, since viscoelastic media are important in morphogenesis (for example, mesenchyme is best characterized as viscoelastic [9, p. 98][10]), we expect to be dealing with materials of differing viscosity and elasticity. Finally, some of these materials may be anisotropic (e.g., epithelium), and our formalism must accommodate that possibility.

The formalism must be capable of describing active elements, such as living cells and microrobots, as well as passive elements, such as diffusing chemicals, fluid media, and solid substrates. In particular, the formalism should be applicable to living agents as well as to nonliving ones, for embryological morphogenesis, which is the inspiration for this technology, is based on living cells, and we also want to address artificial morphogenetic processes based on genetically engineered organisms and other products of synthetic biology.

Energetic issues are critical to embodied computational systems, which must be maintained in a nonequilibrium state either for a definite duration or indefinitely. Therefore the formalism needs to be able to describe and define the flow of matter and energy through the system. Dissipation of energy is especially critical for microscopic nonequilibrium systems, and should be addressed by the formalism. Similarly embodied computation must address the disposal of other waste products, such as unrecyclable chemical reaction products.

Descriptions of morphogenetic and developmental processes in terms of partial differential equations are naturally expressed relative to a fixed, external three-dimensional reference frame. There are two problems with this. First, the natural reference frame is the developing body itself, which might not have any significant relationship to a fixed frame. Second, since active elements (such as migrating cells) are responding to their local environments, it it natural to describe their behavior in terms of their intrinsic coordinates or their immediately local frame. Therefore tensors seem to be the best way to describe the properties and behaviors of elements.

Biologists often find it convenient to express regulatory networks qualitatively, using *influence models* to indicate that one gene product enhances or represses the expression of another gene [36]. These qualitative regulatory relationships are an important tool for conceptualizing control processes in which the quantitative relationships are unknown or are not considered critical. Therefore, the formalism should permit the expression of qualitative control relationships.

In embodied computation, especially as it pushes towards the nanoscale, noise, error, uncertainty, defects, faults, and other sources of indeterminacy and unpredictability are unavoidable [17]. Therefore a formalism for embodied computation should be oriented toward the description of stochastic processes. Further, since many self-organizing processes depend on stochasticity and sampling effects for symmetry breaking, the formalism should admit them.

3 Approach

Having outlined the requirements for a formalism for embodied computation, we present a preliminary definition.

Substances. One of the central concepts in the proposed formalism is a *substance*, which refers to a class of phenomenological continua with similar properties, especially from a computational perspective (e.g., cohesion, viscosity, behavioral repertoire). They are defined by (perhaps fuzzy) ranges of parameter values, ratios of parameters, etc. Substances are naturally organized in a hierarchy, from the most general classes (e.g., solid, liquid, gas) to specific physical substances (e.g., liquid water, oxygen gas, mesenchyme, a specific extracellular matrix). The more generic classes are more useful from a computational perspective, since they have the potential of being realized by a greater variety of specific substances. For example, for the purposes of embodied computation it may be sufficient that a substance diffuse and degrade at certain relative rates, and be producible and detectable by other particles, but the choice of specific substance might be otherwise unconstrained.

Bodies. In object-oriented programming the instances (members) of classes are discrete atomic objects, which frequently act as software agents. In our formalism for embodied computing, in contrast, instances of substances are called *bodies* or *tissues*, which are phenomenological continua of elements (discrete *particles* or infinitesimal *material points*). Several bodies may occupy the same region of space and interact with each other. For example, the same region may be occupied by a volume of diffusing chemical and by a mass of cells following the chemical gradient. The kind and degree of interpenetrability possible, as well as other interactions, are determined by the definitions of the substances. Some bodies may be quite diffuse (e.g., disconnected cells migrating through mesenchyme).

Mathematically, a body is defined to be an indefinitely large set $\mathcal{B}$ of elements (particles or material points). At any given time t each element $P \in \mathcal{B}$ has a location in a region of three-dimensional Euclidean space $\mathcal{E}$ defined by a vector $\mathbf{p} = C_t(P)$, where C_t is a continuous function that maps $\mathcal{B}$ into this region. As in continuum mechanics, C_t defines the *configuration* of $\mathcal{B}$ at time t, and therefore C_t reflects the *deformation* of the body as a consequence of its own internal dynamics and its interaction with other bodies.

An embodied computation system comprises a finite and fixed number of bodies (or tissues), each having properties that allow it to be classified as one substance or another. The behavioral dynamics of these bodies, in mutual interaction, defines the dynamics of the embodied computation system, but it must be prepared in some appropriate initial state. This is accomplished by specifying that a particular body of a defined substance occupies a specified region of Euclidean space and by specifying particular values for the variable properties of the substance throughout that region. While the full generality of mathematics may be used to define the initial bodies, we are most interested in physically feasible preparations, but this depends on the specifics of the embodied computation system.

Elements. A substance is defined in terms of the properties and behaviors of its elements, and so we need to consider how they may be defined consistently with the requirements of Sec. 2. For most purposes the formalism makes use of *material* (*Lagrangian, reference*) descriptions of these properties and behaviors, rather than a *spatial* (*Eulerian*) reference frame. This means that we consider a physical quantity Q as a time-varying function of a fixed particle $P \in \mathcal{B}$ as it moves through space, $Q(P, t)$, rather than as a time-varying property $q(\mathbf{p}, t)$ at a particular fixed location in space, $\mathbf{p} \in \mathcal{E}$. In effect, the use of material variables in preference to spatial variables is a particle-centered description, and is a more object-oriented or agent-based way of thinking, in that we can think of the particles as carrying their own properties with them and behaving like well-defined entities.

As explained in Sec. 2, in order to maintain independence of the size of the elements, so far as possible all quantities are *intensive*. This is one of the characteristics that distinguishes this model of embodied computing from ordinary mathematical descriptions of physical phenomena. For example, instead of dealing with N, the number of particles corresponding to an element, we deal with its *number density* n, the number of particles per unit volume. Note that if the number density becomes very small, then the dynamics will be subject to small sample effects, which are often important in self-organization.

The requirement for complementary perspectives implies that the properties of elements be treated as bulk quantities, that is, as the collective properties of an indeterminate ensemble of "units" (e.g., cells or molecules). This creates special requirements when the units constituting an element might have differing values for an attribute. For example, cells, molecules, and microrobots have an orientation, which is often critical to their collective behavior. In some cases all the units will have the same orientation, in which case the orientation can be treated as an ordinary intensive property of the element. In most cases, however, we must allow for the fact that the units may have differing orientations (even if a consequence only of defects or thermal agitation). Therefore, we have to consider the *distribution* of orientations; each orientation (represented by a unit vector $\mathbf{u}$) has a corresponding probability density $\Pr\{\mathbf{u}\}$. If n is the number density of an element, then $n\Pr\{\mathbf{u}\}$ is the number density of units with orientation $\mathbf{u}$. Equivalently, the orientation may be interpreted as a vector-valued random variable, $\mathbf{U}$.

Morphogenesis in development sometimes depends on cell shape; for example, a change from a cylindrical shape to a truncated cone may cause a flat tissue to fold [9, pp. 113–16][27]. Generally a shape can be expressed as a vector or matrix of essential parameters that define relevant aspects of the shape. Two complexities arise in the expression of shape in this formalism for embodied computing. First, the shape must be expressed in a coordinate-independent way, which means that we are dealing with a *shape tensor*. Second, since an element represents an indefinite ensemble of units, shape must be treated as a probability distribution defined over shape tensors, or as a tensor-valued random variable.

Behavior. The behavior of the elements (particles, material points) of a body is defined by rules that describe the temporal change of various quantities, primarily intensive tensor quantities. Such changes might be expressed in continuous time, by ordinary differential equations, e.g., $D_t X = F(X, Y)$, or in discrete time by finite difference equations, e.g, $\Delta_t X = F(X, Y)$. Generally, these would be *substantial* or *material* (as opposed to *spatial*) time derivatives, that is, derivatives evaluated with respect to a fixed particle: $D_t X = \partial X(P, t)/\partial t|_{P \text{ fixed}}$.

In order to maintain complementarity between discrete and continuous time, the proposed formalism expresses such relationships in a neutral notation:

$$\mathrm{D}X = F(X, Y).$$

This *change equation* may be read, "the change in X is given by $F(X, Y)$." The rules of manipulation for the Ð operator respect its complementary discrete- and continuous-time interpretations.

A particle-oriented description of behavior implies that in most cases active particles (e.g., cells, microrobots) will not have direct control over their position or velocity. Rather, particles will act by controlling local forces (e.g., adhesion, stress) between themselves and other particles in the same body or in other bodies. Therefore, "programming" active substances will involve implementing change equations that govern stress tensors and other motive forces associated with the particles.

As discussed in Sec. 2, noise, uncertainty, error, faults, defects, and other sources of randomness are unavoidable in embodied computation, especially when applied in nanotechnology. Indeed, such randomness can often be exploited as a computational resource in embodied computing. Therefore, the behavior of elements will often be described by stochastic differential/difference equations. To facilitate the complementary continuous/discrete interpretations, it is most convenient to express these equations in the Langevin form:

$$ĐX = F(X, Y, \ldots) + G_1(X, Y, \ldots)\nu_1(t) + \cdots + G_n(X, Y, \ldots)\nu_n(t),$$

where the ν_j are noise terms, but we must be careful about the interpretation of equations of this kind.

To see why, consider a stochastic integral, $X_t = \int_0^t H_s dW_s$, where W is a Wiener process (Brownian motion). In differential form this is $dX_t = H_t dW_t$. To maintain complementarity, this should be consistent in the limit with the difference equation: $\Delta X_t = H_t \Delta W_t$. This implies that the stochastic integral is interpreted in accord with the Itō calculus. The corresponding stochastic change equation in Langevin form is $ĐX_t = H_t ĐW_t$. Interpreted as a finite difference equation it is $\Delta_t X_t = H_t \Delta_t W_t$, which makes sense. However the corresponding differential equation, $D_t X_t = H_t D_t W_t$, is problematic, since a Wiener process is nowhere differentiable. Fortunately we can treat $D_t W_t$ purely formally, as follows. First observe that $\Delta W_t = W_{t+\Delta t} - W_t$ is normally distributed with zero mean and variance Δt, $\Delta W_t \sim \mathcal{N}(0, \Delta t)$. Therefore $\Delta_t W_t = \Delta W_t / \Delta t$ is normally distributed with unit variance, $\Delta_t W_t \sim \mathcal{N}(0, 1)$, and $\Delta_t W_t$ can be treated as a random variable with this distribution. To extend this to the continuous case, we treat $ĐW_t$ as a random variable, $ĐW_t \sim \mathcal{N}(0, 1)$. Therefore the stochastic change equation $ĐX = HĐW$ has consistent complementary interpretations. Similarly, for an n-dimensional Wiener process $\mathbf{W}^n$, we interpret $Đ\mathbf{W}^n$ to be an n-dimensional random vector distributed $\mathcal{N}(0, 1)$ in each dimension.

As explained above (Sec. 2), one requirement for the formalism is that it be able to express qualitative behavioral rules corresponding to the influence diagrams widely used in biology. Therefore, we define the notation $ĐX \sim -X, Y, Z$ (for example) to mean that the change of X is "repressed" (negatively regulated) by X and "enhanced" (positively regulated) by Y and Z. We have been calling such a relationship a *regulation* and read it, "the change in X is negatively regulated by X and positively regulated by Y and Z." Formally, $ĐX \sim -X, Y, Z$ is interpreted as a change equation $ĐX = F(-X, Y, Z)$ in which F is a function that is unspecified except that it is monotonically non-decreasing in each of its arguments. (Thus the signs of the arguments express positive or negative regulation.)

4   Example

The following example, which illustrates our formalism, extends the model developed by Frederico Bussolino and his colleagues of vasculogenesis, the early stages of formation of capillary networks from dispersed endothelial cells [2,11,33] (cf. [26]). Aggregation is governed by a morphogen that is released by the cells and that diffuses and degrades.

> **substance** morphogen:
>> **scalar fields:**
>>
>>| | |
>>|---|---|
>>| C | ‖ concentration |
>>| S | ‖ source |
>>| **order-2 field D** | ‖ diffusion tensor |
>>| **scalar** τ | ‖ degradation time constant |
>>
>> **behavior:**
>>
>> $$ÐC = \triangle(\mathbf{D}C) + S - C/\tau \quad ‖ \text{ diffusion + release − degradation}$$

($\triangle = \nabla \cdot \nabla$ is the Laplacian for tensor fields.) The cells produce morphogen at a rate α and follow the gradient at a rate governed by attraction strength β. Cell motion is impeded by dissipative interactions with the substrate, which are measured by an order-2 tensor field γ. Since cells are filled with water, they are nearly incompressible, and so they have a maximum density n_0, which influences cell motion; to accommodate this property (which will also apply to microrobots), the model uses an arbitrary function, ϕ, that increases very rapidly as the density exceeds n_0.

> **substance** cell-mass **is** morphogen **with:**
>> **scalar fields:**
>>
>>| | |
>>|---|---|
>>| n | ‖ number density of cell mass |
>>| ϕ | ‖ cell compression force |
>>| **vector field v** | ‖ cell velocity |
>>
>> **scalars:**
>>
>>| | |
>>|---|---|
>>| n_0 | ‖ maximum cell density |
>>| α | ‖ rate of morphogen release |
>>| β | ‖ strength of morphogen attraction |
>>| **order-2 field** γ | ‖ dissipative interaction |
>>
>> **behavior:**
>>
>> $$S = \alpha n \quad ‖ \text{ production of morphogen}$$
>> $$‖ \text{ follow morphogen gradient, subject to drag and compression:}$$
>> $$Ð\mathbf{v} = \beta\nabla C - \gamma \cdot \mathbf{v} - n^{-1}\nabla\phi$$
>> $$Ðn = -\nabla \cdot (n \cdot \mathbf{v}) + \mathbf{v} \cdot \nabla n \quad ‖ \text{ change of density in material frame}$$
>> $$\phi = [(n - n_0)^+]^3 \quad ‖ \text{ arbitrary penalty function}$$

References

1. Adamatzky, A., Costello, B.D.L., Asai, T.: Reaction-Diffusion Computers. Elsevier, Amsterdam (2005)
2. Ambrosi, D., Gamba, A., Serini, G.: Cell directional persistence and chemotaxis in vascular morphogenesis. Bulletin of Mathematical Biology 67, 1851–1873 (2004)
3. Barish, R.D., Rothemund, P.W.K., Winfree, E.: Two computational primitives for algorithmic self-assembly: Copying and counting. Nano Letters 5, 2586–2592 (2005)
4. Benzi, R., Parisi, G., Sutera, A., Vulpiani, A.: Stochastic resonance in climatic change. Tellus 34(10-16) (1982)
5. Beysens, D.A., Forgacs, G., Glazier, J.A.: Cell sorting is analogous to phase ordering in fluids. Proc. Nat. Acad. Sci. USA 97, 9467–9471 (2000)
6. Brooks, R.: Intelligence without representation. Artificial Intelligence 47, 139–159 (1991)
7. Clark, A.: Being There: Putting Brain, Body, and World Together Again. MIT Press, Cambridge (1997)
8. de Gennes, P.G.: Soft matter. Science 256, 495–497 (1992)
9. Forgacs, G., Newman, S.A.: Biological Physics of the Developing Embryo. Cambridge University Press, Cambridge (2005)
10. Fung, Y.C.: Biomechanics: Mechanical Properties of Living Tissues. Springer, New York (1993)
11. Gamba, A., Ambrosi, D., Coniglio, A., de Candia, A., Di Talia, S., Giraudo, E., Serini, G., Preziosi, L., Bussolino, F.: Percolation, morphogenesis, and Burgers dynamics in blood vessels formation. Physical Review Letters 90(11), 118101-1 – 118101-4 (2003)
12. Haykin, S.: Neural Networks: A Comprehensive Foundation, 2nd edn. Prentice Hall, Upper Saddle River (1999)
13. Johnson, M., Rohrer, T.: We are live creatures: Embodiment, American pragmatism, and the cognitive organism. In: Zlatev, J., Ziemke, T., Frank, R., Dirven, R. (eds.) Body, Language, and Mind, vol. 1, pp. 17–54. Mouton de Gruyter, Berlin (2007)
14. Kirkpatrick, S., Gelatt Jr., C.D., Vecchi, M.P.: Optimization by simulated annealing. Science 220, 671–680 (1983)
15. MacLennan, B.J.: Universally programmable intelligent matter: Summary. In: IEEE Nano 2002 Proceedings, pp. 405–408. IEEE Press, Los Alamitos (2002)
16. MacLennan, B.J.: Molecular combinatory computing for nanostructure synthesis and control. In: IEEE Nano 2003 Proceedings, p. 13_01. IEEE Press, Los Alamitos (2003)
17. MacLennan, B.J.: Natural computation and non-Turing models of computation. Theoretical Computer Science 317(1-3), 115–145 (2004)
18. MacLennan, B.J.: Radical reconfiguration by universally programmable intelligent matter. In: Air Force Reconfigurable Electronics Workshop, Santa Fe, NM, May 17-18. Air Force Research Laboratory (2004)
19. MacLennan, B.J.: Radical reconfiguration of computers by programmable matter: Progress on universally programmable intelligent matter — UPIM report 9. Technical Report CS-04-531, Dept. of Computer Science, University of Tennessee, Knoxville (2004)
20. MacLennan, B.J.: Highly programmable matter and generalized computation: Research in reconfigurable analog and digital computation in bulk materials. In: 1st AFRL Reconfigurable Systems Workshop, Albuquerque, NM, Feburary 14-15, pp. 14–15. Air Force Research Laboratory (2007)

21. MacLennan, B.J.: Embodiment and non-Turing computation. In: 2008 North American Computing and Philosophy Conference: The Limits of Computation, Bloomington, IN, July 10-12. The International Association for Computing and Philosophy (2008), http://www.cs.utk.edu/~mclennan/papers/AEC-TR.pdf

22. MacLennan, B.J.: Computation and nanotechnology (editorial preface). International Journal of Nanotechnology and Molecular Computation 1(1), i–ix (2009)

23. MacLennan, B.J.: Configuration and reconfiguration of complex systems by artificial morphogenesis. In: Reconfigurable Systems Workshop 2009: Discovery Challenge Thrust, Santa Fe, NM, July 20-22. Air Force Office of Scientific Research (2009)

24. MacLennan, B.J.: Super-Turing or non-Turing? Extending the concept of computation. International Journal of Unconventional Computing 5(3-4), 369–387 (2009)

25. MacLennan, B.J.: Bodies — both informed and transformed: Embodied computation and information processing. In: Dodig-Crnkovic, G., Burgin, M. (eds.) Information and Computation. World Scientific, Singapore (in press)

26. Merks, R.M.H., Brodsky, S.V., Goligorksy, M.S., Newman, S.A., Glazier, J.A.: Cell elongation is key to in silico replication of in vitro vasculogenesis and subsequent remodeling. Developmental Biology 289, 44–54 (2006)

27. Odell, G.M., Oster, G., Alberch, P., Burnside, B.: The mechanical basis of morphogenesis. I. Epithial folding and invagination. Developmental Biology 85, 446–462 (1981)

28. Pfeifer, R., Bongard, J.C.: How the Body Shapes the Way We Think — A New View of Intelligence. MIT Press, Cambridge (2007)

29. Pfeifer, R., Lungarella, M., Iida, F.: Self-organization, embodiment, and biologically inspired robotics. Science 318, 1088–1093 (2007)

30. Rothemund, P.W.K., Papadakis, N., Winfree, E.: Algorithmic self-assembly of DNA Sierpinski triangles. PLoS Biology 2(12), 2041–2053 (2004)

31. Rothemund, P.W.K., Winfree, E.: The program-size complexity of self-assembled squares. In: Symposium on Theory of Computing (STOC), New York, pp. 459–468. Association for Computing Machinery (2000)

32. Salazar-Ciudad, I., Jernvall, J., Newman, S.A.: Mechanisms of pattern formation in development and evolution. Development 130, 2027–2037 (2003)

33. Serini, G., Ambrosi, D., Giraudo, E., Gamba, A., Preziosi, L., Bussolino, F.: Modeling the early stages of vascular network assembly. The EMBO Journal 22(8), 1771–1779 (2003)

34. Steinberg, M.S., Poole, T.J.: Liquid behavior of embryonic tissues. In: Bellairs, R., Curtis, A.S.G. (eds.) Cell Behavior, pp. 583–607. Cambridge University Press, Cambridge (1982)

35. Taber, L.A.: Nonlinear Theory of Elasticity: Applications in Biomechanics. World Scientific, Singapore (2004)

36. Tomlin, C.J., Axelrod, J.D.: Biology by numbers: Mathematical modelling in developmental biology. Nature Reviews Genetics 8, 331–340 (2007)

Biologically–Inspired Network Architecture for Future Networks

Masayuki Murata

Graduate School of Information Science and Technology,
Osaka University Suita, Osaka 565-0871, Japan
murata@ist.osaka-u.ac.jp
http://www.anarg.jp/

Abstract. An architecture for future networks (often referred to as new–generation networks, or the future Internet), is now actively discussed in a "clean–slate" fashion. That is, the future network architecture could be differently from the current Internet architecture to overcome its problems. We first discuss why such an approach is necessary, and how we can reach a new era of information networks. Then, we introduce our approach towards the new network architecture. It is a biologically–inspired self–organizing network. Its robustness and adaptiveness attained by the bio–inspired approach is quite useful for satisfying the requirement of the future networks.

Keywords: Future networks, Self–organizing networks, Bio–inspired approach.

1 Introduction

The Internet is being now conceived as a commodity of industrial and social activities. The Internet was originally designed to have a thin waist of IP, in order to allow an convergence of a wide variety of link–layer technologies such as Ethernet, optical networks, wireless networks and so on. Also, it can allow to deploy new emerging applications and services on IP, leading to assure an interoperability. It has been an actual source of innovation by introducing such a design principle.

However, assumptions behind the Internet design methodology are now being changed. For example, the wireless technology is utilized for the last hop more and more, which poses a need of supporting mobility in a more efficient way. Commercialization and civilization of the Internet has introduced deep concerns about security and trust. Use of the Internet as a communication infrastructure has raised the need of much improved resilience and fault–tolerance through better control and management. Technologies and related protocols have not been designed without energy efficiency. It is true that in the literature, such needs have been tackled, but those often fail to satisfy a deployable requirement from the current Internet. Or, the effect is quite limited if those are based on the current Internet architecture. That is, it has not been possible to introduce any

F. Peper et al. (Eds.): IWNC 2009, PICT 2, pp. 34–41, 2010.

major changes to the deployed base of the Internet. Also, an incremental change for solving some specific problem has a limited effect and even introduce other problems. Then, it is now convinced that it is time to redesigning the Internet for resolving the problems while ensuring enough flexibility to incorporate future possible requirements. It is called as a "clean slate" approach for realizing a new paradigm of architectural design against traditional approaches of the incremental design.

To tackle problems of fragile networks, we have been researching on a future network architecture based on self–organizing control. For this purpose, we are learning from biology with the knowledge gained through analyzing the behavior of various living organisms. After we started the project, we soon noticed that a direct application of those studies into ours is insufficient. We decided to use well–established, biologically inspired mathematical models and apply those to the control methods for information networks, especially for recently emerging networks like peer-to-peer (P2P) networks, mobile wireless ad hoc networks and sensor networks. More recently, we are applying biologically–inspired approaches to the existing optical networks, and convince that the bio–inspired approach is a useful and powerful tool for building the infrastructure of the current and even future networks.

In what follows, we first describe why we take biologically inspired approaches in Section 2. We then describe our the recent achievement based on a so-called "attractor selection principle" in Section 3. This article concludes with future perspectives of the bio–inspired approaches.

2   Why Bio-inspired Approaches?

Historically, in information networks including the Internet, it has been assumed that static nodes (i.e., routers) serve packet forwarding based on the pre-defined routing protocol. The adequate operation of those networks requires careful network planning in order to satisfy the objectives of those networks, e.g., QoS (Quality of Service) support. On the other hand, the new emerging networking technologies have a quite different structure from the traditional networks; the node itself may move, and on–demand search is necessary for finding the shared information in P2P networks and the peer location in ad hoc networks. Also, in sensor networks, topological control like clustering control without precise knowledge on network configuration becomes crucial in order to obtain scalability and power saving control.

In future networks, we consider that the following three characteristics are mandatory for network control & management:

1. Scalability: We need to cope with the growing number of nodes and end users, and a wide variety of devices attached to the network. More importantly, the number of nodes and terminals has never been predicted in advance. This means that the conventional network planning method becomes meaningless.

2. Mobility: In addition to the users' mobility, we should also consider the mobility of network nodes in ad hoc and sensor networks. It implies that stable packet forwarding by nodes is no longer expected.
3. Diversity: We need to support a wide variety of network devices generating traffic in a quite different nature from network applications and services.

In our view, a possible solution to meeting above characteristics is that end hosts must be equipped with adaptability to the current network condition for finding communicating peers and/or for controlling congestion. At the same time, network control should satisfy the requirements of end hosts in a real–time fashion. Since the network is becoming complex and large–scaled, it becomes more difficult to utilize a conventional "well–managed" approach.

In many conventional approaches, we have been targeted the new protocol or network control methods in order to tackle current problems or to improve existing proposals. Then, we show significance or validity of new proposals. In order to do so, we explicitly or implicitly assume that

1. processing capability and topology of the network are known in advance,
2. traffic characteristics and required QoS level are given by upper layers, and
3. lower layers are stable enough so that obtained performance level is not so changed.

We then obtain performance metrics in concern by means of mathematical methods (including probabilistic arguments or traffic theory/queueing theory). When the analysis is difficult to apply to the current problem, the simulation technique is utilized. The problem is apparently that such an approach is just an optimization of one protocol within a specific layer. It is quite far from designing the whole system.

Another obvious problem is that in the conventional system, we have been considered the throughput and/or packet delays as the performance metrics. Those are "average" metrics, and validation of using those metrics is that we implicitly/explicitly consider the stable system, and those are meaningful only when the target system is sufficiently stable. However, as described in the above, the future network is very dynamic in the sense that

1. Topology often changes. It is typical in overlay networks and mobile networks. However, as the network size becomes large, it would be common even in the other networks that the failure always happens somewhere within the network. Then, we can never use a "single–failure" assumption that we see at most single failure at the same time within the network.
2. Traffic often changes. Due to the topological changes, the traffic flow changes if the adaptable traffic control method is applied. Also, diversified applications and services would generate a wide variety of traffic characteristics.

We have another reason that average performance metrics like throughput and/ or packet delays are no longer useful. In the future networks, we have to consider the other "quality"–related metrics more in order to argue about the quality level of the network. Those include manageability, availability, reliability, adaptability,

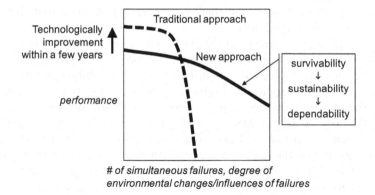

Fig. 1. Performance curve for the future networks

dependability, sustainability, reconfigurability, and others. Once we satisfy those in defined metrics, then, we would have the "quality–rich" network instead of the current "quantity–rich" network.

Then, what kind of the performance–curve is preferred in the future networks? It is a next question. In Fig. 1, we plot the performance. Here, we suppose the throughput averaged over some time duration for simplicity, but we can consider the other metrics as the vertical axis. The horizontal axis shows the number of simultaneous failures within the network, the degree of environmental changes, the influence of failures or whatever like those. In the traditional approach, we aim at maximizing the performance when we have no failures. Or, the performance is still good when the single–failure assumption is considered to treat the reliability issue.

However, in those systems, it is likely to happen when two or more numbers of simultaneous failures occur, the performance of the target system is suddenly degraded. On the other hand, in the future network, we want to have the performance curve labeled as "new approach" in the figure. Even if the number of simultaneous failures is increased, the performance is not so degraded. When we have no failures, the performance may not be optimized when compared with the "traditional approach." However, we claim that the performance will be compensated soon by the technological improvement in the near future. Importantly, by the new approach, we would have a sustainable network system, and only by such an approach, we would achieve the dependability of the target system. That is, as is widely recognized now, the word "dependability" includes some psychological flavor, and the users can rely on the network only when the users know that the system continues working on.

To realize such a new feature, we are taking the self–organizing approach for the network control and management. Generally, self–organization is a set of dynamical mechanisms whereby structures appear at the global level of a system from interactions among its lower–level components [1]. The rules specifying the interactions among the system's constituent units are executed on the basis of purely logical information, without reference to the global pattern which is an emergent property of the system rather than a property imposed upon the system by an external ordering influence.

Self–organization is a universal behavior that many natural systems exhibit. Among them, we learn from the self–organizing property of biological systems. It is known that the biological system has an excellent feature of adaptability. For this reason, the biologically–inspired approach is expected to be promising to meet the above–mentioned requirements of the future networks. Of course, biologically–inspired approaches for information technologies are not new; those include genetic algorithms and neural networks. However, most of them have been concentrated on the optimization problems in the field of network controls. On the contrary, we are focusing on the adaptability of the biological system, in order to get robustness by virtue of its self–organization property.

One promising approach based on the biological system is so-called swarm intelligence or stigmergy where a group exhibits an intelligent and organized behavior without any centralized control, but with local and mutual interactions among individuals. Then, the behavior would be adaptive to changes in the environment, and a group keeps working even if a part fails. Such behaviors can be found in many biological systems. Those include ant trail (foraging behavior of ants), cemetery organization and brood sorting, colonial closure, division of labor and task allocation, pattern formation, and synchronization in flashing fireflies. A pioneer work is done by [2] where the ant trail behavior is applied to the network routing protocol.

We have successfully applied the above–mentioned mechanisms in the fundamental technologies of the information networks, especially, P2P and sensor/ad hoc networks where it is important to consider adaptability to the changing environments [3,4,5,6,7]. By biologically–inspired approaches, we can establish fully–distributed and self–organizing techniques. An important point is that when we apply the biologically–inspired approach to the information networks, we should refrain from simply mimicking the biology. What we have to learn from the biology is the dynamical system behavior and its mathematical modeling method. Then, we can have mathematical discussion. We can tune the system based on the mathematical model if necessary. We can predict the system behavior even in the experimental phase of the target system.

3   Attractor Selection Principle and Its Application to Networking Problems

Among the bio–inspired approaches, we have been recently engaged in the application of the "attractor selection" principle proposed by the biologists group at Osaka University. The attractor selection principle is based on the study of the cell biology [8]: why the biological system is so adaptive to the environmental changes, and robust eventually? In [8], the authors experimentally studied the effects of two mutually inhibitory operons in E.coli cells reacting to the lack of a nutrient in their exposed medium. A mathematical model was also proposed in [8], which is essentially stochastic N-differential equations.

$$\frac{dx_i}{dt} = f_i(x_1, \ldots, x_N)\alpha + \eta_i, i = 1, \ldots, N \tag{1}$$

The basic dynamic behavior can be described as follows. The system state contains all x_i and is derived from the concentrations of the messenger RNA (mRNA) molecules in the original model. The functions f_i define the attractors to which the dynamic orbit of the system will eventually converge in spite of the existence of an inherent noise term η_i. A key term is α, which is a non-negative function representing the cell's growth rate and is related to its activity. Essentially, this function influences the actual selection by switching between two modes of operation. In the first case, if $\alpha > 0$, the dynamics of Eq. (1) follows a rather deterministic way and the fluctuations introduced by will not influence the convergence to an attractor, under the condition that the noise amplitude is sufficiently small. On the other hand, when α approaches 0, the dynamics is entirely governed by η_i leading to a random walk in the phase space. Last we note that a same idea of the attractor selection is not limited in the cellular system of E.coli. It is in a sense universal in the biological systems from the molecular motor level to the human brain level (see, e.g., [9]). More essential is that those effectively use the fluctuation everywhere.

From the above, one could easily imagine that the attractor selection princioke is applicable to the information network control methods, in which one solution is self–adaptively selected among a set of candidates utilizing the inherent noise in the system. That is, the selection follows the system dynamics embedded in a set of differential equations and the selection itself is performed without explicit rules, as each node simply follows the same dynamical pattern, as shown in it seems well suited for application in ambient network environments. Actually, in [3,9], the above modeling principle is applied to the multi–path routing problem in the overlay network, and we are now extending it to the various networking problems.

Of course, a huge variety of optimization algorithms or quasi–optimization algorithms have been developed for controlling networks. In most of those approaches, however, it is assumed that the current network status is known in advance. The network status may be obtained through real–time network measurements, but even in that case, the network is implicitly assumed to be static at least during the optimization problem is solved. On the other hand, we completely take a different approach: under the condition that the current network status is measured in a real–time fashion, the system is driven by the time–differential equation. Then the adaptability to changing environments, which includes the influences by external source like the traffic changes and internal source like the node/link failures and the measurement errors, is established. Finding the optimal solution is also performed in real–time, while it does not always guarantee the optimality. In the last point, our claim is that the optimality is important, but the degradation might be acceptable and more important is adaptability to the environmental changes.

One important aspect of the self–organization is that it exhibits an emergent behavior without any centralized control, but with only local and mutual

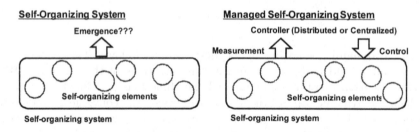

Fig. 2. Self-organizing system with/without control

interactions among individuals as described before. It is a good property for realizing a distributed principle, which is essential for future very large networking systems. At the same time, however, it cause the problem that it often leads to be unpredictable, that is, a difficulty of forecasting the resulting pheromone due to the intrinsic nature of "emergency." (see, e.g., [10] for detailed discussions about the emergence). In the networking area, on the other hand, the managed system is another pursuitment. For realizing it, we need to build the rigorous resource management. That has been a intrinsic obstacle for applying the self–organization principle to the networking system while the good properties of the self–organization is now well understood. In the above attractor selection principle, on the other hand, a whole system is driven by activity α which can be set to be a target performance metric of interest (see Fig. 2). So, our system might be called as a "managed self–organizing system." It provides a range of an operating regime by an external control while still allowing a self–organization property.

4   Concluding Remarks

In this article, we have introduced the biologically inspired networking control and management methods. By extending the attractor selection principle described in Section 3, we are running a project called Global COE Project "Center for Excellence for Founding Ambient Information Society Infrastructure," [11] (or "Ambient Information Environment (AmIE)" project in short), which is supported by MEXT (the Ministry of Education, Culture, Sports, Science & Technology, Japan). See [11] for more detail. Technological discussions on AmIE can also be found in [12].

Finally, we comment on the importance of the interdisciplinary approach like ours. Establishing the new networking environment is of course an engineering problem. However, we have to have a scientific basis for designing the network. Traditionally, queuing and/or traffic theories have been used for this purpose. However, those can no longer become a basis for such purposes because in those approaches, the steady–state system behavior is of primary concern. For the future networks, we want to treat the time–dependent system, and to establish the robust and adaptive system. Fortunately, the biological system essentially has such characteristics. It is the reason why the biologically–inspired approach is so important. To learn from the biology is not just an analogy, but a basis of the interdisciplinary science/engineering approach, which is now widely recognized as a source of the innovation in the next technological era.

References

1. Bonabeau, E., Dorigo, M., Theraulaz, G.: Swarm Intelligence: From Natural to Artificial Systems. Oxford University Press, Oxford (1999)
2. Di Caro, G., Dorigo, M.: AntNet: Distributed Stigmergetic Control for Communications Networks. Journal of Artificial Intelligence Research (JAIR) 9, 317–365 (1998)
3. Leibnitz, K., Wakamiya, N., Murata, M.: Biologically–Inspired Self–Adaptive Multi–Path Routing in Overlay Networks. CACM 49(3), 62–67 (2006)
4. Hasegawa, G., Murata, M.: TCP Symbiosis: Congestion Control Mechanisms of TCP based on Lotka–Volterra Competition Model. In: Proc. of Workshop on Interdisciplinary Systems Approach in Performance Evaluation and Design of Computer & Communications Systems (2006)
5. Koizumi, Y., Miyamura, T., Arakawa, S., Oki, E., Shiomoto, K., Murata, M.: Application of Attractor Selection to Adaptive Virtual Network Topology Control. In: Proc. of ACM 3rd International Conference on Bio–Inspired Models of Network, Information, and Computing Systems (2008)
6. Wakamiya, N., Hyodo, K., Murata, M.: Reaction–diffusion based Topology Self–organization for Periodic Data Gathering in Wireless Sensor Networks. In: Proc. of Second IEEE International Conference on Self–Adaptive and Self–Organizing Systems (2008)
7. Sugano, M., Kiri, Y., Murata, M.: Differences in Robustness of Self–Organized Control and Centralized Control in Sensor Networks Caused by Differences in Control Dependence. In: Proc. of the Third International Conference on Systems and Networks Communications, ICSNC 2008 (2008)
8. Kashiwagi, A., Urabe, I., Kaneko, K., Yomo, T.: Adaptive Response of a Gene Network to Environmental Changes by Fitness–induced Attractor Selection. Plos ONE 1(1), e49 (2006)
9. Leibnitz, K., Wakamiya, N., Murata, M.: A Bio-Inspired Robust Routing Protocol for Mobile Ad Hoc Networks. In: Proc. of 16th International Conference on Computer Communications and Networks (ICCCN 2007), pp. 321–326 (2007)
10. De Wolf, T., Holvoet, T.: Emergence Versus Self-Organisation: Different Concepts but Promising When Combined. In: Brueckner, S.A., Di Marzo Serugendo, G., Karageorgos, A., Nagpal, R. (eds.) ESOA 2005. LNCS (LNAI), vol. 3464,.pp. 1–15. Springer, Heidelberg (2005)
11. Global COE Program, Center for Excellence for Founding Ambient Information Society Infrastructure, http://www.ist.osaka-u.ac.jp/GlobalCOE
12. Murata, M.: Towards Establishing Ambient Network Environment. IEICE Transactions on Communications E92.B(4), 1070–1076 (2009)

Foraging Behaviors and Potential Computational Ability of Problem-Solving in an Amoeba

Toshiyuki Nakagaki[1,2]

[1] Research Institute For Electronic Science,
Hokkaido University, N12 W6, Sapporo 001-0021, Japan
[2] JST, CREST, 5, Sanbancho, Chiyoda-ku, Tokyo, 102-0075, Japan
nakagaki@es.hokudai.ac.jp

Abstract. We study cell behaviors in the complex situations: multiple locations of food were simultaneously given. An amoeba-like organism of true slime mold gathered at the multiple food locations while body shape made of tubular network was totally changed. Then only a few tubes connected all of food locations through a network shape. By taking the network shape of body, the plasmodium could meet its own physiological requirements: as fast absorption of nutrient as possible and sufficient circulation of chemical signals and nutrients through a whole body. Optimality of network shape was evaluated in relation to a combinatorial optimization problem. Here we reviewed the potential computational ability of problem-solving in the amoeba, which was much higher than we'd though. The main message of this article is that we had better to change our stupid opinion that an amoeba is stupid.

Keywords: *Physarum*, sub-cellular computing, primitive intelligence, problem-solving, ethology.

1   Introduction

An amoeba of true slime mold, the plasmodium of *Physarum polycephalum*, has the ability to solve a maze although it consists of only a single cell. Cells are units, which have common nature over all of organisms. People are likely to look at human and other higher animals when considering what makes *intelligence* appear. From an evolutionary point of view, however, any tiny organisms can show *bud* of intelligence in some sense. Otherwise, it may be difficult to succeed to survive for such a long time as a little less than a billion of years. The thought that single celled-organisms may be more intelligent than people normally expected has been claimed repeatedly since one-hundred years ago[1]. It is quite interesting to study how intelligent cells are and what the mechanism of intelligence is in terms of dynamics and in terms of computational algorithm.

To test intelligence in animals other than human, appetite for food in animals plays an important role. Foods are given to animals but it is difficult for them to get the foods, then they try to demonstrate their potential ability of finding a way to get the foods. A typical example is that chimpanzee in the room

F. Peper et al. (Eds.): IWNC 2009, PICT 2, pp. 42–54, 2010.

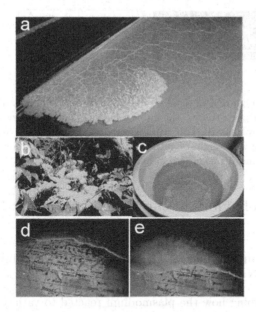

Fig. 1. Beauty and wonder of the plasmodium of *Physarum polycephalum*. a) a large plasmodium, which is single-celled but multinuclear. It is coming downward in the picture. The frontal part is sheet-like and the complicated network of tubular channel develops toward the rear. The frontal part with the semi-circle shape is approximately 3cm in the diameter. b) the plasmodium of *Fuligo*, a different species from *Physarum*, in the wild nature. c) a plasmodium that crawls up along the wall of bucket. d-e) a plasmodium that takes over the globe. Locomotion speed is a few centimeters per hour.

uses a tool to get a banana, which is hung from the ceiling and impossible to reach directly by a hand. A similar method is used in the test for unicellular organisms.

Here we describe a nice model organism of unicellular organism, a giant amoeba of true slime mold (see Fig.1). The plasmodium of true slime mold *Physarum polycephalum* is an amoeba-like organism in which network structure of tubular element developed. Through the network, nutrient and signal are carried through the organism. When multiple small food sources (FSs) were presented to a starved plasmodium that spread over an entire culture dish, it concentrated at each FS. Almost the entire plasmodium accumulated at the FSs and covered each of them in order to absorb nutrient. Only a few thick tubes remained connecting the quasi-separated components of the plasmodium[2,3,4,5]. By forming such a pattern, the organism derives the maximum nutrition in the minimum time. Hence this strategy is rather smart; it implies that the plasmodium can solve a complex problem[6,7,8].

We review here how high the computational ability of problem-sovling is in the amoeba and what the mechanism is in some sense of computational algorithm.[9,10,11,12,13,14,15].

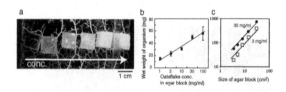

Fig. 2. Dependency of gathering mass of organism on food amount. a) Photograph of a typical plasmodium gathering at the cubic FSs with the same size ($1 \times 1 \times 1$ cm^3) and different concentrations (0, 1, 3, 10, 30 and 100 mg/ml from left to right). b, c) Relationships between wet weight of gathering organism at FS and food amount at the FS: dependency on concentration (b) and size (c) of FS. Refer to [15].

2   Foraging Behaviors in the Case of Spatially Distributed FSs

2.1   Accumulation of Mass of Organism to FS in Relation to Amount of Nutrient at FS

We began by describing how the plasmodium reacted to variations of nutrient amount at FS[15]. Various kinds of FS were given to a large plasmodium (10×60 cm^2) as shown in figure 2a and wet weight of plasmodium gathering on a FS was teared off and measured. An agar block containing powdered oats flake was made as a FS. Concentration of oats powder and size of agar block were changed.

The organism gathered to FS in proportional to the logarithmic concentration of nutrients (1 to 100 mg/ml) at a fixed size of FS ($1 \times 1 \times 1$ cm^3) and in proportional to the surface area of FS (3 to 30 cm^2) at a fixed concentration (3 and 30 mg/ml), as shown in figure 2b and 2c. More organism stayed at FS as food amount at FS was larger.

The amoeba absorbs the nutrients from the surface of food block by the process of endocytosis. Area of contact to the food block restricts the rate of absorption. After the food block is entirely wrapped by the amoeba, no more portion of amoeboid body comes to FS. The amoeba may find the size of FS by finishing to wrap the food block.

2.2   Time Course of Two Connections between Two FSs in Relation to Food Amount at FS

Next we tested which connection the organism selected from two possibilities, as shown in figure 3a, in response to variations of size of FS at a fixed concentration of oats [15]. Volume of organism initially applied was fixed but food amount at FS was changed.

Figure 3b shows the typical time variations of remaining connections. The grey bars indicate existence of each connection: if space is not painted by grey, it means collapse of connections. When FS has size of 0.4 cm^2 in surface area, both connections remained (the left panel). Only longer one disappeared in the FS 0.7

(the middle). The short connection collapsed after the longer one disappeared in the size 1.7 (the right). Usually, the longer connection was cut earlier than the short. Figure 3c shows time course of mean number of remaining tube at the various amounts of applied food (0.2, 0.4, 0.7, 1.7 and 2.5 cm^2 in surface area of FS). The number decreased as time went by and also as food size was larger. In any cases, the longer connection collapsed earlier than the shorter one. This implies that there might be a simple rule for connection selection: the longer connection collapsed earlier.

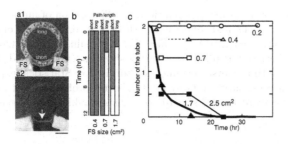

Fig. 3. Tube selection from two possibilities in response to food amount. a) Photographs of typical results 0 (a1) and 12 hours (a2) after FS presentation. The shorter connection remained (marked by the white arrow) though the longer one collapsed. Scale bar: 1 cm. b) Typical time variations of remaining connections. Each gray bar indicates existence of a connection. c) time course of mean number of remaining tube at the various amounts of applied food. See the main text in detail. Refer to [15].

2.3 Order of Tube Collapse in Five Connections with Different Length between Two FSs

To test the selection rule described earlier, we consider the situation of five connections as shown in figure 4a1 [15]. Figure 4b shows typical time courses of connection survival. Final number of surviving connections was reduced as the food size got larger: two, one and zero connection(s) in 7, 10 and 13 cm^2, respectively. But the order of collapsing seemed to be similar: the longer connection died earlier. Exactly speaking, there were some exceptions as indicated by the arrow in the middle figure. Nonetheless statistical analysis clearly shows that the longer tube tended to disappear earlier (fig. 4c). So we can conclude that a simple selection rule works in the different conditions of food amount. Notice that the food size determined when the selection process stops.

2.4 Maze-Solving

From the experiment described below, *Physarum* solves a maze[2,3,9]. The large plasmodium was cut into many small pieces and these pieces were placed in the maze. After several hours, they began to extend and met each other to become a larger organism by fusing spontaneously. Finally the maze was filled with the

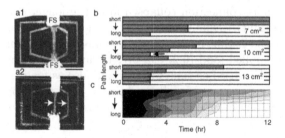

Fig. 4. Path selection from five possibilities. a1) Photograph of initial shape of organism. The five connections had different length of 2, 4, 6, 8 and 10 cm, between two equal FSs. scale bar: 1 cm. a2) A typical result of remaining tubes. Only the shortest and the second shortest remained (marked by the white arrows) and the others have died out. b) Typical time course of connection survival. c) Statistical expression of time variations of connection survival. Gray level indicates probability of tube survival, obtained 12 experiments at the FS size 10 cm^2. Darker is higher in survival probability. Refer to [15].

single organism. Two small food-sources (FSs) were presented at two locations in the maze then. The organism began to move to FSs with drastic changes in the body shape. First the organism pulled back from the dead-end corridors and transported the pulled parts to FSs, and remained the tubular body through all connection routes in the maze. Next the tube in a longer route disappeared earlier. Finally the single tube was left only along the shortest route.

Why does the organism find the shortest route in the maze? One can make an interpretation by considering a physiological desire of organism's own. In the experiment, the organism had a good appetite for food because it was starved. So any parts of organism would like to move to FS in order to absorb nutrients. On the other hand, the organism seemed to try to maintain a large body as it did not like splitting into two smaller individuals. This is another desire.

Maze-solving enable the organism to meet both desires: the volume of organism staying on food-sources was maximized with connecting two major parts of body on FSs, because the volume of body used for connection was minimized. Then the absorption of nutrients was as fast as possible and as much as possible, and moreover intracellular communication of chemical signals was very efficient since flow rate of intracellular viscous materials was higher in a shorter and thicker tube (Please remind the hydrodynamic approximation of Poisuelle flow).

Along the line of understanding as above, one can answer to the question posed at the beginning of this section: the organism tried to maximize its own desires in the maze. This organism's ability can be used to solve human's problem. What has done in the maze-solving experiment is to make our problem of maze-solving come to agreement with satisfaction of organism's desire. Once we know much more about behavioral characters of organism, chances to use the characters are expanded wide-open.

2.5   Risk-Minimum Path in the Inhomogeneous Field of Risk

Two separate FSs were presented to the organism, which was illuminated by an inhomogeneous light field. Because the plasmodium is photophobic, tubes connecting the FSs did not follow the simple shortest path but reacted to the illumination inhomogeneity. We reported that the behavior of the organism under these conditions and proposed a mathematical model for the cell dynamics[11]. Here we describe how *Physarum* solver is used for optimization problem in inhomogeneous field. This is a new type of task involving optimization by *Physarum*.

2.6   Networking of Three and More FSs

The slime mold constructs a network appropriate for maximizing nutrient uptake from multiple separate FSs (more than three). When FSs were presented at different points on the plasmodium it accumulated at each FS with a few tubes connecting the plasmodial concentrations. The geometry of the network depended on the positions of the FSs. Statistical analysis showed that the network geometry met the multiple requirements of a smart network: short total length of tubes, close connections among all the branches (a small number of transit FSs between any two FSs) and tolerance of accidental disconnection of the tubes[4,5,14,12]. These findings indicate that the plasmodium can achieve a better solution to the problem of network configuration than is provided by the shortest connection of Steiner minimum tree (or SMT).

2.7   Not Only the Final Answer But Also Transient

Figure 5a shows a typical time course of tube network in three FSs[4,5,15,12]. Tubes collapsed in time after three FSs were given at the vertices of equilateral triangle to a circular organism and the final network looked like the shortest connection of Steiner's minimum spanning tree.

In order to evaluate the transient shapes of the tube network, we introduced the criteria[5]: *fault tolerance (FT)* and *total length (TL) of the network*. FT_n

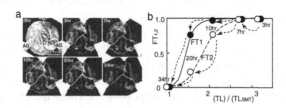

Fig. 5. Tube network in 3 FSs. a) photographs of network at 0, 3, 7, 10, 20, 34 hours after presentation of FSs (AG), located at three vertices of an equilateral triangle. scale bar: 1cm. b) FT-TL relationship of tube network. The black line and the dark line indicates optimal FT value at a specific value of TL for all of TL values (See the main text in detail). Refer to [15].

is the probability that the organism is not fragmented if n times accidental disconnections occur at random points along the tubes. Since the probability of disconnection of a tube is proportional to its length, a longer tube has a higher risk of disconnection.

TL and FT decreased as tubes collapsed in time as shown in figure 5b. The Data points traces the black line (for FT_1) and the dark line (for FT_2). These lines are the master curves of FT-TL relationship which was obtained in the previously published paper[5] and indicates the optimal value of FT under the constraints of that specific TL for all of the TL values. This means that the organism reduced TL with keeping FT optimal. This is a hidden rule for tube selection in the complex dynamics of functional network.

3   Mathematical Modeling for the Foraging Behaviors to Connect the Spatially Distributed FSs through a Smart Network

3.1   Mathematical Formulation of Network Dynamics Adaptable to Streaming

The processes of maze-solving are based on morphogenesis of tubular structure in a body. A key phenomenological mechanism has been reported: thickness of tube became thicker (or thinner) if flow through the tube was large (or small) enough[7,16]. This character is adaptability of tube to the volume of flow. By incorporating the adaptability, a simple mathematical model was proposed. Many realistic things like elasticity of tube etc were neglected and the adaptability to flow was focused. A tube network of *Physarum* was represented by a network of water pipeline. Flow through the network was first calculated and dynamics of tube thickness was next determined based on the flow through the tube.

The mathematical model for some cases of two food-sources (or FSs) was already proposed[9], which consisted of two kinds of equations for flow of protoplasm and adaptability of tube thickness. This model was based on biological mechanism experimentally obtained but took a form which was mathematically simplified and tractable. Outline of model is described here as details were already described in the previous papers[9].

In the model, the shape of the cell body is represented by a graph: an edge corresponds to a plasmodial tube and a node corresponds to a junction between tubes. The two nodes with food-source are labeled N_1 and N_2 and the other nodes are numbered $N_3, N_4, N_5, \cdots$. An edge between node i and j is labeled M_{ij}, and if there are multiple edges between these nodes, they are labeled $M_{ij}^1, M_{ij}^2, \cdots$.

Suppose that the pressures at nodes i and j are p_i and p_j, respectively, and that the two nodes are connected by a cylinder of length L_{ij} and radius r_{ij}. Assuming a Poiseuille flow, the flux through the tube is

$$Q_{ij} = \frac{\pi r^4 (p_i - p_j)}{8\xi L_{ij}} = \frac{D_{ij}}{L_{ij}}(p_i - p_j), \tag{1}$$

where ξ is the viscosity of the fluid, and

$$D_{ij} = \frac{\pi r^4}{8\xi} \tag{2}$$

is a measure of the conductivity of the tube. Although the tube walls are not rigid and the radius changes over time, the dynamics of tube adaptation are slow enough (10-20 minutes) that the flow can be taken in steady state. The state of the network is described by fluxes, Q_{ij}, and the conductivities, D_{ij}, of the edges.

The amount of fluid must be conserved, so that at each internal node i $(i \neq 1, 2)$

$$\Sigma_j Q_{ij} = 0. \tag{3}$$

The nodes that correspond to FSs drive the flow through the network by changing their volume, so that at FSs eq.3 is modified by a prescribed source (or sink) term

$$\sum_i Q_{ij} = \begin{cases} -Q_0 & \text{for } j = 1, \\ Q_0 & \text{for } j = 2. \end{cases} \tag{4}$$

The total volume of fluid in the network does not change, so that Q_0 is a constant. These source terms could be periodic in time and drive shuttle streaming through the network. However, because the time scale of network adaptation is an order of magnitude longer than the time scale of shuttle streaming, the sources are taken to be constant.

For a given set of D_{ij} and r_{ij} and source and sink, the flux through each of network edges can be computed. In *Physarum* the radii r_{ij} of tube change in response to this flux[9,18]. In the model, the conductivities D_{ij} evolve according to the equation

$$\frac{dD_{ij}}{dt} = f(|Q_{ij}|) - aD_{ij}. \tag{5}$$

The first term on the right hand side describes the expansion of tubes in response to the flux. The function f is a monotonically increasing function which satisfies $f(0) = 0$. The second term represents a constant rate of tube constriction, so that in the absence of flow the tubes will disappear. Each tube interacts with one another because the total amount of fluid in the network must be conserved. If the flux through a tube changes, it affects all the other tubes in the network.

It is instructive to consider the analogy of an electrical circuit. An edge of the network is regarded as a dynamic resistor, with resistance proportional to L_{ij}^{-1} and r_{ij}^4. The shape of the organism is represented as a network of resistors. A flux through an edge is analogous to currents through the resistor, and the source/sink terms at FSs correspond to input of electrical current. The pressure at the node corresponds to voltage in the circuit. If the current through a resistor is large enough, the resistance decreases and the current through it increases. If the current through the resistor is low, the resistance may go to infinitely small, which corresponds to the collapse of the tube.

The model simulation $(f(|Q_{ij}|) = |Q_{ij}|^\mu$, $\mu = 1$, for the sake of simplicity) can reproduce the maze-solving, based on adaptive dynamics of flow. We want to extract an essence of how the model solve the maze. Each pipe changes its thickness in time, depending on the flow through the tube itself only. This means that information used for dynamics in each tube is not global but local. As, in a practical sense of numerical scheme in the model simulation, the model simulation is not parallel processing. But an actual organism can perform a way of parallel computing.

What is to be emphasized on is that the globally optimal path is obtained from a collective motion of similar element of pipe. Here we point out a hidden effect of global interaction among the pipes in the model. Notice that total flow through a whole network is conserved at Q_0. This means that a flow through a pipe is large and the rest of pipe recieves an influence from the flow. Any pipe interacts with all pipes via conserved quantity of total flow. The conserved quantity play a key role for natural computing in living system of *Physarum*.

3.2   *Physarum* Solver: A Biologically Inspired Method for Path-Finding

The function f is taken to be of the form

$$f(|Q_{ij}|) = |Q_{ij}|^{\mu}. \tag{6}$$

The model network exhibits three different behaviors, depending on the value of μ. The results for $\mu = 1$, $\mu > 1$, and $\mu < 1$ are summarized in the previous paper[9]. Here we always assume $\mu = 1$ for the sake of simplicity and tractability in a mathematical sense.

The model of $\mu = 1$ reproduces the maze-solving in *Physarum* [9]. The way of finding the shortest connection path was tested in the much more complex maze[8]. In fact, it is mathematically proved that the model system can always show the shortest path in some cases of maze[19,20,21]. So we named the model *Physarum* solver for path-finding problem[10]. In this section, algorithmic characters and capacity of *Physarum* solver are considered.

3.3   Risk-Minimum Path in an Inhomogeneous Field

Physarum solver is capable of solving the risk-minimum path in inhomogeneous field of risk[11]. The equation 5 implies that the conductivity tends to vanish exponentially according to the second term $-aD_{ij}$, while it is enhanced by the flux along an edge according to the first term $f(|Q_{ij}|) = |Q_{ij}|$ (assumed $\mu = 1$ here). Note that a is a kinetic constant in the process of tube thinning. Here we focus on the effects of parameter a on the model behavior in order to develop an understanding of the effects of inhomogeneity in the experimental system. We next consider the relationship of the process of tube thinning (second term in Eq.(5)) to the Relative level of risk in bright field, which was experimentally estimated by measuring the locomotion speed α_1 and transport rate of body mass α_2 (see detail in [11]). Equation (5) can also be expressed in terms of the thickness (or diameter) of tube R_{ij} in the form $\frac{dR_{ij}}{dT} = -aR_{ij}$, where $D_{ij} = (\pi r_{ij}^4)/(8\xi)$,

$r_{ij} = R_{ij}/2$ and $t = 4T$, if the process of tube-thickening $|Q_{ij}|$ is neglected. The constant a expresses how rapidly the thickness decreases and is related to the experimentally measured α_1 and α_2. In the simulation, the constant a was therefore set to $a_{dark} = 1$ in the dark field and to $a_{bright} = \alpha_1$ or α_2 for the illuminated field.

The final connecting path in the numerical simulation was determined by the ratio a_{bright}/a_{dark}, corresponding to α_1. The simulation results for the final connecting path were robust to the assumption of Poiseuille flow $D \propto R^4$; the qualitative behavior is insensitive to the exponent (4, 3 or 2) and only affects the time constant for convergence to the final path. *Physarum* solver can minimize not only the actual physical length of path but also effective length expressed by a-value in the model although the space is not homogeneous in a-value.

3.4   Effect of Food Amount on Tube Selection between Two FSs

Here we introduced the variations of food amount at $FS_{1,2}$[15]. We assumed that total volume of protoplasm (TV) was divided to two parts: part for nutrient absorption (PA) and part for tube formation and force generation (PT). The next assumption was that PA was larger as food amount was larger so that applied food amount was expressed by PA in the model. Only PT affected the following dynamics. Two FS generated pressure as $P_1 = \beta(w_1 - s_1)$ and $P_2 = \beta(w_2 - s_2)$ where w was volume of protoplasmic sol at FS and s was basal volume of protoplasmic sol (see the paper[17] on the physiological meaning of these equations). And β is stiffness constant for protoplasmic gel. These volume variables were given by $s_1 = \frac{PT}{2}(1 + sin2\pi\omega t)$, $s_2 = \frac{PT}{2}(1 - sin2\pi\omega t)$, $\dot{w}_1 = -\sum_i Q_i$, $\dot{w}_2 = \sum_i Q_i$, where ω was frequency of rhythmic contraction in the organism.

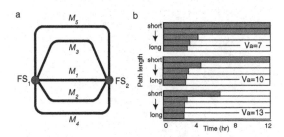

Fig. 6. Simulation results for path selection in the plasmodial tube network. a) graphical representation of initial shape of the organism: five connections, M_i,(i =1 to 5), with the length of 2, 4, 6, 8 and 10 cm, between two equal FSs, FS_1 and FS_2. b) time course of connection survival in the food size (PA) of 5, 10 and 14. Parameter values: $PA(PT) = 5(15), 10(10), 14(6)$(for each simulation), $L_1 = 2.0, L_2 = 4.0, L_3 = 6.0, L_4 = 8.0, L_5 = 10.0, \beta = 1.0, \omega = 36.0$ ($\Rightarrow$ Oscillation Period=100s). Initial Conditions: $D_1(0) = D_2(0) = D_3(0) = D_4(0) = D_5(0) = 6.0, w_1(0) = w_2(0) = PT/2$. Refer to [15].

Figure 6 shows time courses of connection survival in the different food sizes (three different PA values). Clearly, the longer connections collapsed earlier and final number of remaining connections was reduced as food size was larger: two, one and zero connections in the food size (PA) of 5, 10 and 14.

The simple rule for connection selection was found in two equal FSs: longer one collapsed earlier than shorter one and food amount determined the time of ending the selection process. The mathematical model reproduced the experimental observation and told us a possible underlying mechanism: large food reduced volume of flowing sol and it led to tube collapsing.

4   Concluding Remarks: Possibility of *Physarum* Computing

We described the ability of *Physarum* to solve some geometrical puzzles of combinatorial optimization. The potential ability of problem-solving may be used for a new type of problem in the future[8]. One possible problem is dynamic optimization: a problem in which constraints and boundary conditions vary in space and time. A typical example is to find the time-minimum route in car-navigation through complicate road-network, under the traffic congestion varying in space and time[10]. In fact, *Physarum* solver can find a solution. Similarly to solving the risk-minimum path in the inhomogeneous field of risk, the rate constant of thinning process of tube conductance plays a key role. The rate constant should depend on space and time in the car-navigation problem. This application of *Physarum solver* is just beginning and we need studies in the future.

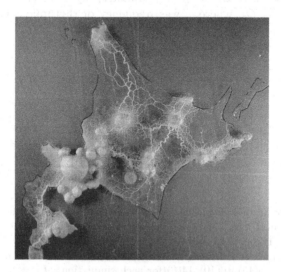

Fig. 7. Networking of some cities in Hokkaido region (a north country of Japan) through *Physarum* tube-network. FSs were placed at the geographical positions of cities. Size of FS was proportional to population of city.

One more possible problem is to solve multi-objective optimization problem[14]. An example was already shown in the formation of multi-functional network (short total-length of network, fault-tolerance of global connectivity against an accidental connection, efficiency of short connection distance) in the case of many FSs. Such multi-functionality can be requested for public transportation network. A real network may be evaluated in comparison to *Physarum* solution. Figure 7 shows a preliminary result of network in Hokkaido region, obtained by a real *Physarum*.

As we described so far in this article, the amoeba has high ability of solving a geometrical puzzle. The other aspects of smartness are also interesting. The following has been reported. *Physarum* can anticipate periodic environmental events and can show *indecisive* behavior when encountering a weak poison [22,23]. Although behavioral smartness in brainless organisms like unicellular organism and green plants have been reported since one-hundred years ago[1,24,25], studies on information processing in a simple primitive organism is still an exciting frontier in modern science. The simple organism may help us to understand a secret of biological information processing.

Acknowledgements

This work was supported MEXT KAKENHI No.18650054 and No. 20300105, Human Frontier Science Program Grant RGP51/2007.

References

1. Bray, D.: Wetware. Oxford University Press, Oxford (2009)
2. Nakagaki, T., Yamada, H., Tóth, Á.: Maze-solving by an amoeboid organism. Nature 407, 470 (2000)
3. Nakagaki, T., Yamada, H., Tóth, Á.: Path finding by tube morphogenesis in an amoeboid organism. Biophys. Chem. 92, 47–52 (2001)
4. Nakagaki, T., Yamada, H., Hara, M.: Smart network solutions in an amoeboid organism. Biophys. Chem. 107, 1–5 (2004)
5. Nakagaki, T., Kobayashi, R., Ueda, T., Nishiura, Y.: Obtaining multiple separate food sources: behavioral intelligence in the *Physarum* plasmodium. Proc. R. Soc. Lond. B 271, 2305–2310 (2004)
6. Nakagaki, T.: Smart behavior of true slime mold in labyrinth. Res. Microbiol. 152, 767–770 (2001)
7. Nakagaki, T., Guy, R.: Intelligent behaviors of amoeboid movement based on complex dynamics of soft matter. Soft Matter 4, 1–12 (2008)
8. Nakagaki, T., Tero, A., Kobayashi, R., Onishi, I., Miyaji, T.: Computational ability of cells based on cell dynamics and adaptability. New Generation Computing 27, 57–81 (2008)
9. Tero, A., Kobayashi, R., Nakagaki, T.: Mathematical model for adaptive transport network in path finding by true slime mold. J. Theor. Biol. 244, 553–564 (2007)
10. Tero, A., Kobayashi, R., Nakagaki, T.: *Physarum* solver -A biologically inspired method for road-network navigation. Physica A363, 115 (2006)

11. Nakagaki, T., Iima, M., Ueda, T., Nishiura, Y., Saigusa, T., Tero, A., Kobayashi, R., Showalter, K.: Minimum-risk path finding by an adaptive amoebal network. Phys. Rev. Lett. 99, 068104 (2007)
12. Tero, A., Yumiki, K., Kobayashi, R., Saigusa, T., Nakagaki, T.: Flow-network adaptation in *Physarum amoebae*. Theory in Biosciences 127, 89–94 (2008)
13. Tero, A., Nakagaki, T., Toyabe, K., Yumiki, K., Kobayashi, R.: A method inspired by *Physarum* for solving the Steiner problem. International Journal of Unconventional Computing (2009) (in press)
14. Tero, A., Takagi, T., Saigusa, T., Ito, K., Bebber, D.P., Fricker, M.D., Yumiki, Y., Kobayashi, R., Nakagaki, T.: Rules for biologically-inspired adaptive network design (submitted)
15. Nakagaki, T., Saigusa, T., Tero, A., Kobayashi, R.: Effects of food amount on path selection in transport network of an amoeboid organism. Topological Aspects of Critical Systems and Networks, 94–100 (2007)
16. Nakagaki, T., Yamada, H., Ueda, T.: Interaction between cell shape and contraction pattern. Biophys. Chem. 84, 195–204 (2000)
17. Kobayashi, R., Tero, A., Nakagaki, T.: Mathematical model for rhythmic amoeboid movement in the true slime mold. Journal of Mathematical Biology 53, 273–286 (2006)
18. Nakagaki, T., Yamada, H., Ueda, T.: Interaction between cell shape and contraction pattern. Biophys. Chem. 84, 195–204 (2000)
19. Miyaji, T., Ohnishi, I.: Mathematical analysis to an adaptive network of the Plasmodium system. Hokkaido Mathematical Journal 36, 445–465 (2007)
20. Miyaji, T., Ohnishi, I.: Physarum can solve the shortest path decision problem mathematically rigorously. International Journal of Pure and Applied Mathematics (in press)
21. Miyaji, T., Ohnishi, I., Tero, A., Nakagaki, T.: Failure to the shortest path decision of an adaptive transport network with double edges in Plasmodium system. International Journal of Dynamical Systems and Differential Equations 1, 210–219 (2008)
22. Saigusa, T., Tero, A., Nakagaki, T., Kuramoto, Y.: Amoebae anticipate periodic events. Physical Review Letters 100, 018101 (2008)
23. Takagi, S., Nishiura, Y., Nakagaki, T., Ueda, T., Ueda, K.: Indecisive behavior of amoeba crossing an environmental barrier. In: Proceedings of Int. Symp. on Topological Aspects of Critical Systems and Networks, pp. 86–93. World Scientific Publishing Co., Singapore (2007)
24. Trewavas, A.: Green plants as intelligent organisms. Trends Plant Sci. 10, 413–419 (2005)
25. Trewavas, A.: Aspects of plant intelligence. Annals Bot. 92, 1–20 (2003)

Two Molecular Information Processing Systems Based on Catalytic Nucleic Acids

Milan Stojanovic

Columbia University, USA

1 Introduction

Mixtures of molecules are capable of powerful information processing [1]. This statement is in the following way self-evident: it is a hierarchically organized complex mixture of molecules that is formulating it to other similarly organized mixtures of molecules. By making such a statement I am not endorsing the extreme forms of reductionism; rather, I am making what I think is a small first step towards harnessing information processing prowess of molecules and, hopefully, overcoming some limitations of more traditional computing paradigms. There are different ideas on how to understand and use molecular information processing abilities and I will list some below. My list is far from inclusive, and delineations are far from clear-cut; whenever available, I will provide examples from our research efforts. I should stress, for a computer science audience that I am a chemist. Thus, my approach may have much different focus and mathematical rigor, then if it would be taken by a computer scientist.

I will start by mentioning the biomimetic approach that we have not pursued in our published research. Here, one looks at what molecules do to process information in Nature, and one tries to organize other (or same) molecules into artificial systems. This approach may help us understand information processing in Nature, may facilitate the design process in synthetic biology, and may lead to new molecular devices useful when only rudimentary biocompatible computing is needed. The most important consequences of this research may come if it would be in the future tightly coupled with systems biology; the end result may be formalization and mathematical understanding of the biological processes. But, from the perspective of what devices can actually be implemented, the approach usually neglects that the most powerful and non-trivial aspects of information bioprocessing are result of multiple hierarchical transitions. It seems then unlikely that any approach based on molecules will be able to compete with cell-based computing paradigm (e.g., artificial neuronal networks) when it comes to overcoming the most serious limitations of high-power silicon-based computing.

2 Systems Based on Deoxyribozyme-Based Logic Gates

As an alternative, one can look at some well-known (and usually elementary) computing models, design new or find old molecules that behave in a way that is somewhat analogous to elementary components used in these models, and try to build more complex structures from such molecular components. In the past, this approach was the most successful [2,3,4] and I will use our own molecular logic gates and automata [4]

F. Peper et al. (Eds.): IWNC 2009, PICT 2, pp. 55–63, 2010.
© Springer 2010

to illustrate it. The strength of such an approach is that once we recognize basic building blocks in our molecules (or a group of molecules) the route forward is laid for us by the well-developed theory. However, the approach may miss the key strengths of molecules; instead of asking "What are molecules intrinsically good at?", we impose behaviors on molecules that are not natural to them.

Let us have a set of enzymes that can take allosteric effectors as inputs. We can write a truth table (look-up table) describing presence or absence of these allosteric effectors and their effect on an output (enzymatic activity). Thus, we can say that this enzyme is processing information presented to it as a presence or absence of a set of effectors in a way that is similar (but not identical) to logic gates that are basis of digital computing. This allows us to use digital computing paradigm to organize molecules in mixtures performing complex logical (Boolean) calculations. I will now illustrate some complex molecular behaviors obtained by mixing logic gates based on nucleic acid enzymes [5]. These gates use oligonucleotides as inputs and outputs.

In order to achieve a logic gate-like behavior in deoxyribozymes we use modular design [6] and combine two modules: (1) a catalytic module [6] based on DNA-based nucleic acid catalysts cleaving other oligonucleotides (cleavases or phosphodiesterases, Figure 1a); and (2) a recognition module, a stem-loop oligonucleotide inspired by molecular beacons [7] (Figure 1b). We often take advantage of fluorogenic labeling of substrates, in order to monitor reactions of these cleavases. If substrate is labeled with a 5' fluorophore and a 3' quencher (Figure 1a), after its cleavage, the fluorophore is separated from the quencher and increase in fluorescence signal is observed.

The stem-loop will establish allosteric control over the activity of the deoxyribozyme, as long as the conformational changes in the stem-loop region influence the catalytic activity of the deoxyribozymes. For example, if we use a stem-loop region to block access of the deoxyribozyme substrate to the substrate recognition region (Figure 1c) we obtain "catalytic molecular beacon" [8] (a detector or a sensor gate, sometimes called by chemists a YES gate), which uses positive allosteric regulation by an oligonucleotide to sense inputs. Upon binding of the complementary oligonucleotide to the loop, the stem will open, allowing substrate to bind and the catalytic reaction to proceed.

Introduction of a negative allosteric regulation leads to a NOT gate-like behavior [9] Figure 2a): a stem-loop region is embedded within so called catalytic core of deoxyribozymes, and the binding of a complementary oligonucleotide distorts this core, turning off enzymatic activity. Multiple-input gates are constructed by the addition of two or three controlling modules to one deoxyribozyme. For instance, placement of two stem-loop controlling regions on the opposite ends of the substrate recognition region will create an AND gate (Figure 2b). This gate requires the presence of both inputs for enzymatic activity.

Combination of sensor gate and NOT gate gives an ANDNOT gate (not shown) is active in the presence of one input for as long as a second input is not present. Furthermore, if a NOT gate-like element is added to an AND gate one obtains a three input ANDANDNOT gate (in Figure 12c [4]). The action of input oligonucleotides can be reversed by adding their complements, and we took advantage of this property to turn ANDANDNOT gate into ANDAND and ANDNOTNOT gates [10].

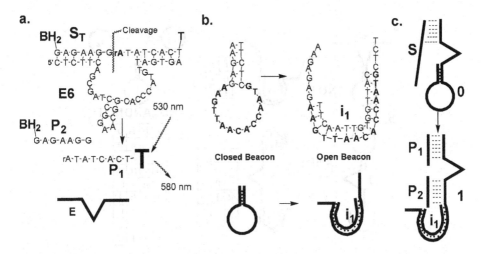

Fig. 1. a. E6 deoxyribozyme shown here in the complex with its substrate. Substrate is labeled for fluorogenic cleavage with a fluorophore (e.g., TAMRA, or T) and a quencher (e.g., Black Hole 2, or BH2). Upon cleavage there is an increase in fluorescence of TAMRA.   **b.** Molecular beacon stem-loop: Closed beacon has a stem-loop conformation, but the addition of complementary input i_1 opens the stem.   **c.** Catalytic molecular beacon or YES or sensor gate is constructed by attaching a beacon module to one of the substrate-recognition regions of the deoxyribozyme module. Upon the addition of an input (i_1), the gate is turned into its active form. The reaction can be monitored fluorogenically, and the graph in the insert shows increase in fluorescence with time upon addition of increasing amount of input.

Once we have basic molecular logic gates, we can arrange them together in solution by simply mixing increasingly large numbers of them around the same substrates in what is called implicit OR arrangement (implicit OR — if any of the gates that cleave the same substrate is active, the output would be 1). Our ability to obtain increasingly complex logic is limited by our ability to combine only up to three inputs in conjunctions in a single logic gate — for more we need cascades that are not the topic of the current review. But what do we use to demonstrate this newly obtained power? If we simply put large number of arbitrary gates in solution we would be open to criticism that we selected their logic for convenience, and that some other logic might not work. Thus, we must have an unbiased test of the complexity that can be achieved by our information processing medium and for this purpose we have chosen to study game-playing by molecular automata (the other approach, not discussed here, being some standard circuits from electrical engineering).

We can define molecular automata in several ways, the most general being as sets of molecules that change states according to a series of inputs and a set of rules that determine the changes between each state. In a molecular world, we approached the definition of automata a bit more subjectively, as to differentiate them from ordinary circuits: Molecular automata are circuits that leave an observer with an impression of a two way-communication when exposed to a series of consecutive inputs. Essentially, we translate our own challenges to automata in a "language" molecules can accept,

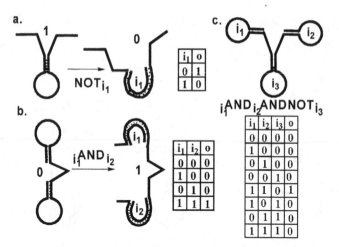

Fig. 2. A set of deoxyribozyme-based logic gates: **a.** NOT gate is constructed when a stem-loop is added to a catalytic core. Opening of the stem will distort the catalytic core and inhibit the reaction.   **b.** AND gate is contructed when two stem-loops are added to both substrate recognition regions. Both stems have to open for a substrate to be cleaved, and both inputs have to be present. **d.** A three input gate (ANDANDNOT), which is active when two inputs (i_1 and i_2) are present, but not the third one (i_3), is a combination of an AND gate and a NOT gate.

which are in our case input oligonucleotides, and we obtain a response interpretable as a game-playing over adding a sequence of inputs.

We have focused on the construction of molecular automata that play simple games of perfect information, such as tic-tac-toe. There is an element of tradition in our choice, because it was one of the first games played by computers, almost fifty years ago. Our first automaton, MAYA (from Molecular Array of YES and AND gates, now renamed MAYA-I, Figure 3a) [4], played a restricted (symmetry-pruned) game of tic-tac-toe. Our second automaton, MAYA-II [4], played an unrestricted game, but required different encoding of inputs.

Games are played in 3x3 wells of 384 well plates (mimicking the nine squares of a tic-tac-toe game board), with individual wells sequentially numbered with #1-#9. In MAYA (shown in Figure 4) the automaton always goes first in the middle well, and the human player is allowed to respond in only one of the corner squares or side squares (Figure 3A). This simplification reduces our engineering effort to programming the automata to play a total of 19 legal games (out of 76 possible, after the automaton's first move), 18 of which end in the automaton winning, because human is not playing perfectly, and one ending in draw, if both players play perfectly.

The automaton is turned on by adding Mg^{2+} to all wells, activating a constitutively active deoxyribozyme in the middle well (#5). Human intentions are communicated to automaton through oligonucleotide inputs, which have to be added to all the wells. These inputs trigger response from gates, and the automaton's response is observed by monitoring a fluorescence increase in a particular well. In MAYA, eight inputs are used, keyed to the human move into wells #1-#4 and #6-#9 (#5 is already used by the

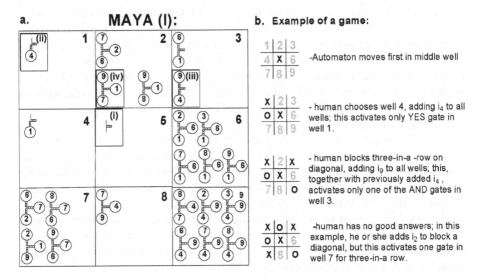

Fig. 3. MAYA-I, automaton that plays symmetry-pruned tic-tac-toe game: **a.** Distribution of gates in wells. Central well (#5) contains constitutively active deoxyribozyme, while the other wells contain logic gates. Gates used in our example game are boxed.   **b.** An example of a game, in which human does not play perfectly and looses. There are total of 19 games encoded in this distribution of logic gates.

automaton in the opening move). In order to move into well 1 the human player will add input i_1, to move into well 2 the human player will add input i_2, and so on. An array of 23 logic gates distributed in these eight wells calculates a response to the human player's input causing only one well to display an increase in fluorescence indicating the automaton's chosen move (of note, there is usually some non-digital behavior in other wells, sometimes due to conformational changes, sometimes due to imperfect inhibition of enzymes). The cycle of human player input followed by automaton response continues until there is a draw or a victory for the automaton. The perfect game playing ability of our initial MAYA was demonstrated in over 100 test games.

After the successful demonstration of MAYA, we decided to build a larger tic-tac-toe playing automaton, MAYA-II, in order to test the limits of integration of individual logic gates. While MAYA-II always played first in the middle well, the human player was free to choose the first move in the any of the remaining wells. Thus, the automaton plays all 76 permissible games. MAYA-II uses an array of 96 logic gates based on E6 deoxyribozymes. Translating a human move for MAYA-II required an increase from 8 to 32 input sequences, because these new inputs encoded both the well position and the order of the human move (their first, second, third, or fourth move into wells 1-8). The automaton is also "user-friendly", with a two-color fluorogenic output system displaying both human and automaton moves. Human moves are displayed in a green channel (fluorescein) and automaton moves are displayed in a red channel (TAMRA), similar to our adders. Human moves are displayed through an action of 32 YES gates based on the 8-17 deoxyribozymes.

In all approaches to molecular computing, it is inevitable that discrepancies between idealized and desirable molecular behaviors and real molecules catch up, leading to the increasingly difficult struggle to contain device failures (in our case cross-interactions between elements) as we scale up and target non-trivial behaviors. With MAYA-II this struggle was significantly more difficult than with MAYA-I; we observed multiple examples of unpredictable failures that required us to individually address emerging issues, change gates and inputs, or optimize concentrations, before all legal games were demonstrated successfully.

This also points to the possibility that, instead of just mixing more and more molecules in solution, a hybrid approach, introducing concepts from biology (e.g., using compartmentalization), may lead to the more successful information processing architectures. Nevertheless, and without trying to discourage anyone, in the absence of multiple truly breakthrough ideas the task of accomplishing something that silicon-based computers cannot do much better seems to me daunting; as with biomimetic approach it is most likely that practical applications will be limited to the situation in which required computing power is negligible and some form of biocompatibility is needed (e.g., autonomous therapeutic devices).

3   Spider-Based System

Aside from biomimetic and "silicomimetic" approaches, we can also recognize or design in some molecule an efficient form of information processing to begin with, and then try to build more complex behaviors taking advantage of what the molecule is actually intrinsically good at. Indeed, many biological molecules (e.g., a ribosome or any of polymerases) are proficient in information processing, but they are evolutionary locked into the larger context of cellular operations, and the scale-up problem may be similar as discussed above. It seems advantageous, therefore, to take a molecule that has not been highly contextualized by the evolution, and that can be redesigned per will, without loosing much of its information processing power. Of course, if this molecule or some of its variations, readily fits into an existing computing paradigm, that would allow rapid development of systems with complex behaviors. The cumulative nature of scientific progress ensures that even if we hit some barrier with one system, the next generation approach (by us or others) will take into account this limitation from the very beginning, and the progress toward more complex systems will be ensured. As an example of such approach we discuss now our initial results with spider systems (NICK from Nanoassembly Incorporating Catalytic Kinesis, [11]).

We start by looking at a single constitutively active deoxyribozyme (Figure 1a). This enzyme (let's call it a leg for the reasons that will next become obvious) recognizes all other even partially complementary oligonucleotides, but we are mostly interested in interactions with substrates and products deposited on some surface. When leg binds to the product on the surfaces it dissociates at some rate. When leg binds to substrate it dissociates more slowly than from products and mostly only after cleaving it. This information processing reaction can be described as a random walk, but of a particular kind — a random walk with a memory; once visited sites are recognized through the change of behavior of our hopping leg (residence time). The process is difficult to control

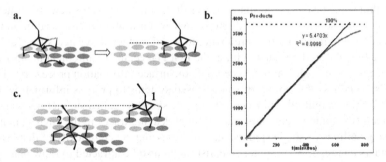

Fig. 4. a. A spider with four deoxyribozyme legs interacts with surface covered by substrates (dark gray ellipses) and products (light gray ellipses). As a result, it moves towards new substrates, instead of "eating" itself up in the field of products. **b.** Release profile (# of substrates vs. time) of products by four legged spider from the matrix displaying substrates [11]. The measurement was performed by surface-plasmon resonance. One can observe nearly linear release for up to 400 minutes (at 1:3800 ratio of spiders to substrates). **c.** One example of two spiders communicating through the surface (cf. stigmergic interaction). Spider **1** passed first, directing spider **2** to take a turn preferentially towards the only direction in which there are substrates left.

spatially, because there is nothing to stop the leg from moving away from the surface, and randomly binding anywhere else. There is an interesting consequence of this property, if we put together multiple legs together on an inert body in a spider (Figure 4a). One might think that such a multivalent species will quickly capture itself in the field of grazed products, moving very slowly towards new substrates, at the rate that is limited by the rate of diffusion on the product surface. However, as pointed out by Krapivsky and Antal in their paper on "molecular spiders with memory" [12], such spider would move rapidly toward new substrates or would follow the line of substrates, if exposed to them, due to an interface effect. Namely, a multivalent species, with some legs on substrates, and some legs on products, would shift towards substrates preferentially, because legs on products would dissociate more readily than legs on substrates.

The proposed behavior, moving towards new substrates in multivalent species, was indeed observed in bulk (Figure 4b). This Figure shows release of products, that is, substrate cleavage from the matrix displaying substrates by a four-legged spider (four deoxyribozymes coupled to streptavidin). The release is nearly constant over prolonged periods of time, at the rate that is between 3-4 times of the rate at which one leg cleaves substrates in solution. This indicates that the spider is at 100, 200, or 300 minutes in very similar environments, in the presence of an excess of substrates. Yet, substrates are attached to the surface, so it must be the spider that is replenishing them by moving constantly in their direction. Importantly, because of the multiple legs, spiders do not leave surface (matrix) easily, and we estimated that they can undergo several hundreds of cleavages, before leaving the surface (jumping). Similar results were obtained more recently on 2-D surfaces on which spider was observed using single molecule imaging techniques (unpublished, with several collaborators).

So, let's recapitulate what has happened here: a molecule processes information displayed on the surface by "touching" elements distributed over the surface and differentially interacting with them; as a result of this process and imposed by the body keeping the legs within some distance, the spider moves directionally (towards new substrates), processively (over several hundreds of steps, that is, on a multimicron scale), and it "remembers" where it was before, preferring to avoid previously visited sites. Reformulated like this, the spider seems to fit, at the first glance, the requirement mentioned above: it is a molecule efficient in "some kind" of surface information processing. The next goal is to harness this ability of spider-based systems to process information, by coupling activities of multiple spiders (cf. Figure 4c shows so called stigmergic interaction, in which two spiders "communicate" through leaving traces in the environment), by increasing abilities of single spiders, and by increasing the complexity of displayed signals spiders can read. A computer scientist might also be attracted to draw analogy between this process and a Turing machine. Yet, to a chemist like me, the question of a dramatically decreasing yield in linearly coupled multi-stepped processes indicates the need to find an alternative approach, perhaps one based on some type of convergent or hierarchical organization of groups of spiders. Such studies are now in progress with our multidisciplinary group of collaborators.

4   Conclusions

In this brief review, I focused on two models of information processing, I personally find very interesting, and that we pursued in some details. The first, based on deoxyribozyme-logic gates, is well advanced and resulted in some of the most complex in vitro molecular computing devices to date. The second, based on a molecule that moves over surface, changing it, has yet to be harnessed in devices that will perform recognizable tasks (beyond just moving). Hopefully, my account will act as a catalyst for others to look at widely different chemical systems from an information processing perspective.

References

1. Credi, A.: Molecules That Make Decisions. Angew Chem. Int. Ed. Engl. 46(29), 5472–5475 (2007)
2. Rothemund, P.W.K., Papadakis, N., Winfree, E.: Algorithmic Self-assembly on DNA Sierpinski Triangles. PLoS Biology 2, e424 (2004)
3. Benenson, Y., Gil, B., Ben-Dor, U., Adar, R., Shapiro, E.: An Autonomous Molecular Computer for Logical Control of Gene Expression. Nature 429, 423–429 (2004)
4. Macdonald, J., Stefanovic, D., Stojanovic, M.N.: DNA computing for Play and Work. Scientific American (November 2008)
5. Breaker, R.R., Joyce, G.F.: A DNA enzyme with Mg(2+)-dependent RNA phosphoesterase activity. Chem. Biol. 2(10), 655–660 (1995)
6. Breaker, R.R.: Engineered allosteric ribozymes as biosensor components. Current Opinions in Biotechnology 13, 31–39 (2002)
7. Tyagi, S., Krammer, F.R.: Molecular beacons: Probes that fluoresce upon hybridization. Nature Biotechnoogy 14, 303–309 (1996)
8. Stojanovic, M.N., de Prada, P., Landry, D.W.: Catalytic Molecular Beacons. Chem. BioChem. 2(6), 411–415 (2001)
9. Stojanovic, M.N., Mitchell, T.E., et al.: Deoxyribozyme-based logic gates. J. Am. Chem. Soc. 124(14), 3555–3561 (2002)

10. Lederman, H.S., Macdonald, J.J., Stefanovic, D., Stojanovic, M.N.: Deoxyribozyme-Based Three-Input Logic Gates and Construction of a Molecular Full Adder. Biochemistry 45(4), 1194–1199 (2006)

11. Pei, R., Taylor, S., Rudchenko, S., Stefanovic, D., Mitchell, T.E., Stojanovic, M.N.: Behavior of polycatalytic nanoassemblies on substrate-displaying matrices. J. Am. Chem. Soc. 128, 12693–12697 (2006)

12. Antal, T., Krapivsky, P.L.: Molecular spiders with memory. Phys. Rev. E 76, 021121 (2007)

The Effect of Community on Distributed Bio-inspired Service Composition

Raymond Carroll, Sasitharan Balasubramaniam,
Dmitri Botvich, and William Donnelly

TSSG, Waterford Institute of Technology,
Waterford, Ireland
{rcarroll,sasib,dbotvich,wdonnelly}@tssg.org

Abstract. The Future Internet is expected to cater for both a larger number and variety of services, which in turn will make basic tasks such as service lifecycle management increasingly important and difficult. At the same time, the ability for users to efficiently discover and compose these services will become a key factor for service providers to differentiate themselves in a competitive market. In previous work, we examined the effect adding biological mechanisms to services had on service management and discovery. In this paper we examine the effects of community on services, specifically in terms of composing services in a distributed fashion. By introducing aspects of community we aim to demonstrate that services can further improve their sustainability and indeed their efficiency.

1 Introduction

Adding biological mechanisms such as migration, replication, death and gradient emission to services has proved beneficial [1] to service management and discovery. In this paper our search becomes more complex, as now we are attempting to form composed services in a distributed fashion. Distributed service composition relies heavily on service discovery, as each composition would need to perform a large number of searches. In a service network with a large number of nodes (e.g. Data centres) this search can become inefficient, even with a gradient-based approach, as the number of messages required becomes large. Also, as distributed composition is so complex, limiting the set of services available for selection to those within a restricted proximity can lead to reduced composition completion.

With this in mind we adapt ideas from social communities to reduce the search overhead and improve the overall success of composition. More successful composition leads to higher service utilisation and improved sustainability of services (since services gain extra revenue from fulfilling requests). The benefit of community is that it provides a common pool of information and resources that members can utilise. One application of community is in learning the capabilities and relationships of its members by monitoring their interactions. By monitoring the interactions of services within the network, we aim to provide more directed/focussed service recommendations when attempting service composition. Although numerous research works have addressed service composition [2,3,4,5], including bio-inspired approaches [6,7,8], our solution focuses on autonomous behaviours of these services to support dynamic environments

F. Peper et al. (Eds.): IWNC 2009, PICT 2, pp. 64–71, 2010.

(e.g. users changing requirements, service functionalities changes). We believe that applying biologically inspired approaches allows services to adapt and change, mimicking the way living organisms are able to collectively and individually survive through harsh environments.

This paper is organised as follows. Section 2 gives an overview of our biologically inspired service management solution. Section 3 then describes the community concepts we applied to services. In section 4 we discuss the simulations carried out and their results. Finally, section 5 presents our conclusions.

2   Bio-inspired Service Management

In this section we will briefly outline the basic elements of our biologically inspired service management solution, where more details can be found in [1]. Services (or Service Agents, note we will use the terms *agent* or *service* interchangeably) are augmented with biological behaviour including replication, migration, death and gradient emission. Services provide energy to the nodes on which they run for utilising resources (e.g. cpu) and receive energy from users in return for serving requests. If an agent is not being utilised it runs out of energy and dies, hence removing redundant services from the environment. When service demand is high, agents can replicate in order to meet the increased demand. Also, when the load on a given node becomes too high agents can migrate to other, less loaded nodes.

Another feature of services is gradient emission. Agents create a gradient field by *diffusing* service advertisements to nodes in close proximity, which attracts requests towards the advertising agent (see Fig. 1a for illustration of gradient field formed by SA_1). When a service is utilised it gains energy, and uses some of this energy to maintain and extend the gradient field. The more popular a service is, the larger its gradient field and the more visible it is within the network. If a service agent is searching for another service to compose with, the gradient mechanism floods the network with search messages looking for the gradient of the required agent. By looking for the gradient, as opposed to the agent itself, the search is more efficient.

2.1   Distributed Service Composition

In our previous work the creation of the composition plan was done in a centralised manner and then passed to the service network to discover specific services. In this paper however, we adopt a distributed service composition approach. Attempting service composition in a distributed fashion makes finding the optimal composition even more difficult, as we do not have complete knowledge of the services in our network. In this paper we provide only basic optimisation of our composition plan, with more sophisticated optimisation to be addressed in future work.

Most service composition approaches view a service in terms of its input and output parameters, and/or state information such as pre- and post-conditions (e.g. [4,5]). For example, a travel booking composed service might take a date as an input and return a flight and hotel reservation as the output, where the actual service was provided by two individual services. For the sake or our simulations we model services and service

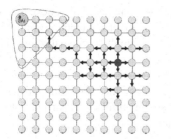

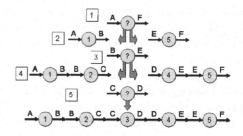

Fig. 1a. Non-Community Structure   **Fig. 1b.** Service Composition Process

requests only in terms of input and output parameters. Service requests describe the basic input provided and output required from the service, and from this we attempt to determine a service chain that provides the required functionality. Fig. 1b depicts the iterative composition search process.

The service request arrives, describing the required service in terms of its input and output (line 1). If no atomic service is found then we try to compose a service to fulfil the request. We do this by searching for two separate services which provide the required input and the required output. Note, however, that it is not mandatory that both the input and output service are found, the process will continue if just one of these is discovered. When these services are found (line 2, service 1 and 5) we examine if they can be joined directly, if not then our next iteration must search for a link service to join the two (line 3). If no atomic link service is found then we again break the service into separate input and output parts and search for these individually (line 4). This process continues until a suitable link service is found (line 5) and both ends of the chain can be joined, thus forming a composed service.

3   Community

The overall goal of the community is to increase the survivability of individual services by increasing their ability to serve requests. Just as living organism communities serve to enhance the lives of their members, service communities support service members by providing a source of group knowledge that every member can utilise. In effect, when trying to satisfy some request, a community provides a facility where community members can get advice about how best to satisfy a request. The more communities a service agent is part of, the more sources of information the agent has at its disposal.

In society more than one type of community exists, and these are typically defined by what the community's goal or purpose is. While in future work we will address multiple community types, for this paper we limit our solution to just one community type. This we refer to as the interaction community. In this community services that interact regularly to fulfil tasks form communities. As an analogy, imagine a work place where a large number of employees, each with different expertise, interact to carry out some task. Finding and determining who has the required expertise for the task takes time, so remembering member interactions and capabilities for this task might lead to quicker job completion in the future. We can apply the same logic to services, where services that constitute a composed service remember their interactions, thereby forming an interaction community. This information can be queried later to determine suitable services for a given service request. An underlying assumption for this behaviour is that some services will be requested frequently, so as the same (or similar) requests re-occur, recording the interactions becomes beneficial.

3.1 Community Structure and Formation

The structure of a community can take many forms. These can vary from hierarchical communities with many levels, to completely flat, peer-based communities [9] (Fig. 2a). These different structures determine how knowledge is distributed throughout the community and ultimately how a search of the community is performed. If the community is completely flat (no hierarchy) then community knowledge is distributed among the services themselves, with each individual service maintaining its own view of the community. Therefore, querying the community involves querying individual services. As such, searching for a composed service would require querying individual services in a hop by hop fashion, following a chain of recommended interactions through many nodes without a broader view of the overall process. Also, since a service can only see other services within its gradient field or within a searchable distance, then the services community is also limited to this scope. This problem could be alleviated by maintaining a high gradient value to somewhat broaden the community perspective, but higher gradients require higher energy to maintain the gradient field.

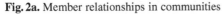
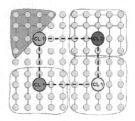

Fig. 2a. Member relationships in communities   **Fig. 2b.** Community Structure

If a hierarchy is employed, the *community leader* (CL) retains a broader view of the community interactions and so is in a better position to make recommendations. Each agent records its own interactions and informs the CL of these. By maintaining a broader view of the community the CL can attempt to form composed services (or sections of a composed service) itself, without having to query multiple agents. In our approach we have adopted a 2-tiered hierarchy, where each community member has a supervising CL. We create multiple CLs within the network in order to increase redundancy, where each is aware of the other community leaders within the network. The community structure employed in this work can be seen in Fig. 2b.

Since the community is based on service interactions, before community information can be utilised the community must be formed. If a community does not exist when a service request is received then the node attempts to form the composed service itself, using the normal gradient-based approach (Fig. 1a). Every time an agent interacts with another agent, as part of a composed service, this is recorded by each agent. Over time, if this interaction is used repeatedly and a certain threshold is met, the agents inform the CL of this interaction. The aim of informing the CL is to build up a more 'complete' view of all the interactions taking place within the community, thereby improving its ability to make recommendations. Once the community has been established, a service request that enters the network at a given node is forwarded to the nearest CL. If this community leader cannot satisfy the request itself, then it broadcasts the request to the other leaders in the network.

4   Simulation and Results

Our simulations evaluate two basic scenarios, employing the community and non-community based algorithms. In the non-community scenario, requests arrive at a node who attempts to compose the service by searching its list of known agents. If this is unsuccessful, it then employs the gradient-based search to look further into the service network. In the community-based model, when a request arrives, it is immediately sent to the nearest CL. In our simulation CLs are pre-selected. Future research will look at models for selecting these leaders through more 'natural' processes, where agents emerge as leaders [9]. The basic parameters used for the simulation are described in Table 1.

Table 1. Parameter Table

Parameter	Values
Arrival Rate (per sec)	15
Number of Nodes	100
IO Permutations	9-36
Community Threshold	5-20
Agents Density (per node)	1-5
Max CS Size	3-6
Gradient Size	3-5
Gradient Search TTL	5s

For our simulation each service has only one input and one output. The number of possible types available for input or output is variable and range from 9–36. Hence, the number of possible unique service combinations range from 72–1260. There are a number of reasons that requests may fail. For example, if the search fails to find any part of a candidate link service then the search fails. If the composed service grows to a greater length than the defined maximum length (*csMax*), this also leads to failure. Our search technique also employs a mechanism to avoid loops found in the service composition.

In Fig. 3a we show the success rate of forming service compositions for the community scenario versus non-community. For each scenario we show a number of simulations, each at a different gradient size. What we can see is that the community-based algorithm greatly out-performs the non-community. As the gradient varies we see minor variations in the completion rate for each scenario. In the non-community scenario, this is because as the gradient size drops we compensate by performing more gradient searches. Hence, as we see in Fig. 3b, the number of messages increases dramatically. In the community scenario the completion rate starts low, but steadily increases. This is a result of the time taken to populate the community and is also reflected in Fig. 3b, where the messaging rate for community is initially high since initial population utilises gradient-based search. Once this is complete the message rate drops-off dramatically. For the community case, gradient size does not greatly affect the search since search is primarily done via the CL. The varying gradient only has an effect when community is unable to recommend a service and so a gradient search is required.

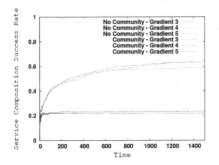
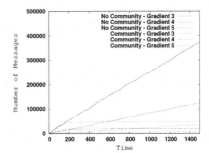

Fig. 3a. Community v Non-Community Composition Completion at varying gradients

Fig. 3b. Community v Non-Community Message Volume at varying gradients

In Fig. 3b we also see that the gradient size does affect the number of messages greatly, specifically in the non-community simulation. As the gradient size drops, more gradient searches are performed which greatly increases the number of messages used. For the community simulation, since it primarily relies on the CL, the number of messages used is much lower. At the same time, the gradient size only has a minor affect when gradient searches are performed in the community simulation.

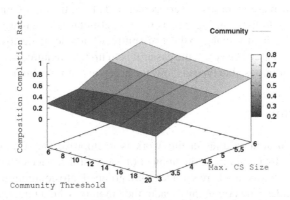

Fig. 4. Composition Completion v Community Threshold v Max. Composed Service Size

In Fig. 4 we compare the effect of the community threshold and the maximum composed service size parameters on the service completion rate using the community algorithm. As we can see from the graph, as the maximum composed service size is increased the completion rate increases dramatically. Increasing the size of the composed service from 3 to 6 increases the completion rate from approx. 25% to 70% (depending on the community threshold value). With regards to the community threshold we can see that altering this from 5 to 10 only has a mild effect on the completion rate. In general, the completion rate is higher when the threshold value is lower. This is as expected since a lower barrier to entering the community will lead to a community of higher population and hence more recommendations.

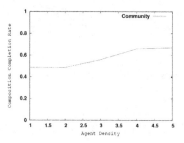 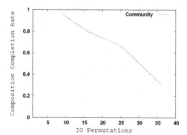

Fig. 5a. Composition Completion v Agent Density

Fig. 5b. Composition Completion v IO Parameters

In Fig. 5a and 5b we examine how the composed service completion rate is affected in the community simulation by varying the agent density and IO parameters. For both these simulations the other parameters remain stable with the following values: $csMax = 6; CommunityThreshold = 15; GradientSize = 5, AgentDensity = 5$. The agent density is measured by the number of agents on each node. Fig. 5a shows that low agent density results in a low completion. This is due to less services (and hence service variation) being available for selection/discovery. As shown in the graph, as the density increase from just 1 service per node to 5, the completion rate increase from approx 48% to 67%. In Fig. 5b, as the IO permutations increase the completion rate decreases. Again this relates to the availability of suitable agents to match a request. The less variance in service types, the easier it is to find the required service. Therefore, when the number of possible types is low, the acceptance rate is high, and vice versa.

5   Conclusion

The main aim of communities in this work is to improve the sustainability of bio-inspired services by maximising the number of successful service compositions. To do this we proposed a system where services were grouped into communities and assigned a community leader. The community leader then monitors the interactions taking place between services and records those that occur repeatedly. By maintaining this interaction knowledge, the community leader can recommend part, or even complete solutions for a given service request. As such, by utilising community we provide trusted advice through community experience. In essence, based on the community's experience of services working together, we can more quickly determine services that are a suitable match for a given request.

From the simulations carried out, we can see that our community-based solution out performs the basic non-community solution in terms of service completion. By increasing the number of successful compositions found we increase the profitability of the bio-inspired services, thereby increasing their ability to survive. At the same time, the hierarchical structure of the community ensures that the improvements made do not come at a cost, in terms of messaging overhead. This is particularly important for large topologies where flood-based searches can cause significant overheads. The results show that, by utilising community leaders to localise search, we reduce the overall cost on the network.

Acknowledgements

This work has received support from the Higher Education Authority in Ireland under the PRTLI Cycle 4 programme, in a project Serving Society: Management of Future Communications Networks and Services.

References

1. Balasubramaniam, S., Botvich, D., Carroll, R., Mineraud, J., Nakano, T., Suda, T., Donnelly, W.: Adaptive Dynamic Routing Supporting Service Management for Future Internet. Accepted for publication in proceedings of the Global Communication Conference (Globecom) 2009, Hawaii (2009)
2. Herborn, S., Lopez, Y., Seneviratne, A.: A Distributed Scheme for Autonomous Service Composition. In: Proceedings of 1st ACM International Workshop on Multimedia Service Composition, Singapore (2005)
3. Gu, X., Nahrstedt, K.: Distributed Multimedia Service Composition with Statistical QoS Assurances. IEEE Transactions on Multimedia (2005)
4. Hu, S., Muthusamy, V., Li, G., Jacobsen, H.: Distributed automatic service composition in large-scale systems. In: Proceedings of the Second International Conference on Distributed Event-Based Systems, DEBS 2008, Rome, Italy, July 1-4 (2008)
5. Fujii, K., Suda, T.: Dynamic Service Composition Using Semantic Information. In: Proceedings of 2nd international conference on Service oriented computing, NewYork (2004)
6. Miorandi, D., Yamamoto, L., Dini, P.: Service Evolution in a Bio-inspired Communication System. Service Evolution in a Bio-inspired Communication System 2(6), 573–587 (2006)
7. Suzuki, J., Suda, T.: A Middleware Platform for a Biologically Inspired Network Architecture Supporting Autonomous and Adaptive Applications. IEEE Journal on selected areas in Communications 23(2) (February 2005)
8. Nakano, T., Suda, T.: Self-organizing Network Services with Evolutionary Adaptation. IEEE Transaction on Neural Networks 16(5) (September 2005)
9. Hui, P., Crowcroft, J., Yoneki, E.: BUBBLE Rap: Social-based Forwarding in Delay Tolerant Networks. In: Proceeding of the 9th ACM International Symposium on Mobile Ad Hoc Networking and Computing (MobiHoc), HongKong (May 2008)

Efficient Computation in Brownian Cellular Automata

Jia Lee[1,2] and Ferdinand Peper[2]

[1] College of Computer Science, Chongqing University, Chongqing, China
[2] National Institute of Information and Communications Technology (NiCT),
Nano ICT Group, Kobe, Japan

Abstract. A Brownian cellular automaton is a kind of asynchronous cellular automaton, in which certain local configurations—like signals—propagate randomly in the cellular space, resembling Brownian motion. The Brownian-like behavior is driven by three kinds of local transition rules, two of which are locally reversible and rotation symmetric, thus mapping a rule's left-hand side into a right-hand side is equivalent modulus rotations of multiples of 90 degrees. As a result, any update of cells using these rules can always be followed by a reversed update undoing it; this resembles the reversal of chemical reactions or other molecular processes. The third transition rule is not reversible and is merely used for diffusive purposes, so that signals can fluctuate forward and backward on wires like with random walks of molecules. The use of only these three rules is sufficient for embedding arbitrary asynchronous circuits on the cellular automaton, thus making it computationally universal. Key to this universality is the straightforward implementation of signal propagation as well as of the active backtracking of cell updates, which enables an effective realization of arbitration and choice – a functionality that is essential for asynchronous circuits but usually hard to implement efficiently on non-Brownian cellular automata. We show how to speed up the operation of circuits embedded in our Brownian cellular automaton. One method focuses on the design scheme of the circuits, by confining all necessary Brownian motions to local configurations representing primitive elements of circuits, such that a wire connecting two elements no longer needs backward propagation of signals on it, thus allowing the use of conventional design schemes of asynchronous circuits without change. Another method is to implement ratchets on the input and output lines by using various configurations in the cellular space, so as to further increase the speeds of signals on the wires.

1 Introduction

Cellular automaton models are increasingly attracting interest as architectures for computers based on nanometer-scale electronic devices [6,17,14]. Already considered a problem in the current pre-nanocomputer era, noise and fluctuations are expected to become a major factor interfering with the operation of nanometer-scale electronic devices, to the extent that they cannot be coped

F. Peper et al. (Eds.): IWNC 2009, PICT 2, pp. 72–81, 2010.

with by traditional techniques, which are limited to the suppression of noise and fluctuations and the correction of errors caused by them [2,7,3]. One effective approach to this problem is an asynchronously timed cellular automata in which computation is realized through cell configurations that implement circuits in which signals fluctuate randomly in forward and backward directions, as if they were subject to Brownian motion [13,12].

The fluctuations are employed in a search process through computational state space that eventually culminates into signals appearing at designated output terminals, which then indicates the termination of the computation. Though the randomness of the fluctuations looks just like that-randomness-it actually forms a powerful resource that can be employed to backtrack circuits out of deadlocks and to equip the circuits with arbitration ability. The fluctuations add to the functionality of the circuits to the extend that only two very simple primitive modules suffice as the basis from which more complicated circuits can be constructed [13]. As a result, our cellular automaton requires merely three states and three transition rules, which is far less than achieved thus far in literature for computationally universal asynchronous cellular automaton models [1,9,10,16].

The circuits to be embedded in the Brownian cellular automaton are called Brownian circuits [18,13], which allows signals running on the lines randomly in either direction, and modules switching their functionalities between their input/output lines freely, except the Ratchet module. As shown in [13], two kinds of simple primitives, called CJoin and Hub respectively, can be used to construct any Brownian circuits. Though fluctuation of signals can make the computational path of the circuits to be free of deadlock states, it will usually slow down the operation of the circuits because of the overhead of their Brownian nature. To improve the efficiency, we propose two methods, with one relying on the design scheme of circuits to be embedded into the Brownian cellular automaton. That is, CJoin and Hub, along with the Ratchet, can be used to construct any primitive modules of delay-insensitive circuits (e.g. [5]). As a result, we can embed networks of primitive delay-insensitive modules into the Brownian cellular space, where fluctuation of signals never need between any two primitives. Thus we restrict the space of Brownian behavior merely inside the relatively small configurations representing the primitives. The other method is to use various configurations to implement Ratchets along wires of the circuits.

This paper is organized as follows: Section 2 gives an overview on our Brownian cellular automaton. Section 3 shows how to embed an arbitrary asynchronous circuit into the Brownian cellular automaton in an effective way. Section 4 illustrates more variety of configurations of Ratchets which can be used to significantly speed up the circuits' operations, followed by conclusions in Section 5.

2   A Brownian Cellular Automaton Model

A *cellular automaton* is a discrete dynamical system consisting of an 2-dimensional array of identical finite-state automata (cells). Each cell is connected uniformly to a neighborhood of a finite number of cells, and has a state from a finite state set. It updates its state according to a transition function, which determines a cell's state based on the states of the cells in its neighborhood. In most CA models, there is a special state called *quiescent* state, in which a cell will never change its state as long as all the cells in its neighborhood are quiescent.

We assume a cellular automaton model where each cell takes a state from the state set $\{0, 1, 2\}$, denoted by white, gray, and black, respectively. Each cell does state transitions based upon the state of itself, along with the states of the four adjacent cells in non-diagonal direction (the von Neumann neighborhood). Also, transition rules are given in the form:

such that they not only change the state of the central cell, but also its four neighbors, so our cellular automaton is similar to a block cellular automaton. Moreover, updates of the cells take place in a random order in the way that at any time step only a randomly selected cell from the cellular space is allowed to undergo a state transition. Thus, our cellular automaton is asynchronous.

Cells are updated in accordance with three kinds of transition rules: R_1, R_2, R_3, as listed in Fig. 1, together with their rotational equivalences. Transition rules R_1 and R_2 are reversible, and whose righthand sides are just the same as their lefthand sides after being rotated in 90- and 180-degrees, respectively. Thus, the rule R_1 or R_2, as well as any of their rotated symmetries, will always have a reversed rule within its rotational equivalences, as illustrated in Fig. 1. As a result, any transition of cells according to rule R_1 or R_2 can always be immediately followed by a reversal transition to cancel the former transition, which actually resembles the reversal possibility in chemical reactions, or other molecular processes. On the other hand, the rule R_3 is obviously irreversible, but it can be used to transmit a signal on a line forward and backward in a random fashion, resembling the random walk of a free particle. This is illustrated in Fig. 2.

In addition to the configurations representing line and signal, two essential configurations that can be used as logic devices to process signals are given in Fig. 3 and Fig. 4, respectively. The configuration in Fig. 3 can be used to transfer a signal fluctuating on a horizontal line to a vertical line, and vice versa. Thus, this configuration can be used as the so-called Hub primitive module[13] (see Fig. 6(b)) in Brownian circuits. Additionally, the configuration in Fig. 4 can be used to join two signals coming from horizontal lines and produce two signals each on one of the vertical lines, and vice versa. Thus, it serves as the so-called CJoin primitive[13] in Fig. 6(a). Another useful configuration is shown in Fig 5, which represents the Ratchet in Fig. 6(c), and is used to restrict the random fluctuation of signals so as to run into a specified direction.

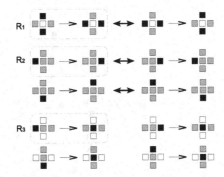

Fig. 1. Transition rules of Brownian cellular automaton. The dashed boxes indicate the basic forms of each rule.

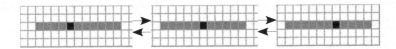

Fig. 2. Random propagation of a signal on a straight line. Here a line is represented by a continuous sequence of cells in state 1 (gray), and a signal is denoted by a cell in state 2 (black) on a line. Since a signal has no direction, and due to R_3 it can fluctuate forward and backward.

Based upon the configurations in Figs.2—5, it is possible to construct arbitrary asynchronous circuits in our Brownian cellular automaton. Since fluctuation of signals as well as reversal behavior of modules will usually reduce the efficiency of circuits' operation, it is essential to adopt some methods to speed up the operation of the circuits. In the following sections, we exploit two methods to improve the computing efficiency of our Brownian cellular automaton. One method depends on the design scheme of the circuits, by which all necessary Brownian motions are confined within small local configurations, such that lines

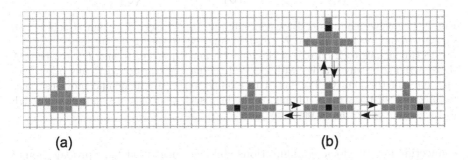

(a)   (b)

Fig. 3. (a) Local configuration representing Hub. (b)Propagation of a signal within the configuration due to transition rule R_2.

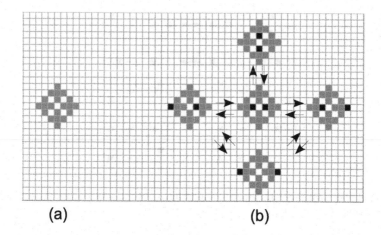

(a)   **(b)**

Fig. 4. (a) Local configuration representing CJoin. (b)Propagation of two signals within the configuration due to transition rules R_1 and R_2.

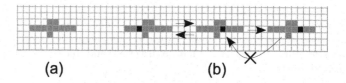

(a)   **(b)**

Fig. 5. (a) Basic configuration of Ratchet on the line. (b)Propagation of a signal before and after passing through the configuration due to transition rules R_2 and R_3.

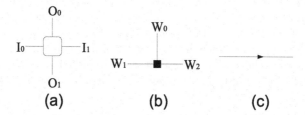

(a)   **(b)**   **(c)**

Fig. 6. Primitive modules of Brownian circuits. (a) Conservative Join (CJoin): Two signals arriving on lines I_0 and I_1 (resp. O_0 and O_1) respectively are joined and give rise to two signals each on one of the lines O_0 and O_1 (resp. I_0 and I_1). (b) Hub: A signal arriving on any line W_i with $i \in \{0, 1, 2\}$ may be transferred to any other lines due to its fluctuation. (C) Ratchet: This module works as a diode and provides a bias to directions of the signals running on the line, such that once a signal pass through it, the signal can not move back any more.

connecting any two local configurations no longer need backward propagation of signals on them, and hence, conventional design schemes of asynchronous circuits can be exploited without change. Moreover, another method is to place ratchets on the input and output lines by using various configurations in the cellular space, so as to further increase the speeds of signals running on the lines.

3   Embedding Delay-Insensitive Circuits into Brownian Cellular Automaton

A delay-insensitive circuit is a kind of asynchronous circuits, whose correct operation is robust to arbitrary delays involved in lines or operators. Any delay-insensitive circuit can be constructed from a fixed set of primitive modules. An example of universal primitive elements for delay-insensitive circuits is the set consisting of the so-called Merge, Conservative 2x2-Join and Conservative Sequencer (CSequencer) (see Fig. 7) [15]. These primitive elements have in common that they conserve signals. In other words, the number of input signals equals the number of output signals in each primitive. The class of delay-insensitive circuits that conserve signals is called conservative delay- insensitive.

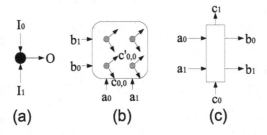

Fig. 7. Primitive modules for conservative delay-insensitive circuits. (a) Merge: A signal arriving on input lines I_0 or I_1 is transferred to output line O. (b) Conservative 2x2-Join: two signals arriving on input line a_i and b_j, respectively, are processed and result in two signals each on one of the output lines $c_{i,j}$ and $c'_{i,j}$, where $i, j \in \{0, 1\}$. (C) Conservative Sequencer: An input signal on line a_0 (resp. b_0) together with an input signal on line c_0 are assimilated, resulting in an output signal on line b_0 (resp. b_1) and on line c_1. If there are input signals on both a_0 and b_0 at the same time as well as an input signal on c_0 , then only one of the signals on a_0 and b_0 (possibly chosen arbitrarily) is assimilated together with the signal on c_0. The remaining input signal will be processed at a later time, when a new signal is available on line c_0.

The primitives of Brownian circuits given in Fig.6 can be used to construct any arbitrary conservative delay-insensitive circuits. In particular, as shown in Fig.8, each primitive element in Fig.7 can be realized by networks of CJoin and Hub. Thus, more complex modules, such as the Half-Adder or Full-Adder in Fig.8 can be constructed in accordance with the design scheme of conventional circuits, which in turn implies that fluctuations of signals along the lines between any primitives of delay-insensitive circuits do not need any more. As a result, it is possible to provide Ratchets on those connection lines to improve the computational efficiency of the entire circuits, as illustrated in Fig. 8.

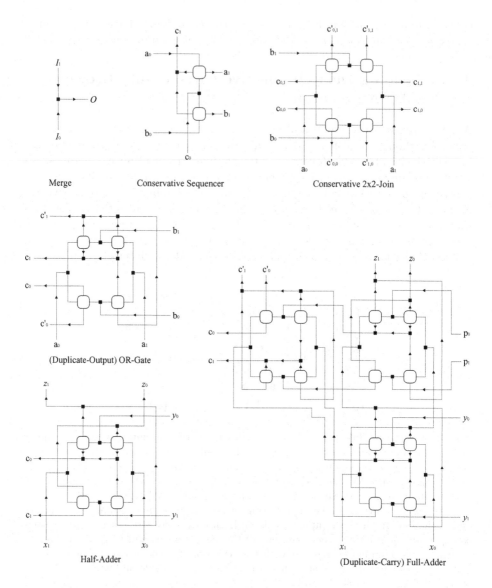

Merge

Conservative Sequencer

Conservative 2x2-Join

(Duplicate-Output) OR-Gate

Half-Adder

(Duplicate-Carry) Full-Adder

Fig. 8. Constructions of Merge, CSequencer, and Conservative 2x2-Join, as well as other more complex circuits from CJoin, Hub, and Ratchet. As demonstrated by the construction of CSequencer, the random fluctuation of signals along with the reversal behavior of modules provide a kind of active backtracking mechanism that enables direct and effective realization of the arbitrating or choice functionality of the Sequencer, whereas the arbitrating behavior is usually difficult to realize without backtracking.

Figure 9 shows the configurations of the Merge, CSequencer, and Conservative 2x2-Join respectively in the cellular space, according to the construction scheme in Fig. 8. Thus, it is possible to embed any arbitrary delay-insensitive circuit in

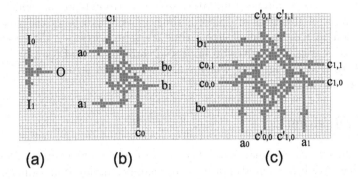

Fig. 9. Configurations of (a) Merge, (b) CSequencer, and (c) Conservative 2x2-Join

our Brownian cellular automaton, by means of duplicating and locating these configurations of primitives at appropriate positions in the cellular space, and after that connecting their input and output ports by lines with each other.

4   Implementing Ratchets with Various Configurations

Once a connection line between two elements of delay-insensitive circuits no need backward propagation of signals, we can place Ratchet on the line to further improve the efficiency of the circuit's operation. In addition to the basic configuration in Fig. 5, Fig. 10 demonstrates various configurations of Ratchets on the input/output-lines of modules that can be effectively implemented in the

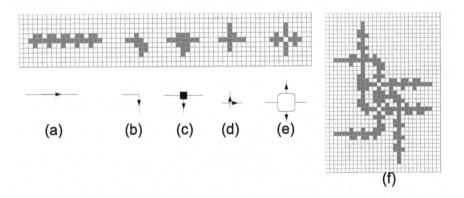

Fig. 10. (a) A straight line with a maximum number of Ratchets placed on it. (b) A left/right turn element with a Ratchet at its output line. (c) A Hub with a Ratchet at one of its lines. (d) two crossed lines with a Ratchet at the crossing point. (e) A CJoin with Ratchets at a pair of its lines, i.e., these two lines serve as the output lines of the CJoin. (f) Configuration of CSequencer with more placements of Ratchets than those in Fig. 9(b).

80   J. Lee and F. Peper

cellular space. For example, Fig. 10(a) shows a straight line on which a maximum number of Ratchets are placed, thus a signal can be transferred on the
line from left to right in almost linear time steps proportion to the length of the
line. Moreover, Fig. 10(f) gives an improved configuration of the CSequencer as
compared to that in Fig. 9(b), which can process input signals nearly five times
faster than the latter on average.

5   Conclusions

Random fluctuation and noise are effective and powerful resource for biological
systems [19]. We investigated that they also can be employed in a similar way as
those in biological systems in the search processes associated with computation,
based upon a cellular automaton model. The resulted Brownian cellular automaton has the potential to become the basis for architectures of nanocomputers.
Not only are there a very small number of transition rules which may possibly
contribute to physical realizations, transitions of configurations taking place in
the cellular space also behave somewhat similar to natural phenomena. This is
due to the local reversiblity of transition rules R_1 and R_2 in Fig. 1, which thus
can be used to emulate the invertible forward and backward actions occurring
simultaneously in most physical processes (see also [8,11]). In addition, random
propagation of signals is resulted from the transition rule R_3 which, along with
the random update nature of our cellular automaton, can actually emulate the
random walk process of molecules.

Acknowledgements

we are grateful to Prof. Kenichi Morita at Hiroshima University, Prof. Nobuyuki
Matsui, and Prof. Teijiro Isokawa at University of Hyogo, Japan, for the fruitful
discussions.

References

1. Adachi, S., Peper, F., Lee, J.: Computation by asynchronously updating cellular
   automata. Journal of Statistical Physics 114(1/2), 261–289 (2004)
2. Bennett, C.H.: The thermodynamics of computation–a review. International Journal of Theoretical Physics 21(12), 905–940 (1982)
3. Dasmahapatra, S., Werner, J., Zauner, K.P.: Noise as a computational resource.
   Int. J. of Unconventional Computing 2(4), 305–319 (2006)
4. Frank, M., Vieri, C., Ammer, M.J., Love, N., Margolus, N.H., Knight Jr., T.: A
   scalable reversible computer in silicon. In: Calude, C.S., Casti, J., Dinneen, M.J.
   (eds.) Unconventional Models of Computation, pp. 183–200. Springer, Singapore
   (1998)
5. Hauck, S.: Asynchronous design methodologies: an overview. Proc. IEEE 83(1),
   69–93 (1995)
6. Heinrich, A.J., Lutz, C.P., Gupta, J.A., Eigler, D.M.: Molecule cascades. Science 298, 1381–1387 (2002)

7. Kish, L.B.: Thermal noise driven computing. Applied Physics Letters 89(14), 144104-1–3 (2006)

8. Lee, J., Peper, F., Adachi, S., Morita, K., Mashiko, S.: Reversible computation in asynchronous cellular automata. In: Calude, C.S., Dinneen, M.J., Peper, F. (eds.) UMC 2002. LNCS, vol. 2509, pp. 220–229. Springer, Heidelberg (2002)

9. Lee, J., Adachi, S., Peper, F., Morita, K.: Embedding universal delay-insensitive circuits in asynchronous cellular spaces. Fundamenta Informaticae 58(3-4), 295–320 (2003)

10. Lee, J., Adachi, S., Peper, F., Mashiko, S.: Delay-insensitive computation in asynchronous cellular automata. Journal of Computer and System Sciences 70, 201–220 (2005)

11. Lee, J., Peper, F., Adachi, S., Morita, K.: An Asynchronous Cellular Automaton Implementing 2-State 2-Input 2-Output Reversed-Twin Reversible Elements. In: Umeo, H., Morishita, S., Nishinari, K., Komatsuzaki, T., Bandini, S. (eds.) ACRI 2008. LNCS, vol. 5191, pp. 67–76. Springer, Heidelberg (2008)

12. Lee, J., Peper, F.: On Brownian cellular automata. In: Theory and Applications of Cellular Automata, p. 278. Luniver Press (2008)

13. Lee, J., Peper, F., et al.: Brownian circuits—part II (in preparation)

14. MacLennan, B.J.: Computation and nanotechnology. Int. J. of Nanotechnology and Molecular Computation 1 (2009)

15. Patra, P., Fussell, D.S.: Conservative delay-insensitive circuits. In: Workshop on Physics and Computation, pp. 248–259 (1996)

16. Peper, F., Lee, J., Adachi, S., Mashiko, S.: Laying out circuits on asynchronous cellular arrays: a step towards feasible nanocomputers? Nanotechnology 14(4), 469–485 (2003)

17. Peper, F., Lee, J., Isokawa, T.: Cellular nanocomputers: a focused review. Int. J. of Nanotechnology and Molecular Computation 1, 33–49 (2009)

18. Peper, F., Lee, J., et al.: Brownian circuits—Part I (in preparation)

19. Yanagida, T., Ueda, M., Murata, T., Esaki, S., Ishii, Y.: Brownian motion, fluctuation and life. Biosystems 88(3), 228–242 (2007)

A Molecular Communication System

Yuki Moritani[1], Satoshi Hiyama[1], and Tatsuya Suda[2]

[1] Research Laboratories, NTT DOCOMO, Inc.
3-5 Hikarinooka, Yokosuka-shi, Kanagawa 239-8536 Japan
[2] Information and Computer Science, University of California,
Irvine, CA 92697-3425 USA
{moritani,hiyama}@nttdocomo.co.jp, suda@ics.uci.edu

Abstract. Molecular communication uses molecules (i.e., chemical signals) as an information carrier and allows biologically- and artificially-created nano- or cell-scale entities to communicate over a short distance. It is a new communication paradigm and is different from the traditional communication that uses electromagnetic waves (i.e., electronic and optical signals) as an information carrier. This paper focuses on system design and experimental results of molecular communication.

1 Introduction

Molecular communication [1] is inspired by the biological communication mechanisms (e.g., cell-cell communication using hormones) [2] and artificially creates a biochemically-engineered communication system in which communication processes are controllable. Molecular communication uses molecules (i.e., chemical signals) as an information carrier and allows biologically- and artificially-created nano- or cell-scale entities (e.g., biohybrid devices) to communicate over a short distance. It is a new communication paradigm and is different from the traditional communication that uses electromagnetic waves (i.e., electronic and optical signals) as an information carrier.

Molecular communication is a new and interdisciplinary research area that spans the nanotechnology, biotechnology, and communication technology, and as such, it requires research into a number of key components. This paper focuses on system design and experimental results of molecular communication components.

The rest of this paper is organized in the following manner. Section 2 presents an overview of a molecular communication system. Section 3 explains system design and initial experimental results of key components in a molecular communication system. Section 4 concludes the paper.

2 Overview of a Molecular Communication System

Molecular communication is a new communication paradigm and is different from the traditional communication (Table. 1). Unlike the traditional communication that utilizes electromagnetic waves as an information carrier, molecular

F. Peper et al. (Eds.): IWNC 2009, PICT 2, pp. 82–89, 2010.

communication utilizes molecules as an information carrier. In addition, unlike in the traditional communication where encoded information such as voice, text, and video is decoded and regenerated at a receiver, in molecular communication, information molecules cause some biochemical reactions at a receiver and recreate phenomena and/or chemical status that a sender transmits.

Although the communication speed/distance of molecular communication is slower/shorter than that of the traditional communication, molecular communication may carry information that is not feasible to carry with the traditional communication (such as biochemical status of a living organism) between the entities that the traditional communication does not apply (such as biological entities). Molecular communication has unique features that are not seen in the traditional communication and is not competitive but complementary to the traditional communication.

Figure 1 depicts an overview of a molecular communication system that includes senders, molecular communication interfaces, molecular propagation systems, and receivers.

A sender generates molecules, encodes information onto the molecules, and emits the information encoded molecules (called information molecules) into a propagation environment. The sender may encode information on the type of the information molecules used or the concentration of the information molecules used. Possible approaches to create a sender include genetically modifying eukaryotic cells and artificially constructing biological devices that are capable of performing the encoding. A molecular communication interface acts as a molecular container that encapsulates information molecules to hide the characteristics of the information molecules during the propagation from the sender to a receiver to allow a generic transport of information molecules independent of their biochemical/physical characteristics. Using a lipid bilayer vesicle [3] is a promising approach to encapsulate the information molecules. Encapsulated information molecules are decapsulated at a receiver. A molecular propagation system passively or actively transports information molecules (or vesicles that encapsulate information molecules) from a sender to an appropriate receiver through the propagation environment. The propagation environment is aqueous solution that is typically found within and between cells. Using biological motor systems (motor proteins and cytoskeletal filaments) [4] are a promising approach

Table 1. Comparisons of key features between the traditional communication and molecular communication

Key features	Traditional communication	Molecular communication
Information carrier	Electromagnetic waves	Molecules
Signal type	Electronic and optical signals	Chemical signals
Propagation speed	Light speed (3×10^5 km/sec)	Slow speed (a few μm/sec)
Propagation distance	Long (ranging from m to km)	Short (ranging from nm to m)
Propagation environment	Airborne and cable medium	Aqueous medium
Encoded information	Voice, text, and video	Phenomena and chemical status
Behavior of receivers	Decoding of digital info	Biochemical reaction

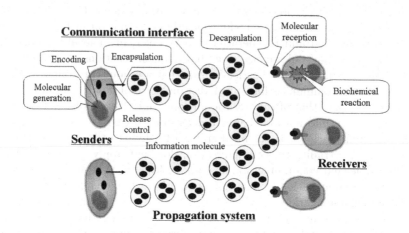

Fig. 1. An overview of a molecular communication system

to actively transport information molecules. A receiver selectively receives transported and decapsulated information molecules, and biochemically reacts to the received information molecules. Possible approaches to create a receiver are to genetically modify eukaryotic cells and to artificially construct biological devices as to control the biochemical reaction.

3   System Design and Initial Experimental Results

This section describes system design and initial experimental results of key components in a molecular communication system.

3.1   Molecular Communication Interface

A vesicle-based communication interface provides a mechanism to transport different types of information molecules in diverse propagation environments [5]. This is because the vesicle structure (i.e., a lipid bilayer membrane) provides a generic architecture that compartmentalizes and transports diverse types of information molecules independent of their biochemical/physical characteristics. Key research issues in implementing the vesicle-based communication interface include how vesicles encapsulate information molecules at a sender and how vesicles decapsulate the information molecules at a receiver.

The authors of this paper have proposed a molecular communication interface that uses a vesicle embedded with gap junction proteins (Fig. 2) [6]. A gap junction is an inter-cellular communication channel formed between neighboring two cells, and it consists of two docked hemichannels (connexons) constructed from self-assembled six gap junction proteins (connexins) [7]. When a gap junction is open, molecules whose molecular masses are less than 1.5 kDa can directly

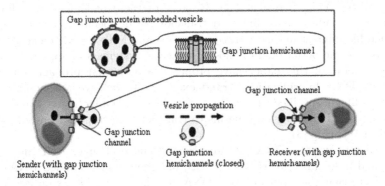

Fig. 2. A schematic diagram of a molecular communication interface using a vesicle embedded with gap junction proteins

propagate through the gap junction channel connecting two cells according to the molecular concentration gradient. A gap junction hemichannel is closed unless two hemichannels are docked.

In this system, a sender has gap junction hemichannels and stores information molecules inside itself. When a vesicle with gap junction hemichannels physically contacts the sender, gap junction channels are formed between the sender and the vesicle, and the information molecules are transferred from the sender to the vesicle according to the molecular concentration gradient. When the vesicle detaches from the sender, the gap junction hemichannels at the sender and at the vesicle close, and the information molecules transferred from the sender to the vesicle are encapsulated in the vesicle. The vesicle is then transported to a receiver by a molecular propagation system. A receiver also has gap junction hemichannels, and when the transported vesicle physically contacts the receiver, gap junction channels are formed between the vesicle and the receiver, and the information molecules in the vesicle are transferred into the receiver according to the molecular concentration gradient.

In order to investigate the feasibility of the designed communication interface, the authors of this paper used connexin-43 (one of the gap junction proteins) embedded vesicles [8]. Microscopic observations confirmed that calceins (hydrophilic dyes used as model information molecules) were transferred between connexin-43 embedded vesicles and the transferred calceins were encapsulated into the vesicles [9]. This result indicates that the proposed molecular communication interface may encapsulate information molecules and receive/transfer information molecules from/into a sender/receiver through gap junctions.

3.2   Molecular Propagation System

In eukaryotic cells, biological motors (e.g., kinesins) load/unload particular types of cargoes (e.g., vesicles) without external stimuli and transport them along cytoskeletal filaments (e.g., microtubules (MTs)) using the energy of adenosine triphosphate (ATP) hydrolysis [4]. Because of these biological capabilities of autonomous loading/unloading and transporting of specified cargoes, there is considerable interest in incorporating kinesins and MTs into artificially-created transporters and actuators in nano- or cell-scale systems and applications [10].

The authors of this paper have proposed a molecular propagation system that uses DNA hybridization/strand exchange to achieve autonomous loading/unloading of specified cargoes (e.g., vesicles) at a sender/receiver and reverse geometry of MT motility on kinesins to transport the cargoes from a sender to a receiver [11] (Fig. 3).

In order to use the DNA hybridization/strand exchange, each gliding MT, cargo, and unloading site is labeled with different single-stranded DNAs (ssDNAs). Note that the length of an ssDNA attached to an MT is designed to be shorter than that of the cargo, and the length of an ssDNA attached to a cargo is designed to be as long as that of the unloading site. Cargoes are pooled at a given loading site (a given sender) (Fig. 3a) and the ssDNA for the cargo is designed to be complementary to that of the MT. When an MT labeled with an ssDNA passes through a given loading site, a cargo is selectively loaded onto the gliding MT through DNA hybridization without external stimuli (Fig. 3b). The cargo loaded onto the MT (i.e., an MT-cargo complex) is transported by MT motility on kinesins toward a given unloading site (a given receiver) (Fig. 3c). To achieve autonomous unloading at a given unloading site, the ssDNA attached to each unloading site is designed to be complementary to that attached to the cargo. When the MT-cargo complex passes through an unloading site, the cargo is unloaded from the gliding MT through DNA strand exchange without external stimuli (Fig. 3d).

In order to investigate the feasibility of the designed propagation system, the authors labeled MTs with ssDNAs using a chemical linkage that cross-links thiolated ssDNAs and amino groups of MTs, while maintaining smooth gliding of labeled MTs on kinesins [12]. Microscopic observations confirmed that 23-bases ssDNA labeled cargo-microbeads (used as model vesicles in which information

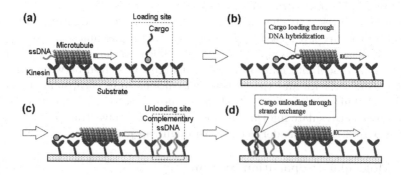

Fig. 3. A schematic diagram of a molecular propagation system using the reverse geometry of MT motility on kinesins and DNA hybridization/strand exchange

molecules were encapsulated) were selectively loaded onto gliding MTs labeled with complementary 15-bases ssDNAs [12]. Microscopic observation also confirmed that loaded cargoes were selectively unloaded from the gliding MTs at a micro-patterned unloading site where complementary 23-bases ssDNAs were immobilized [13]. These results indicate that gliding MTs may load/unload cargo-vesicles at a sender/receiver through the DNA hybridization/strand exchange.

3.3   Sender and Receiver

Researchers at Nara Institute of Science and Technology (NAIST) and the authors of this paper have proposed a receiver that uses a giant liposome embedded with gemini-peptide lipids [14][15]. A liposome is an artificially created vesicle that has the lipid bilayer membrane structure similar to vesicles. The gemini-peptide lipids are composed of two amino acid residues, each having a hydrophobic double-tail and a functional spacer unit connecting to the polar heads of the lipid. The liposomes embedded with the same type of gemini-peptide lipids assemble in response to an external stimulus (e.g., light, ions, and temperature) [16][17]. This allows a selective reception of information molecules at a receiver. The assembled liposomes with gemini-peptide lipids are also dissociated reversibly by applying a complementary external stimulus (e.g., applying UV-light for liposome assembly and applying visible light for liposome dissociation), and the selective dissociation mechanism may be applied to the selective transmission mechanism at a sender (Fig. 4).

The gemini-peptide lipids are used as a molecular tag. A small liposome embedded with molecular tags acts as a molecular capsule of information molecules and a giant liposome embedded with a molecular tag act as a sender/receiver. A receiver is embedded with a specific molecular tag and a molecular capsule whose destination is the receiver is also embedded with the same type of molecular tag.

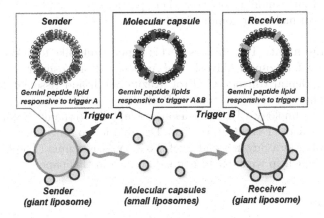

Fig. 4. A schematic diagram of senders and receivers using giant liposomes with gemini-peptide lipids

When an external stimulus is applied to the senders and the molecular capsules, molecular capsules embedded with a molecular tag responsive to the applied external stimulus are transmitted. When another external stimulus is applied to the receivers and the molecular capsules, a receiver embedded with a molecular tag that is responsive to the applied external stimulus receives molecular capsules embedded with the same type of molecular tag.

In order to investigate the feasibility of the designed receiver, researchers at NAIST created molecular capsules (liposomes with a diameter $\approx$100 nm) and receivers (liposomes with a diameter >2 μm) both containing zinc-ion responsive molecular tags in their lipid bilayer membranes. When zinc ions were added to the aqueous environment where both the molecular capsules and receivers exist, the selective binding of the molecular capsules to the receivers was observed [14]. The researchers at NAIST also created molecular capsules and receivers both containing photo-responsive molecular tags in their lipid bilayer membranes to control reception of molecular capsules to a receiver [15]. These results indicate that a receiver may selectively receive tagged molecular capsules (encapsulating information molecules) by applying an external stimulus.

4   Conclusions

This paper described basic concepts and key system components of molecular communication. This paper also discussed in detail system design of a communication interface that uses a vesicle embedded with gap junction proteins, a propagation system that uses MT motility on kinesins and DNA hybridization/strand exchange, and a sender/receiver that uses a giant liposome with gemini-peptide lipids. The feasibility of the designed system components was confirmed through the biochemical experiments.

Molecular communication is an emerging research area. The authors of this paper hope that a number of researchers participate in and contribute to the development of molecular communication.

Acknowledgement. The authors of this paper would like to acknowledge collaborators; Prof. Kazunari Akiyoshi and Associate Prof. Yoshihiro Sasaki (Tokyo Medical and Dental University), Prof. Kazuo Sutoh and Associate Prof. Shoji Takeuchi (The University of Tokyo), Prof. Jun-ichi Kikuchi (Nara Institute of Science and Technology). The authors also would like to thank Prof. Ikuo Morita (Tokyo Medical and Dental University), Dr. Shin-ichiro M. Nomura (Kyoto University), Prof. Akira Suyama and Prof. Yoko. Y. Toyoshima (The University of Tokyo) for their help with experiments described in this paper.

References

1. Hiyama, S., Moritani, Y., Suda, T., Egashira, R., Enomoto, A., Moore, M., Nakano, T.: Molecular Communication. In: Proc. NSTI Nanotechnology Conference and Trade Show, Anaheim, May 2005, vol. 3, pp. 391–394 (2005)

2. Alberts, B., Bray, D., Johnson, A., Lewis, J., Raff, M., Roberts, K., Walter, P.: Essential Cell Biology - An Introduction to the Molecular Biology of the Cell. Garland Publishing, New York (1998)
3. Luisi, P.L., Walde, P.: Giant Vesicles. John Wiley & Sons Inc., Chichester (2000)
4. Vale, R.D.: The Molecular Motor Toolbox for Intracellular. Transport. Cell. 112, 467–480 (2003)
5. Moritani, Y., Hiyama, S., Suda, T.: Molecular Communication among Nanomachines Using Vesicles. In: Proc. NSTI Nanotechnology Conference and Trade Show, Boston, May 2006, vol. 2, pp. 705–708 (2006)
6. Moritani, Y., Nomura, S.-M., Hiyama, S., Akiyoshi, K., Suda, T.: A Molecular Communication Interface Using Liposomes with Gap Junction Proteins. In: Proc. Bio Inspired Models of Network, Information and Computing Systems, Cavalese (December 2006)
7. Kumar, N.M., Gilula, N.B.: The Gap Junction Communication Channel. Cell 84, 381–388 (1996)
8. Kaneda, M., Nomura, S.-M., Ichinose, S., Kondo, S., Nakahama, K., Akiyoshi, K., Morita, I.: Direct formation of proteo-liposomes by in vitro synthesis and cellular cytosolic delivery with connexin-expressing liposomes. Biomaterials 30, 3971–3977 (2009)
9. Moritani, Y., Nomura, S.-M., Hiyama, S., Suda, T., Akiyoshi, K.: A Communication Interface Using Vesicles Embedded with Channel Forming Proteins in Molecular Communication. In: Proc. Bio Inspired Models of Network, Information and Computing Systems, Budapest (December 2007)
10. Van den Heuvel, M.G.L., Dekker, C.: Motor Proteins at Work for Nanotechnology. Science 317, 333–336 (2007)
11. Hiyama, S., Isogawa, Y., Suda, T., Moritani, Y., Sutoh, K.: A Design of an Autonomous Molecule Loading/Transporting/Unloading System Using DNA Hybridization and Biomolecular Linear Motors. In: Proc. European Nano Systems, Paris, December 2005, pp. 75–80 (2005)
12. Hiyama, S., Inoue, T., Shima, T., Moritani, Y., Suda, T., Sutoh, K.: Autonomous Loading/Unloading and Transport of Specified Cargoes by Using DNA Hybridization and Biological Motor-Based Motility. Small 4, 410–415 (2008)
13. Hiyama, S., Gojo, R., Shima, T., Takeuchi, S., Sutoh, K.: Biomolecular-Motor-Based Nano- or Microscale Particle Translocations on DNA Microarrays. Nano Letters 9(6), 2407–2413 (2009)
14. Sasaki, Y., Hashizume, M., Maruo, K., Yamasaki, N., Kikuchi, J., Moritani, Y., Hiyama, S., Suda, T.: Controlled Propagation in Molecular Communication Using Tagged Liposome Containers. In: Proc. Bio Inspired Models of Network, Information and Computing Systems, Cavalese (December 2006)
15. Mukai, M., Maruo, K., Kikuchi, J., Sasaki, Y., Hiyama, S., Moritani, Y., Suda, T.: Propagation and amplification of molecular information using a photoresponsive molecular switch. Supramolecular Chemistry 21, 284–291 (2009)
16. Iwamoto, S., Otsuki, M., Sasaki, Y., Ikeda, A., Kikuchi, J.: Gemini peptide lipids with ditopic ion-recognition site. Preparation and functions as an inducer for assembling of liposomal membranes. Tetrahedron 60, 9841–9847 (2004)
17. Otsuki, M., Sasaki, Y., Iwamoto, S., Kikuchi, J.: Liposomal sorting onto substrate through ion recognition by gemini peptide lipids. Chemistry Letters 35, 206–207 (2006)

Properties of Threshold Coupled Chaotic Neuronal Maps

Manish Dev Shrimali

The LNM Institute of Information Technology, Jaipur 302 031, India

Abstract. We study coupled chaotic neuronal maps. Chaos in chaotic neuronal maps is controlled by threshold activated coupling and the system yields synchronized temporally periodic states. A threshold is applied to all sites of a lattice and excess is transferred to other sites at varying degree of: relaxation time, dynamical updating, and connection topology. In each different realization, a transition from spatiotemporal chaos to ordered state is obtained. We also propose a scheme to obtain dynamical logic cell, using a threshold coupled chaotic neuronal drive–response systems. The study of threshold coupled chaotic dynamical systems suggest its potential application in computation and information processing.

Keywords: Chaotic dynamics, Control and Synchronization.

1 Introduction

One of the chaos control strategy based on threshold mechanism was introduced by Sinha *et al.* [1], where the control is triggered whenever the variable under observation exceeds the critical threshold. This mechanism works in marked contrast to the OGY method [2]. In the OGY method the chaotic trajectories in the vicinity of unstable fixed points are controlled onto these points. In threshold control, on the other hand, the system does not have to be close to any particular fixed point before implementing the control. Here the trajectory merely has to exceed the prescribed threshold. So the control transience is typically very short. It has been shown analytically and verified experimentally that a wide range of regular behaviour can be obtained from a threshold controlled chaotic system [1,3,4]. Such threshold controlled chaotic systems has potential utility for chaos-based applications, such as chaos computing [5,6,7].

The threshold-activated coupling of chaotic systems are relevant for certain mechanical systems like chains of nonlinear springs, as also for some biological systems, such as synaptic transmissions among neurons [8]. It is also relevant to population migrations as it is reasonable to model the population of an area as a nonlinear map and when this population exceeds a certain critical amount the excess population moves to a neighboring area. The threshold mechanism is also reminiscent of the Bak-Tang-Wiesenfeld cellular automata algorithm [9], or the sandpile model, which gives rise to self organized criticality (SOC). The

F. Peper et al. (Eds.): IWNC 2009, PICT 2, pp. 90–98, 2010.

model system studied here is however significantly different, the most important difference being that the threshold mechanism now occurs on a *nonlinearly
evolving substrate*, i.e. there is an *intrinsic deterministic dynamics* at each site.
So the local chaos here is like an *internal* driving or perturbation, as opposed
to *external* perturbation or driving in the sandpile model, which is effected by
dropping sand from outside.

In this review paper, we report the study of a lattice of chaotic model neurons incorporating *threshold activated coupling* under different connection topology, varying relaxation time, and varying dynamical updating scheme. We also
propose a scheme to obtain dynamical logic, using a drive–response system of
threshold coupled chaotic model neurons. Here, we focus on the lattice of chaotic
neuron model that has been proposed in studies of the squid giant axon and the
Hodgkin–Huxley equation [10]. We have applied the threshold activated coupling
to chaotic neurons to control chaos [1,4]. The chaotic neurons are controlled with
the threshold activated coupling and spatially synchronized temporal patterns
with different periods are obtained. We have investigated the effect of the variation of the relaxation time, different dynamical updating, and local–nonlocal
coupling on the spatiotemporal characteristics of the threshold coupled chaotic
neurons.

The paper is organized as follows. In next section, a model of chaotic neuron is
described and threshold activated coupling is introduced as a control mechanism
for both local and non–local coupling. The system under varying relaxation time
and different degree of synchronicity in the dynamical updating is studied followed by the application to chaos based computing. The conclusion are presented
in the last section.

2   The Model System: Neuronal Maps

We study the following lattice of uni–directionally coupled neuron model maps
proposed by Aihara *et al.* in studies of the squid giant axon and the Hodgkin–
Huxley equation [10].

$$y_{t+1}(i) = ky_t(i) - \alpha f(y_t(i)) + a$$
$$x_{t+1}(i) = f(y_{t+1}(i)) \tag{1}$$

where, $f(x) = 1/[1 + exp(-x/\varepsilon)]$ The internal state of the ith neuron is $y_t(i)$ at
time t, $x_t(i)$ is the output of the ith neuron at time t, k is the decay parameter of
the refractoriness, and α is the refractory scaling parameter. The sigmoid function, $f(.)$, is the output function of the neuron, and ϵ is the steepness parameter
of the sigmoid function.

A threshold activated coupling is incorporated on the lattice of model neurons
[1]. The coupling is triggered when the internal state of neuron at certain site
in the lattice exceeds the critical value y^* i.e. $y_t(i) > y^*$. The super critical site
then relaxes by transporting the excess $\delta = (y_t(i) - y^*)$ to its neighbour:

$$y_t(i) \rightarrow y^*$$
$$y_t(i+1) \rightarrow y_t(i+1) + \delta \tag{2}$$

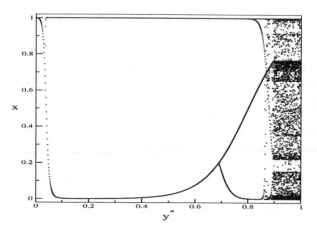

Fig. 1. The output of threshold coupled neurons x as a function of threshold y^* for the fixed value of $\alpha = 1.4$

The process above occurs in parallel, i.e. all supercritical sites at any instant relax simultaneously, according to Eqs. 2, and this constitutes one relaxation time step. After r such relaxation steps, the system undergoes the next chaotic update. In some sense then, time t associated with the chaotic dynamics is measured in units of r. The boundaries are open so that the excess may be transported out of the system. We denote as r_c the critical value of r after which the spatiotemporal behaviour of the system is effectively quite indistinguishable from the very long relaxation time limit (namely the asymptotic quasistationary limit). So the behviour of system with $r \geq r_c$ is essentially the same as for $r \to \infty$.

In Fig. 1, we have shown the output x of each chaotic neuron as a function of threshold y^* for the fixed value of $\alpha = 1.4$. There are two period–doubling bifurcations near $y^* \sim 0$ and ~ 0.7. The network of chaotic neurons is synchronized [11] with period two for $y^* < 0.7$ and $\alpha = 1.4$, where the system is chaotic in the absence of threshold activated coupling [12].

2.1   Relaxation Time

The dynamical outcome crucially depends the relaxation time r, i.e. on the timescales for autonomously updating each site and propagating the threshold-activated coupling between sites. For sufficiently large value of relaxation time, i.e. in the limit $r \to \infty$, the system is fully relaxed (sub-critical) before the subsequent dynamical update. So the time scales of the two processes, the intrinsic chaotic dynamics of each site and the threshold-activated relaxation, are separable. Here the relaxation mechanism is much faster than the chaotic evolution, enabling the system to relax completely before the next chaotic iteration. At the other end of the spectrum is the limit of very small r where the local dynamics and the coupling take place simultaneously. It is evident that lowering r essentially

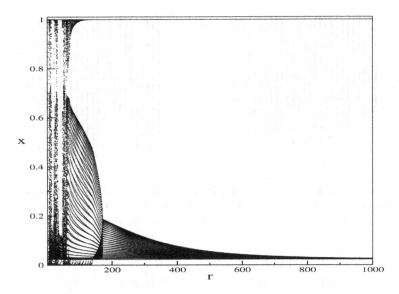

Fig. 2. Bifurcation diagram of the output state x of each neuron with respect to the relaxation time r for an array of threshold–coupled chaotic neurons with threshold $y^* = 0.5$ and system size $N = 50$

allows us to move from a picture of separated relaxation processes to one where the relaxation processes can overlap, and disturbances do not die out before the subsequent chaotic update. It was observed in [12,13] that for short relaxation times the system is driven to spatiotemporal chaos. This is due to the fast driving rate of the system which does not provide enough time to spatially distribute the perturbations and allow the excess to propagate to the open boundaries. However large r gives the system enough time to relax, and allows the excess to be transported out of the system through the open ends. So for large r the system displays very regular behavior for a large range of threshold values.

Fig. 2 shows a example of dynamical transition of threshold coupled chaotic neurons from a fixed temporal behavior to spatiotemporal chaos as r becomes smaller. There is a transition from the spatiotemporal fixed point to spatiotemporal chaos as we decrease the relaxation time [12]. These transitions result from the competition between the rate of intrinsic driving arising from the local chaotic dynamics and the time required to propagate the threshold–activated coupling.

2.2 Nonlocal Coupling

Let us consider the uni–directionally coupled system given by Eqs. 1 & 2, with its coupling connections rewired randomly, and try to determine what dynamical properties are significantly affected by the way connections are made between the elements. In our study, we transport the excess δ with probability p to randomly chosen sites j in the network and with probability $(1-p)$ to the nearest neighbour [14,15].

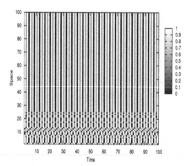 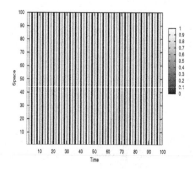

Fig. 3. Space-time density plots of an array of threshold-coupled chaotic neurons, with threshold value $y^* = 0.8$, size $N = 100$, relaxation time $r = 1000$, and re-wiring fraction (a) p=0.0, (b) p=0.1. The x-axis denotes the time and the y-axis denotes the site index.

$$y_t(i) \rightarrow y^*$$
$$y_t(i+1) \rightarrow y_t(i+1) + \delta \qquad with \ \ prob \ \ (1-p)$$
$$y_t(j) \rightarrow y_t(j) + \delta \qquad with \ \ prob \qquad p \qquad (3)$$

The basin of attraction of the synchronized state is not 1 for certain threshold values which becomes 1 with small fraction of rewiring in the threshold coupling. The synchronized fraction does not always increase with increasing random rewiring fraction. There exists an optimum degree of nonlocality yielding the largest basin of attraction for the synchronized state [14,15]. In threshold regimes where the network is unsynchronized under regular nearest neighbour coupling, the random rewiring yield synchronized network (see Figure 3). This provides a basis for understanding the robust synchronization mechanisms in threshold activated systems. Chaotic neuronal maps with threshold activated coupling at selected pinning sites is also studied. In pinning control method, there is an optimal fraction of sites where it is necessary to apply the control algorithm in order to effectively suppress chaotic dynamics [16].

2.3   Asynchronous Updating

There are many studies for the time evolution of coupled nonlinear systems under parallel or synchronous updating. In such models all individual local maps of the lattice are iterated forward simultaneously. Here on the other hand, we focus on *asynchronous evolution*, that is, one in which the *updates are not concurrent* [17,18]. The asynchronous updating can be closer to physical reality than synchronous updates in certain situations. For instance, there exist specific physical situations where an extended system is comprised of a collection of elemental dynamical units which evolve asynchronously (like in neuromorphology where the neurons, neuronal groups and functional layers of the brain, are believed to be usually asynchronous [19]).

It has been shown that the asynchronous updating of coupled chaotic logistic map induce more order and synchronization [18]. We study the network of chaotic

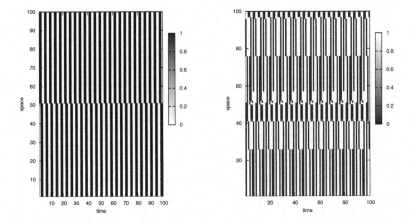

Fig. 4. Space-time density plots of an array of threshold-coupled neurons for asynchronous updating (a) $n_{sync} = N/2 = 50$ and (b) $n_{sync} = 3N/4 = 75$, with threshold value $y^* = 0.5$, size $N = 100$ and relaxation time $r = 1000$

neuron map models under threshold activated coupling with varying degree of synchronicity in the updating scheme with parameter values $k = 0.5$, $\alpha = 1.4$, $a = 1.0$, and $\epsilon = 0.04$ [20]. In the case of sequential updating with one-way coupling, where one site is updated in a sequential manner, the individual sites in the network of chaotic neurons shifts from the value 1 to 0 with time [20]. When, we update few sites between 1 and N, the pattern of the output $x_t(i)$, changes with the degree of synchronicity $p_{sync} = n_{sync}/N$, where n_{sync} is number of sites updated simultaneously. In Fig. 4 an examples with $p_{sync} = 0.5$ and 0.75 are shown.

2.4 Dynamic Logic Cell

Threshold-coupled chaotic systems have the capacity to directly and flexibly implement fundamental logic and arithmetic operations [5]. The one-way threshold coupled chaotic maps can be used for the dynamic and flexible implementation of fundamental logic gate NOR (from which all gates may be constructed), and the logic gates AND and XOR (which yield the building block of arithmetic processing: the bit-by-bit addition operation) [5,6,7].

We study the one-way threshold coupled drive-response system of chaotic model neurons [10]. The internal state of the both neurons is relaxed to threshold value $y^*_{1,2}$, drive and response parameters respectively, if it exceeds to it by transporting the excess $\delta_{1,2} = y_t - y^*_{1,2}$. The excess from the drive neuron is used as a coupling to response system as follows:

$$y_t(1) \rightarrow y^*_1$$
$$y_t(2) \rightarrow y_t(2) + \delta_1 \qquad (4)$$

The excess δ_2 from the response neuron is transported out of the system, where $\delta_2 = y_n(2) - y^*_2$. The value of the excess transported out of the system is a output

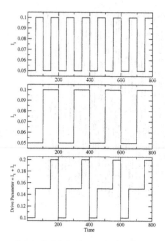 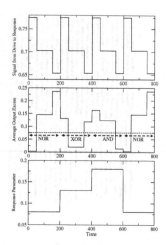

Fig. 5. Left Panel: The input signals I_1, I_2, and the drive parameter $y_1^* = I_1 + I_2$. Right Panel: The signal (excess) transmitted from the drive to response, the average excess out from the response unit (the critical value of excess $e_c = 0.075$ is shown by dotted line), and the Response parameter y_2^* time series.

1 if it is less then the critical value of excess e_c otherwise 0. Here, we study the threshold coupled chaotic neuron map models with parameter values $k = 0.5$, $\alpha = 1.4$, $a = 1.0$, and $\epsilon = 0.04$.

The logic output will be given by the excess emitting out by the response neuron. In this scheme the parameter of the drive neuron, which is the threshold acting on drive neuron y_1^*, is determined by the inputs and the parameter of the response neuron, the threshold acting on response neuron y_2^*, acts as a *logic gate controller*. The output is simply the excess emitted by the response neuron: large excess is output 0, and output 1 corresponds to a very small value of excess (see Figure 5). A significant feature of this scheme is that a single nonlinear drive–response unit can be used to flexibly yield the different desired logical sequences in time by simply varying the time series of the *logic controller*.

3   Conclusion

We have studied the spatiotemporal behaviour of threshold coupled chaotic neuronal maps. We observe that the threshold activated coupling control the chaos in the system and synchronized temporally periodic states are obtained. We have obtained a transition from spatiotemporal chaos to fixed spatiotemporal profiles, by lengthening the relaxation time scale. We have also studied the effect of nonlocal connections in threshold coupled chaotic neurons. The network goes to synchronized state with the rewiring of threshold activated coupling. This suggest a robust synchronization mechanism in threshold activated natural systems. The different updating scheme provide a controlled chaotic coupled neuron maps with different patterns, which can have a possible application in computation and information processing. The drive–response unit of threshold coupled chaotic neuron maps is studied for flexible implementation of logic operations.

Threshold-coupled chaotic systems have the capacity to directly and flexibly implement fundamental logic and arithmetic operations [5]. Such extended dynamical systems offer much scope of parallelism, allowing rapid solutions of certain problems utilizing the collective responses and collective properties of the system [7]. So the varied temporal and spatial responses of the array of model neurons studied here, can also be potentially harnessed to accomplish different computational tasks, and the system may be used for information processing [21].

Acknowledgments

We would like to thank our collaborators Sudeshna Sinha and Kazuyuki Aihara. MDS would also like to thank IWNC 2009 organizing committee and DST, India for financial support.

References

1. Sinha, S., Biswas, D.: Adaptive dynamics on a chaotic lattice. Phys. Rev. Letts. 71, 2010–2013 (1993)
2. Ott, E., Grebogi, C., Yorke, J.A.: Controlling chaos. Phys. Rev. Letts. 64, 1196–1199 (1990)
3. Murali, K., Sinha, S.: Experimental realization of chaos control by thresholding. Phys. Rev. E 68, 016210 (2003)
4. Sinha, S.: Unidirectional adaptive dynamics. Phys. Rev. E 49, 4832–4842 (1994); Sinha, S.: Chaos and Regularity in Adaptive Lattice Dynamics. Int. Journ. Mod. Phys. B 9, 875–931(1995); Sinha, S.: Chaotic Networks under Thresholding. International Journal Modern Physics B 17, 5503–5524 (2003)
5. Sinha, S., Ditto, W.L.: Dynamics based computation. Phys. Rev. Letts. 81, 2156–2159 (1998)
6. Murali, K., Sinha, S.: Using Synchronization to obtain Dynamic Logic Gates. Physical Review E 75, 25201 (2007)
7. Sinha, S., Munakata, T., Ditto, W.L.: Parallel computing with extended dynamical systems. Phys. Rev. E 65, 036214 (2002); Munakata, T., Sinha, S., Ditto, W. L.: IEEE Trans. on Circuits and Systems 49, 1629–1633 (2002)
8. Aihara, K., Matsumoto, G.: Chaotic oscillations and bifurcations in squid giant axons. In: Holden, A.R. (ed.) Chaos, pp. 257–269. Manchester University Press, Manchester (1986)
9. Bak, P., Tang, C., Weisenfeld, K.: Self-organized criticality: An explanation of the 1/f noise. Phys. Rev. Letts. 59, 381–384 (1987)
10. Aihara, K., Takebe, T., Toyoda, M.: Chaotic neural networks. Phys. Lett. A 144, 333–340 (1990)
11. Pikovsky, A., Rosenblum, M., Kurths, J.: Synchronization: A Universal Concept in Nonlinear Sciences. Cambridge University Press, Cambridge (2001)

12. Shrimali, M.D., He, G.G., Sinha, S., Aihara, K.: Control and synchronization of chaotic neurons under threshold activated coupling. In: de Sá, J.M., Alexandre, L.A., Duch, W., Mandic, D.P. (eds.) ICANN 2007. LNCS, vol. 4668, pp. 954–962. Springer, Heidelberg (2007)
13. Mondal, A., Sinha, S.: Spatiotemporal Consequences of Relaxation Timescales in Threshold Coupled Systems. Phys. Rev. E 73, 026215 (2006)
14. Sinha, S.: Consequences of Random Connections in Networks of Chaotic maps under Threshold Activated Coupling. Physical Review E 69, 066209 (2004)
15. Shrimali, M.D.: Chaotic neurons with random threshold coupling. In: Daniel, M., Rajasekar, S. (eds.) Nonlinear Dynamics, p. 389 (2009)
16. Shrimali, M.D.: Pinning control of threshold coupled chaotic neuronal maps. Chaos 19, 033105 (2009)
17. Lumer, E.D., Nicolis, G.: Synchronous versus asynchronous dynamics in spatially distributed systems. Physica D 71, 440 (1994)
18. Mehta, M., Sinha, S.: Asynchronous Updating of Coupled Maps leads to Synchronisation. Chaos 10, 350 (2000); Shrimali, M. D., Sinha, S., Aihara, K.: Asynchronous Updating induces Order in Threshold Coupled Systems. Physical Review E 76, 046212 (2007)
19. Hertz, J., Krogh, A., Palmer, R.G.: Introduction to the Theory of Neural Computation. Addison-Wesley, Reading (1991)
20. Shrimali, M.D., Sinha, S., Aihara, K.: Asynchronous updating of threshold coupled chaotic neurons. Pramana - Journal of Physics 70, 1127 (2008)
21. Aihara, K.: Chaos engineering and its application to parallel distributed processing with chaotic neural networks. Proceedings of the IEEE 90, 919–930 (2002)

Implementation of Rotary Element with Quantum Cellular Automata

Susumu Adachi

Nano ICT Group,
National Institute of Information and Communications Technology,
588-2 Iwaoka, Iwaoka-cho, Nishi-ku, Kobe, 651-2492 Japan
sadachi_nict@yahoo.co.jp

Abstract. In this paper we propose a physical implementation of Rotary Element by using quantum cellular automata. Rotary Element has 4-input and 4-output lines, and rotation bar. Two states of the rotation bar correspond to two states of polarization in quantum cellular automata. Proposed way of implementation is based on the switching due to the Coulomb repulsion between an input charged particle and two electrons in the quantum cellular automata. Dynamical behaviors of the Rotary Element exhibit correct operations. Some ideas of the other modules required for the universal computing are also shown.

1   Introduction

It is predicted that a nanoscale device will have more than 10^{12} elements in it in the near future, and it will become harder to perform correct operations on it due to increased fluctuations per one element. To conquer the problem of increased error rate and thermal noise, fault-tolerant architectures will be required. One candidate is quantum cellular automata.

Quantum cellular automata (QCAs) [1] are one of the CA in which each cell consists of 4 or 5 quantum dots, includes two electrons and takes two polarization states. QCA have some advantages; (1) bistability of the polarization states, (2) no central clock is required in a circuit (edge driven), etc. However, one of disadvantages of QCA is that circuit speed will not be so fast ($\sim$1 MHz) [3,4].

To cover this disadvantage, we propose an implementation way of the Rotary Element (RE) in which the rotation bar consists of the QCA. The RE have been introduced as a reversible logic module that is driven by one signal [2]. In our way, an input signal that consists of a charged particle switches alternatively in accordance with the QCA state due to the Coulomb interaction, while the QCA state changes its state in accordance with where input signal comes from. Implementation of RE with the QCA make it possible to increase the stability of circuit operations.

In this paper, we describe the RE and its operations in section 2. Section 3 describes the QCA, and section 4 shows the implementation of the RE with the QCA. Numerical analysis is shown. Section 5 shows the other modules, Delay Element and C-JOIN.

F. Peper et al. (Eds.): IWNC 2009, PICT 2, pp. 99–106, 2010.

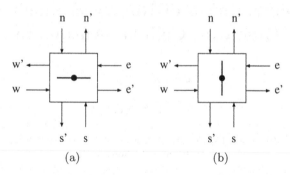

Fig. 1. (a) An RE in horizontal state, and (b) an RE in vertical state

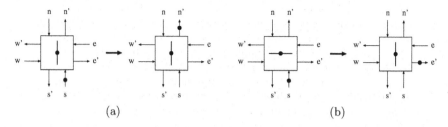

Fig. 2. RE operating on an input signal in (a) parallel case, and (b) perpendicular case. The signal is denoted by a token on a line.

2   Rotary Element

The Rotary Element (RE) introduced by Morita [2] is a reversible logic module. It has four input lines $\{n, e, s, w\}$, four output lines $\{n', e', s', w'\}$, and two states of a rotation bar (see Fig.1), which are displayed as horizontal and vertical bars respectively in the figure. When a signal, which is one-valued, arrives on one of the input lines of an RE, the RE operates on the signal, by possibly changing its state and transmitting the signal to an appropriate output line in the following way: If a signal comes from a direction parallel to the rotating bar of an RE, then it passes straight through to the opposite output line, without changing the direction of the bar (the state of the RE), as in Fig.2 (a); if a signal comes from a direction perpendicular to the rotating bar, then it turns right and rotates the bar by 90° degrees (Fig.2 (b)). Simultaneous signals on any pair of input lines of an RE are not allowed.

The RE is capable of universal computation, in the sense that any reversible Turing machine can be realized by a network of an infinite number of RE modules, in which there is at most one signal moving around in the entire network at any time [2]. Thus, delays in modules or lines do not affect the correct operation of the entire circuit, i.e., this circuit can operate in an asynchronous mode, removing the need of a central clock signal [2,11].

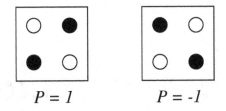

Fig. 3. Quantum cellular automaton consisting of four quantum dots which are occupied by two electrons. It has bistable states $P = \pm 1$ due to the Coulomb interaction between two electrons.

3 Quantum Cellular Automata

The Quantum Cellular Automaton (QCA) offers a novel alternative to the transistor paradigm [1]. A QCA cell (Fig.3) is composed of four quantum dots, two of which can be occupied by electrons. Each electron can move between the dots in a cell due to tunneling effect. Coulomb repulsion between the electrons results in bistable configurations which correspond to two polarization states $P = \pm 1$.

Electron tunneling between cells is not allowed in the standard QCA model. The state of a cell can transmit to neighboring cell due to Coulomb interaction. As nanoscale devices have intrinsic fluctuations, the expectation value of the cell state in the QCA also takes value away from defined value of ± 1, that is non-integer. However, transmission of cell state takes on a stable aspect.

A simple approximation has been investigated by using spin Hamiltonian [5,7,8], reducing 4(5)-dot 2-electron system to one spin. Two ground states of the QCA in a dot correspond to $|\uparrow\rangle$ and $|\downarrow\rangle$. Moreover, it has been shown that the Hamiltonian of the QCA array is equivalent to the transverse Ising model, in which tunneling effect in a dot correspond to the transverse field.

4 Implementation of Rotary Element

We design the RE composed of one QCA and idealized wires as shown in Fig.4 (a), and composed of four QCAs and idealized wires as shown in Fig.4 (b). These have 2-dimensional symmetric structure. In the figure of (a), the polarization state is defined as $P = 1$ if two electrons occupy the dots labeled by 1 and 3, corresponding to the horizontal bar of the RE. And the state is defined as $P = -1$ if two electrons occupy the dots labeled by 2 and 4, corresponding to the vertical bar of the RE. However, the polarization state of only one QCA does not take certain state of $P = \pm 1$. It takes $P = 0$ if there is no external field, i.e., no input signal. Therefore, one QCA can not behave as a memory device. In the figure of (b), in contrast, it is expected that the QCA state becomes robust due to the interaction between the QCA. Then, we adopt this four QCA model.

We analyze the function of the RE by using symplectic integration [9] for an input particle. Assuming that the input particle transports through the wire and

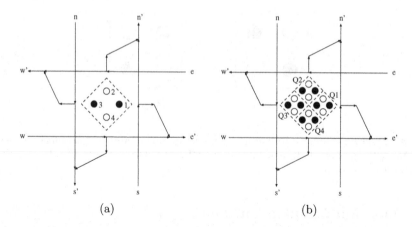

Fig. 4. Rotary element composed of (a) one QCA, and (b) four QCAs. Solid lines surrounding QCAs denote the idealized wires.

is reflected elastically to the boundary of the wire, we adopt classical approximation for the input particle. Let $\mathbf{p}, \mathbf{q}$ be the momentum vector and the position of an input charged particle. Equation of motion of the particle can be described by,

$$\frac{\mathrm{d}\mathbf{p}}{\mathrm{d}t} = -e\mathbf{E} \tag{1}$$

$$\frac{\mathrm{d}\mathbf{q}}{\mathrm{d}t} = \frac{\mathbf{p}}{m} \tag{2}$$

where, e is the charge, m is the mass of the particle, and $\mathbf{E}$ is the electric field affected by the electrons in the QCA. Let $\rho_{n,i}$ ($n = 1, 2, 3, 4$, $i = 1, 2, 3, 4$) be the density of the electrons in the n-th QCA, the electric field $\mathbf{E}$ can be described by,

$$\mathbf{E} = A \sum_{n=1}^{4} \sum_{i=1}^{4} \frac{e\rho_{n,i}}{|\mathbf{q} - \mathbf{q}_{n,i}|^3} (\mathbf{q} - \mathbf{q}_{n,i}) \tag{3}$$

where $q_{n,i}$ ($n = 1, 2, 3, 4$, $i = 1, 2, 3, 4$) is the position of the quantum dot in the n-th QCA, and A is constant.

The Hamiltonian of the QCA is expressed by the extended Hubbard model [1]. Our approach is Rayleigh-Schrödinger perturbation expansion for the transfer (tunneling) term of the model [5,6]. The effective Hamiltonian of one QCA can be expressed by

$$H(t) = -H_z(t)S^z + H_xS^x \tag{4}$$

where, $H_z(t)$ and H_x are the longitudinal and transverse field. $H_z(t)$ depends on the Coulomb interaction between an input particle and the electrons in the QCA. H_x implies spin flip between $P = 1$ and $P = -1$, corresponding to the tunneling effect of the electrons from a dot to the neighboring dot. It is obvious that if there is no input signal, the field $H_z(t)$ is zero, and the ground state of $H(t) = H_x S^x$ takes the linear combination of $|\uparrow\rangle$ and $|\downarrow\rangle$,

$$|\psi_{\pm}\rangle = (|\uparrow\rangle \pm |\downarrow\rangle)/\sqrt{2} \tag{5}$$

This means one QCA without external field takes the neutral state of $P = 0$.

In contrast, the effective Hamiltonian of the four QCA can be expressed by

$$H(t) = -J \sum_n S_n^z S_{n+1}^z - \sum_{n=1}^{4} H_{zn}(t) S_n^z + H_x \sum_{n=1}^{4} S_n^x \tag{6}$$

where, J is exchange interaction between the neighboring QCAs, thus $n+1$ takes 1 if $n = 4$.

There are two reasonable modes of the model dynamics that are synchronous mode and asynchronous mode. These modes correspond to the dynamics without relaxation and with relaxation, respectively. In other words, these are the dynamics in short time coherent mode and in long time decoherent mode, respectively. In short time coherent mode, we solve the time-dependent Schrödinger equation of the QCA,

$$i\hbar|\dot{\psi}(t)\rangle = H(t)|\psi(t)\rangle \tag{7}$$

This differential matrix equation can be solved by Runge-Kutta method.

On the other hand, in long time decoherent mode, the Schrödinger representation of a system dynamics can not describe relaxation term. One of the phenomenological way on this is to start from master equation and the Heisenberg representation for the time-dependent operators. However, this attempt is omitted for simplicity.

The density of each dot can be derived by the expectation value of spin $\langle S_n^z \rangle = \langle \psi(t)|S_n^z|\psi(t)\rangle$. For simplicity, we assume that the densities of the two diagonal dots are same, then the densities are given by,

$$\rho_{n,1} = \rho_{n,3} = (1 - \langle S_n^z \rangle)/2 \tag{8}$$

$$\rho_{n,2} = \rho_{n,4} = (1 + \langle S_n^z \rangle)/2 \tag{9}$$

The horizontal or vertical rotation bar of the RE corresponds to $\langle S_n^z \rangle = -1$ or $\langle S_n^z \rangle = 1$.

The numerical analysis of the RE can be done by solving (1), (2), (3), and (6). The simulation results of the RE are shown in Fig.5 and Fig.6. The filled circles in the quantum dots denote the electrons, and the radius of each of them corresponds to the density. The parameters we use are $e = \hbar = m = 1$ for the constant, the width of the wires is 0.4, $\boldsymbol{p} = (0, -0.87)$ and $\boldsymbol{q} = (1.6, 3.8)$ for the initial values of input particle, $\boldsymbol{q_{1,1}} = (1.0, 0.0)$, $\boldsymbol{q_{1,2}} = (0.7, -0.3)$, $\boldsymbol{q_{1,3}} = (0.4, 0.0)$, and $\boldsymbol{q_{1,4}} = (0.7, 0.3)$ for the position of the quantum dot i in the left QCA (labeled by Q1), and $J = 1.0$, $H_x = 0.6$ for the QCA. The constant of the electric field is $A = 0.2$.

In Fig.5, input particle turns to the right due to the interaction, and the QCA state changes to the vertical state. In Fig.6, input particle transports straight reflecting to the wall, and the QCA state does not change.

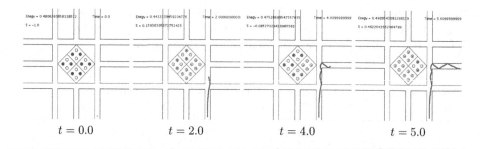

Fig. 5. Dynamical behavior of RE. The initial state of the rotation bar is horizontal.

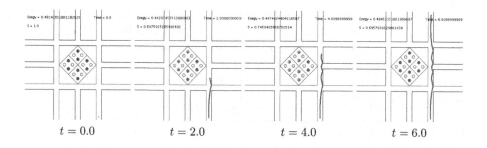

Fig. 6. Dynamical behavior of RE. The initial state of the rotation bar is vertical.

5   Delay Element and C-JOIN

In order to construct a circuit, a set of primitive modules called universal set is required. The investigated sets with the RE are {*RE, Delay*} [2] and {*RE, C-JOIN*} [10].

Delay Element (DE) is shown in Fig.7 (1), that adjusts timing of signals to avoid collision of them in the RE. DE is required for RE circuit which does not need to synchronize signals. The possible implementation of the DE is shown in Fig.7 (2), in which input charged particle is moderated in the electric field.

C-JOIN (Conservative JOIN) is shown in Fig.8 (1), in which signals on both of the two input lines I_1 and I_2 are assimilated and result in two signal output to the lines O_1 and O_2. A signal on only one of the input lines keeps pending until a signal on one of the other input lines is received. C-JOIN is required for building Delay Insensitive circuits. The possible implementation is shown in Fig.8 (2), in which one electron injected from one of the two input line occupies one of the two quantum dots, and if the other electron inputs, both of them output due to the Coulomb interaction. The conditions are that one electron state in 2-dot system is more stable than two-electron state, and no electron state is more stable than one electron state. Assuming one electron state is metastable, two electrons output due to the cascade reaction.

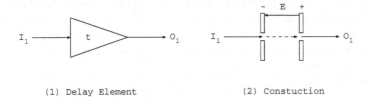

(1) Delay Element   (2) Constuction

Fig. 7. (1) A Delay Element, and (2) its construction with perforated parallel electrode. Delay Element moderate a charged particle in the electric field.

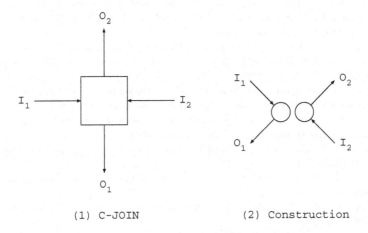

(1) C-JOIN   (2) Construction

Fig. 8. (1) A C-JOIN, and (2) its construction with two quantum dots denoted by circles

6 Concluding Remarks

This paper proposes an implementation way of RE with QCA array. Two states of the QCA correspond to two states of the rotation bar in the RE. Input particle is affected by the Coulomb interaction of the electrons in the QCA, resulting in switching in accordance with the QCA state. At the same time, the QCA state changes in accordance with the input line in which the particle is injected.

To solve this many-body system, we ignore the quantum effect of the input particle, and reduce 4-dot 2-electron system of the QCA to the quantum spin system, for simplicity. As a result of numerical simulations, the model behaves as expected. Moreover, we mention a possibility of other modules, Delay Element and C-JOIN. Either of them is needed to construct a circuit.

Although a charged particle is used as a signal in this paper, it may be replaced by an electron or a few electrons in the future. As reliability of one electron in solid state devices is low due to the screening effect or quantum/thermal fluctuation, it is required to find some alternative.

Our analysis of the RE with QCA in short time coherent mode includes many kind of approximation. An input particle should be also treated as a quantum mechanical wave packet. In this case, the input wave packet and the electrons in the QCA form an entanglement state. This implies that the model can apply to a quantum gate.

References

1. Lent, C.S., Tougaw, P.D.: Quantum Cellular Automata. Nanotechnology 4, 49–57 (1993)
2. Morita, K.: A simple universal logic element and cellular automata for reversible computing. In: Margenstern, M., Rogozhin, Y. (eds.) MCU 2001. LNCS, vol. 2055, pp. 102–113. Springer, Heidelberg (2001)
3. Nikolic, K., Berzon, D., Forshaw, M.: Relative Performance of Three Nanoscale Devices—CMOS, RTDs and QCAs—Against a Standard Computing Task. Nanotechnology 12, 38–43 (2001)
4. Bonci, L., Iannaccone, G., Macucci, M.: Performance Assessment of Adiabatic Quantum Cellular Automata. J. Appl. Phys. 89, 6435–6443 (2001)
5. Adachi, S., Isawa, Y.: Cell Design and Dynamics of Quantum Cellular Automata. Solid-State Electronics 42(7-8), 1361–1366 (1998)
6. Kuramoto, Y., Kitaoka, Y.: Dynamics of heavy electrons. Oxford University Press, Oxford (2000)
7. Cole, T., Lusth, J.C.: Quantum-dot cellular automata. Progress in Quantum Electronics 25(4), 165–189 (2001)
8. McGuigan, M.: Quantum cellular automata from lattice field theories, quant-ph/0307176 (2003)
9. Suzuki, M.: General theory of fractal path integrals with applications to many-body theories and statistical physics. J. Math. Phys. 32, 400–407 (1991)
10. Lee, J., Peper, F., Adachi, S., Mashiko, S.: On Reversible Computation in Asynchronous Systems. In: Quantum Information and Complexity, pp. 296–320. World Scientific, Singapore (2004)
11. Lee, J., Peper, F., Adachi, S., Morita, K.: An Asynchronous Cellular Automaton Implementing 2-State 2-Input 2-Output Reversed-Twin Reversible Elements. In: Umeo, H., Morishita, S., Nishinari, K., Komatsuzaki, T., Bandini, S. (eds.) ACRI 2008. LNCS, vol. 5191, pp. 67–76. Springer, Heidelberg (2008)

Universal 2-State Asynchronous Cellular Automaton with Inner-Independent Transitions

Susumu Adachi[1], Jia Lee[1,2], and Ferdinand Peper[1]

[1] Nano ICT Group,
National Institute of Information and Communications Technology, Japan
[2] College of Computer Science, Chong Qing University, China

Abstract. This paper proposes a computationally universal square lattice asynchronous cellular automaton, in which cells have merely two states. The transition function according to which a cell is updated takes as its arguments the states of the cells at orthogonal or diagonal distances 1 or 2 from the cell. The proposed cellular automaton is *inner-independent*—a property according to which a cell's state does not depend on its previous state, but merely on the states of cells in its neighborhood. Playing a role in classical spin systems, inner-dependence has only been investigated in the context of synchronous cellular automata. The asynchronous update mode used in this paper allows an update of a cell state to take place—but only so with a certain probability—whenever the cell's neighborhood states matches an element of the transition function's domain. Universality of the model is proven through the construction of three circuit primitives on the cell space, which are universal for the class of Delay-Insensitive circuits.

1   Introduction

Cellular Automata (CA) [15,5,19,7] are dynamic systems in which the space is organized in discrete units called cells that assume one of a finite set of states. These cells are updated in discrete time steps according to a transition function, which determines the subsequent state of each cell from the state of the cells inside a certain neighborhood of the cell.

Asynchronous Cellular Automata (ACA) [8] are CA in which each cell is updated at random times. Though ACA are mostly applied to simulations of natural phenomena, there have been efforts to use them for computation, as the lack of a central clock has excellent potential for implementation by nanotechnology. The most recent among these models—and the most efficient in terms of hardware and time resources—use so-called *Delay-Insensitive (DI)* circuits that are embedded on the cell space to implement computation. DI circuits are asynchronous circuits that are robust to delays of signals [6,17,10,1,2,13].

The number of cell states required for achieving computational universality is an important measure for the complexity of a CA model, and it is especially relevant for implementations by nanotechnology. Researchers aim to minimize this number as much as possible, with various degrees of success: the ACA model

F. Peper et al. (Eds.): IWNC 2009, PICT 2, pp. 107–116, 2010.

with a traditional von Neumann neighborhood requires four cell states [11], whereas the model with Moore neighborhood in [1] and the hexagonal model in [2] both require six states. The more recent model in [14] has cells with only three states, whereby the neighborhood is von Neumann, but it requires a special type of transition function in which more than one cell needs to be updated at a time in each transition.

A further reduction in the number of cell states may be very difficult to achieve in ACA, unless a new strategy is adopted. A promising venue in this context is increasing the neighborhood of cells. In [4] this concept is used in a synchronously timed CA, which has a neighborhood defined as the 24 cells lying at orthogonal or diagonal distances 1 or 2 (Moore distance 1 or 2) from a cell. This model's transition function is *totalistic*, meaning that it is only the number of cells in state 1 that determines the next state of a cell, but it is also *inner-independent*—an unusual property in CA according to which a cell's state is irrelevant to its state after update. Inner-independence has a counterpart in nature in the form of a classical spin system, as pointed out in [4]: the orientation of a cell's spin in such a system conforms to those of its neighbors, and is not influenced by its previous spin.

In this paper we propose a square lattice CA that has the same neighborhood as the model in [4], but that differs in that it is asynchronously updated. The model also differs through its use of a non-totalistic transition function, i.e., in an update the state of each individual cell is taken into account, rather than just the sum of the states. The property of inner-independence is included in the proposed model, however, making it a first in the context of asynchronous CA. Computational universality of the model is proven through formulating three primitive modules for DI circuits, and mapping them on the cell space. These modules—the so-called *P-Merge*, *Fork*, and the *R-Counter*—form a universal set for the class of DI circuits, meaning that any arbitrary DI circuit can be constructed from them. The transition function of the proposed CA model can be described in terms of 180 transition rules.

This paper is organized as follows. Section 2 describes the three primitive modules for DI circuits in more detail. These modules are implemented on the ACA cell space in Section 3, which also describes the implementation of signals to exchange information between the modules. This paper finishes with conclusions and a short discussion.

2   On Delay Insensitive Circuits

A DI circuit is an asynchronous circuit in which signals may be subject to arbitrary delays, without these being an obstacle to the circuit's correct operation [9]. Composed of interconnection lines and modules, a DI circuit uses signals—encoded as the change of a line's state—to transfer information from the output side of a module to the input side of another module. The speed of signals is not fixed.

A set of primitive modules from which any DI circuit can be constructed is proposed in Patra [16]. This set, consisting of the so-called *Merge*, *Fork* and

Tria, is universal, but it suffers from the problem that the Tria requires a large number (six) of input and output lines, which is hard to implement on a CA using cells with only four neighbors each. One way around this problem is to relax some of the conditions on DI circuits, like in [18,12], where lines are allowed to carry more than one signal at a time. The advantage of such *buffering lines* is more design freedom, and this translates into simpler structures of circuits and simpler primitive modules. This paper will employ this concept, allowing the use of primitive modules with at most four input or output lines (Fig. 1):

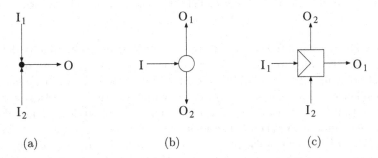

(a)   (b)   (c)

Fig. 1. Primitive modules for the DI circuits. (a) P-Merge, (b) Fork, and (c) R-Counter.

1. **P-Merge (Parallel Merge):** A signal on input line I_1 (I_2) in Fig.1(a) is assimilated and output to O. Simultaneous signals on I_1 and I_2 are assimilated as well, and will be output as two subsequent signals to O.
2. **Fork:** A signal on input line I in Fig.1(b) is assimilated and duplicated on both output lines O_1 and O_2.
3. **R-Counter (Resettable Mod-2 Counter):** Two subsequent signals on I_1 in Fig.1(c) are assimilated and they give rise to one output signal to O_1. This is called *Mod-2 Counter* functionality, because of the double signal required at I_1 to reinstate the initial "zero" state of the module. Alternatively, when there is one signal on each of I_1 and I_2, the module outputs a signal to O_2 after assimilating its inputs; this accounts for the *Reset* operation. A signal on only the input line I_1 keeps pending until a signal on either I_1 or I_2 is received. A signal on only the input line I_2 keeps pending until a signal on I_1 is received.

In the next section these modules will be implemented on the cell space, such that DI circuits can be constructed.

3 Implementing DI-Circuits on Asynchronous CA

The ACA model consists of a 2-dimensional square array of cells, each of which can be in either of the states, 0 and 1. The neighborhood N of a cell C consists of the 24 cells at orthogonal or diagonal distances 1 or 2 from C (*Moore-neighborhood*). As shown in Fig.2, the neighborhood of cell 0 consists of the cells 1, 2,...,24.

9	24	23	22	21
10	1	8	7	20
11	2	0	6	19
12	3	4	5	18
13	14	15	16	17

Fig. 2. Neighborhood of the cell labeled 0 consists of the cells labeled 1 to 24

The cells are updated according to a transition function that is defined by a table of 180 transition rules, shown in the Appendix. The state of the center cell before an update is irrelevant as to what state it becomes after update (inner-independence). The rules are rotation-symmetric and reflection-symmetric, meaning that their equivalents rotated by multiples of 90 degrees are also transition rules, and so are their reflections.

Updates of the cells take place asynchronously, in the way outlined in [1]. According to this scheme, one cell is randomly selected from the cell space at each update step as a candidate to undergo a transition, with a probability lying between 0 and 1. If the states of the neighbors of the selected cell match the states in the Left-Hand-Side of a transition rule, the corresponding transition is carried out.

Implementation of Signals. The propagation of signals over the cell space is governed by transition rules 1 to 6 in the Appendix. The configuration at the left in Fig. 3 will transform through these rules via intermediate configurations into the configuration at the right in the figure, which is the same configuration as the left one, but then shifted one cell to the right.

The transition rules are designed such that they can only be applied in a strictly defined sequential order, even if the update mode is asynchronous. This ensures the reliability of signal propagation or any other operations involved in computation [3]. The design of rules also takes into account the case in which two subsequent signals appear on the same line (not shown here). In this case the two signals will not interfere with each other and keep a distance of at least two cells between them.

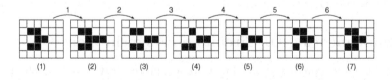

Fig. 3. (1) The basic configuration of a signal is transformed under the direction of transition rules 1 to 6 through steps (2) to (7) to the same configuration shifted one cell to the right

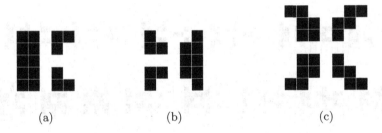

Fig. 4. Configurations of the primitive modules. (a) P-Merge, (b) Fork, and (c) R-Counter.

Implementation of Modules. The three modules introduced in section 2 are represented on the cell space by the configurations in Fig. 4.

The configuration of the P-Merge in Fig.4(a) employs transition rules 7 to 38 in the Appendix for the case when only one signal is input at a time. This case is illustrated in detail by Fig. 5. When the P-Merge receives two input signals at the same time, it uses transition rules 39 to 48 to deal with the resulting complications (not shown here).

The configuration of the Fork in Fig.4(b) employs transition rules 49 to 68 in the Appendix as well as some of the rules for the P-Merge. The processing of a signal by a Fork is illustrated in detail by Fig. 6.

The configuration of the R-Counter in Fig.4(c) employs transition rules 69 to 180 in the Appendix. Fig.7 shows a brief outline of the processing of signals by the R-Counter. Fig.7(a) illustrates the case in which a single signal is input to the left line of the R-Counter. This signal will remain stuck in the R-Counter—we say

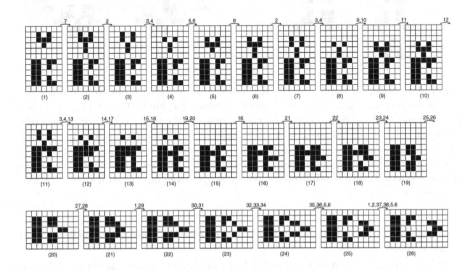

Fig. 5. Processing of one single signal by the P-Merge

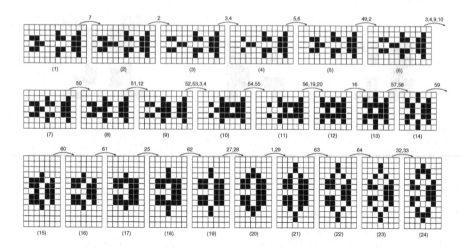

Fig. 6. Processing of a signal by the Fork, giving two output signals

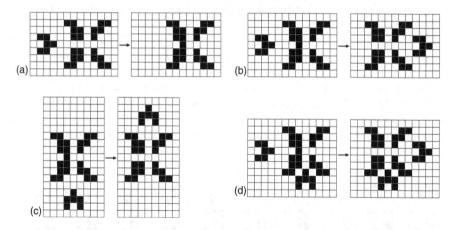

Fig. 7. Processing of signals by the R-Counter. (a) A signal input from the left line becomes a pending signal stuck inside the R-Counter. (b) Mod-2 Counter functionality: a second signal from the left input line results in a signal output to the line at the right. (c) Reset functionality: input from a reset signal to an R-Counter containing a pending signal results in a signal output to the line at the top. (d) When there are two input signals in addition to a pending signal already input from the left, the R-Counter has the choice between two possible operations. The operation illustrated here is the Mod-2 counting operation.

that the signal is pending—while no further processing takes place until one more signal is received.

When one more signal is received from the same input line (Fig. 7(b)) an output signal is produced (at the right), whereby the R-Counter reverts to its initial configuration (Mod-2 Counter functionality).

When a Reset signal is input to the R-Counter in which a signal is pending (Fig. 7(c)), an output signal is produced from the line at the top, whereby the R-Counter reverts to its initial configuration (Reset functionality).

When there are signals at both input lines, while a signal input from the left is already pending, then the R-Counter has the choice to produce either of the outputs, i.e., the output at the right line or the output at the top line. One signal remains pending in both cases, but the line at which it remains pending depends on the choice made by the R-Counter. Fig. 7(d) shows the case in which the R-Counter chooses to conduct the Mod-2 counting operation, leaving the reset signal pending. The choice as to what operation is conducted by the R-Counter is arbitrary. We refer to the ability to make such a choice as *arbitration*.

In addition to the condition that the set of primitive modules is universal, it is necessary to ensure that the primitive modules can be laid out on the cell space such as to form circuits. For this it is necessary to form curves on the cell space to turn a signal left and right. Fortunately, this is an easy task, because it can be implemented by the P-Merge. One more structure that is required to form circuits is a signal crossing. Assuming that signals lack an inherent ability to cross each other, we resort to the design of a circuit specialized for this task. This circuit, shown in Fig. 8, consists of merely one R-Counter, two Forks and two P-Merges. This way of using the arbitration functionality of the R-Counter is much simpler than previously reported in literature [18], making the resulting circuit for signal crossings relatively small.

The configurations of the modules thus allow us to simulate turns to the right and left of signals, as well as crossings of signals. Unlike the synchronous CA model in [4], difficult issues concerning the signal phase and the periodicity of

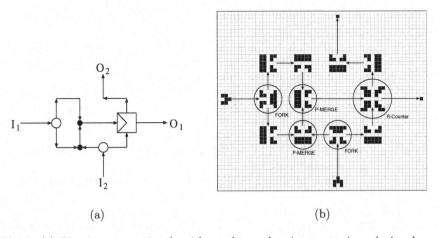

(a)   (b)

Fig. 8. (a) Circuit to cross signals without the need to intersect wires. A signal on I_1 will be directed to O_1, whereas a signal on I_2 will be directed to O_2. Simultaneous signals on I_1 and I_2 will be processed correctly, due to the arbitration functionality of the R-Counter. (b) Implementation of the crossing circuit on the ACA. The circuit is mapped on the cell space, while its topology is preserved.

signals on the cell space do not occur in the proposed ACA model: there is an almost unlimited freedom in laying out modules on the cell space, which is only restricted by the underlying DI circuit topology. In other words, it is quite straightforward to construct any arbitrary DI circuit on the cell space.

4   Conclusions and Discussion

This paper proposes a 24-neighborhood 2-state ACA. The transition rule is inner-independent, meaning that a cell's update does not depend on its own state but only on the states of the neighboring cells. The model is proven computational universal by showing how a universal set of three primitive modules can be embedded on the cell space. Since the primitive modules have at most four input or output lines each, this embedding fits well into the square lattice topology of the cell space.

The number of required transition rules is 180, which is relatively high, when compared to other ACA models. Physical realizations of ACA models tend to require less rules, which is a strong motivation to reduce the number of rules in the model. The transition rules used for the R-Counter are a primary candidate in this context, since they make up more than half of the rules. While a redesign of the rule table is likely to result in a reduced rule set, more efficient implementations may also be achieved through the use of a different set of primitives. This remains a topic for future investigations.

References

1. Adachi, S., Peper, F., Lee, J.: Computation by asynchronously updating cellular automata. J. Stat. Phys. 114(1/2), 261–289 (2004)
2. Adachi, S., Peper, F., Lee, J.: Universality of Hexagonal Asynchronous Totalistic Cellular Automata. In: Sloot, P.M.A., Chopard, B., Hoekstra, A.G. (eds.) ACRI 2004. LNCS, vol. 3305, pp. 91–100. Springer, Heidelberg (2004)
3. Adachi, S., Lee, J., Peper, F.: On signals in asynchronous cellular spaces. IEICE Trans. inf. & syst. E87-D(3), 657–668 (2004)
4. Adachi, S., Lee, J., Peper, F., Umeo, H.: Kaleidoscope of Life: a 24-neighborhood outer-totalistic cellular automaton. Physica D 237, 800–817 (2008)
5. Berlekamp, E.R., Conway, J.H., Guy, R.K.: Wining Ways For Your Mathematical Plays, vol. 2. Academic Press, New York (1982)
6. Hauck, S.: Asynchronous design methodologies: an overview. Proc. IEEE 83(1), 69–93 (1995)
7. Ilachinski, A.: Cellular Automata. World Scientific Publishing, Singapore (2001)
8. Ingerson, T.E., Buvel, R.L.: Structures in asynchronous cellular automata. Physica D 10, 59–68 (1984)
9. Keller, R.M.: Towards a theory of universal speed-independent modules. IEEE Trans. Comput. C-23(1), 21–33 (1974)
10. Lee, J., Adachi, S., Peper, F., Morita, K.: Embedding universal delay-insensitive circuits in asynchronous cellular spaces. Fund. Inform. 58(3/4), 295–320 (2003)

11. Lee, J., Adachi, S., Peper, F., Mashiko, S.: Delay-insensitive computation in asynchronous cellular automata. Journal of Computer and System Sciences 70, 201–220 (2005)
12. Lee, J., Peper, F., Adachi, S., Mashiko, S.: Universal Delay-Insensitive Systems With Buffering Lines. IEEE Trans. Circuits and Systems 52(4), 742–754 (2005)
13. Lee, J., Peper, F., Adachi, S., Morita, K.: An Asynchronous Cellular Automaton Implementing 2-State 2-Input 2-Output Reversed-Twin Reversible Elements. In: Umeo, H., Morishita, S., Nishinari, K., Komatsuzaki, T., Bandini, S. (eds.) ACRI 2008. LNCS, vol. 5191, pp. 67–76. Springer, Heidelberg (2008)
14. Lee, J., Peper, F.: On brownian cellular automata. In: Proc. of Automata 2008, UK, pp. 278–291. Luniver Press (2008)
15. von Neumann, J.: The Theory of Self-Reproducing Automata, edited and completed by A. W. Burks. University of Illinois Press, Urbana (1966)
16. Patra, P., Fussell, D.S.: Efficient building blocks for delay insensitive circuits. In: Proceedings of the International Symposium on Advanced Research in Asynchronous Circuits and Systems, pp. 196–205. IEEE Computer Society Press, Silver Spring (1994)
17. Peper, F., Lee, J., Adachi, S., Mashiko, S.: Laying out circuits on asynchronous cellular arrays: a step towards feasible nanocomputers? Nanotechnology 14(4), 469–485 (2003)
18. Peper, F., Lee, J., Abo, F., Isokawa, T., Adachi, S., Matsui, N., Mashiko, S.: Fault-Tolerance in Nanocomputers: A Cellular Array Approach. IEEE Trans. Nanotech. 3(1), 187–201 (2004)
19. Wolfram, S.: Cellular Automata and Complexity. Addison-Wesley, Reading (1994)

Appendix: Transition Rules

The state of the center cell in the Left-Hand-Side of a rule—though having the color gray here—is irrelevant to the update result (inner-independence property).

Effect of Population Size in Extended Parameter-Free Genetic Algorithm

Susumu Adachi

Nano ICT Group,
National Institute of Information and Communications Technology, Japan
sadachi_nict@yahoo.co.jp

Abstract. We propose an extended parameter-free genetic algorithm. The first step of this study is that each individual includes additional gene whose phenotype indicates a mutation rate. The second step is an extension of the selection rule of the parameter-free genetic algorithm, in which each individual has a characteristic neighborhood radius and the individuals generated near the parents are not selected to avoid trapping a local minimum. The characteristic neighborhood radius of an individual is given by the distance between before mutation and after mutation. As a result of the experiment for function minimization problems, effect of the population size appears and the success rate is improved.

1 Introduction

The genetic algorithm (GA) [1] is an evolutionary computation paradigm inspired by biological evolution. GAs have been successfully used in many practical applications such as functional optimization problems, combinatorial optimization problems, and the optimal design of parameters in machines [2]. However, the design of genetic parameters in a GA has to be determined by trial and error, making optimization by GA ad-hoc.

One of the most important research areas in evolutionary computation is the intelligent search, including self-adaptation [3], parameter-free [4,5], and parameter-less algorithms [6,7]. The parameter-free genetic algorithm (PfGA) and its applications [8,11,10,9] have made clear that it is important to make balance in global search and local search. For difficult problems, having many local minimums, large population size can increase success rate. Our motivation is to establish a mechanism that a GA obtains optimal population size depending on the difficulty of the problem to solve.

This paper proposes a GA based on the PfGA using an extended selection rule depending not only on the fitness but also on the neighborhood radius of individuals. The radius are used for not selecting an individual generated near the parent. To calculate the radius of an individual, we use the distance between an individual before applied mutation and after applied mutation, and determine the mutation rate by using additional gene. The different point of the selection rule from that of the PfGA is that if the fitness of the child is worse than that of the parent, the population size increases, and if the fitness of the child is better

F. Peper et al. (Eds.): IWNC 2009, PICT 2, pp. 117–124, 2010.

than that of the parent, the population size decreases. Due to this mechanism, it is expected that self-adjusting scheme of the mutation rate and the population size realizes.

In this paper, we show the outline of the PfGA in Section 2. The first step of the extension of the GA with additional gene which indicates a mutation rate is shown in Section 3. The second step of the extension of the GA with extended selection rule is shown in Section 4. The experimental results are shown in Section 5. Finally, we conclude and discuss in Section 6.

2   Parameter-Free Genetic Algorithm

PfGA has been proposed to avoid parameter-setting problems [4,5] in which there is no need to set control parameters for genetic operations. The algorithm merely uses random values or probabilities for setting almost all genetic parameters. A 'population' in the PfGA is defined as a sub-group composed of individuals ('sub-population' is used in the original papers). And a population size is the number of the individuals in the population. Outline of the PfGA is shown in Fig. 1, and the procedure of the PfGA is as follows,

- Step 1: The first individual is generated from whole search space randomly and is inserted into the population.
- Step 2: The second individual is generated from whole search space randomly and is inserted into the population.
- Step 3: Two parents P_1 and P_2 are brought out from the population, and two children C_1 and C_2 are generated by multiple-point crossover operation from the parents. The number of the crossover points is determined randomly.
- Step 4: Mutation is applied to one of the children at the probability of $1/2$, in which a randomly chosen portion of the chromosome is inverted (i.e., bit-flipped).
- Step 5: Applying the selection rule (Tab. 1), one to three selected individuals are pushed back to the population. If the population size become one, return back to Step 2. Otherwise, return back to Step 3.

There are four different cases in the selection rule depending on the fitness of the parents and the children as follows,

- Case 1: If both of the fitness values of the children are better than those of the parents, both of the children and the better parent are selected as shown in case 1 of Tab. 1.
- Case 2: If both of the fitness values of the children are worse than those of the parents, the only better parent is selected as shown in case 2 of the table.
- Case 3: If the fitness of the better parent is better than that of the better child, the better parent and the better child are selected as shown in case 3 of the table.
- Case 4: If the fitness of the better child is better than that of the better parent, the only better child is selected as shown in case 4 of the table. In this case, the other individual is generated randomly and is pushed back to the population.

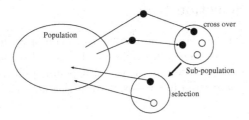

Fig. 1. Outline of PfGA. Two parents indicated by filled circles are chosen from the population, and two children indicated by white circles are generated by the crossover operation. Finally one to three individuals are selected and returned to the population.

Table 1. Selection rule of PfGA. The fitness values are sorted to become $P_1 < P_2$ and $C_1 < C_2$. In case 4, addition $+1$ means that the other individual is generated randomly.

	fitness	selection	addition
case 1	$C_1 < C_2 < P_1 < P_2$	C_1, C_2, P_1	
case 2	$P_1 < P_2 < C_1 < C_2$	P_1	
case 3	$P_1 < C_1 < \min(P_2, C_2)$	P_1, C_1	
case 4	$C_1 < P_1 < \min(P_2, C_2)$	C_1	$+1$

The population size increases in case 1, and decreases in case 2. For the most function to solve, the occurrence rate of the case 1 is less than that of the case 2. Therefore, the population size does not diverge infinitely (always less than $\sim$4). The advantages of the PfGA are compact and fast convergence due to the small population size.

3 Mutation Rate Coding

In the PfGA, the mutation rate takes $1/2$, and the mutation is applied to one of the children. It is not guaranteed that the value of $1/2$ is optimal in general. Then we propose a way of determining the mutation rate by using additional gene. Providing that the length of the gene is L_m and the gene is coded by Gray coding, we obtain the decoded value g ($0 \leq g \leq 2^{L_m} - 1$) and finally the mutation rate can be obtained by,

$$R_m = 2^{-g} \tag{1}$$

Then the mutation rate takes $0 < R_m \leq 1$. This rate is applied to all gene values including the mutation gene itself.

As a result of evolutions in the GA in which each individual has this mutation gene, it is expected that a better individual has a felicitous mutation rate and it survives.

4 Extended Selection Rule

The concept of a neighborhood radius of an individual is shown in Fig. 2. In the figure, r_1 and r_2 are the radii of the neighborhood of parents P_1 and P_2, and there are four cases denoted by (a) to (d) depending on the positions of two children C_1 and C_2 generated by crossover from P_1 and P_2.

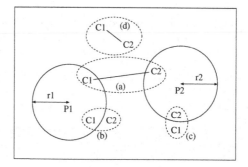

Fig. 2. Concept of neighborhood of each parent. r_1 and r_2 are the radius of the neighborhood of the parents P_1 and P_2. Selection rule is extended according to the positions of the children C_1 and C_2 generated by the crossover from P_1 and P_2.

To calculate the neighborhood radii, the mutation proposed in the previous section need to be applied to all of the individuals generated at first time from whole search space. Let n be the dimension of a real-valued function, the radius r is obtained as that of a hypersphere,

$$r = \sqrt{\sum_{i=1}^{n}(x_i^1 - x_i^0)^2} \tag{2}$$

where x_i^0 and x_i^1 are the i-th coordinates of a function before and after applying mutation. Moreover, the mutation rate and the radius of the child are also calculated when it is produced by the crossover.

The extension of the selection rule follows that of the PfGA basically. Considering the four cases (a) to (d) shown in Fig. 2, the degree of freedom of the selection rule quadruple. Our approach is based on the growth in diversity if the search is failed and the acceleration of the convergence if the search is successful. We propose the extended PfGA (EPfGA) selection rule as shown in Tab. 2, in which the size of the population increases in the case 2-(a), and decreases in the cases 4-(a) and 4-(b).

The algorithm can be rewritten as follows,

- Step 1: The first individual is generated from whole search space randomly. Its mutation rate is calculated and then the point-mutation is applied to obtain the neighborhood radius. Then it is inserted into the population.

Table 2. Selection rule of EPfGA. In case 1-(d), the worst individual in the population is eliminated, then the size does not change in this case.

	fitness	position	selection	addition
case 1	$C_1 < C_2 < P_1 < P_2$	(a),(b),(c)	C_1, C_2	
		(d)	C_1, C_2, P_1	-1
case 2	$P_1 < P_2 < C_1 < C_2$	(a)	P_1	+2
		(b)	P_1	+1
		(c),(d)	P_1, P_2	
case 3	$P_1 < C_1 < \min(P_2, C_2)$	(a),(b)	P_1	+1
		(c),(d)	P_1, C_1	
case 4	$C_1 < P_1 < \min(P_2, C_2)$	(a),(b)	C_1	
		(c),(d)	C_1, P_1	

- Step 2: The second individual is generated from whole search space randomly. Its mutation rate is calculated and then the point-mutation is applied to obtain the neighborhood radius. Then it is inserted into the population.
- Step 3: Two parents P_1 and P_2 are brought out from the population, and two children C_1 and C_2 are generated by multiple-point crossover operation from the parents. The number of the crossover points is determined randomly.
- Step 4: The mutation rates of the children are calculated and then the point-mutations are applied to both of the children. At the same time, the radius of each child is calculated.
- Step 5: Applying the selection rule (Tab. 2), one to three individuals are pushed back to the population. If the population size become one, return back to Step 2. Otherwise, return back to Step 3.

5   Experiments

Experiments were performed using the PfGA and the EPfGA for nine function minimization problems shown in Tab. 3. The functions F1 to F5 have been used in the first ICEO held in 1996.

We set the length L of the gene per one dimension to 24-bit for the functions F1 to F3 and F5 to F9, and 22-bit for the function F4, respectively. And set the length of the mutation gene to $L_m = 8$. Therefore, the total length of the gene is $nL + L_m$.

Fig. 3, 4, and 5 show the results of the experiments in EPfGA for F1, F3, and F5. The number of evaluations is 10000. The population size varies adaptively in Fig. 4 and 5.

Statistical results are shown in Tab. 4, in which the number of trials is 300, comparing with the PfGA. The values of each average population size increase for all functions, especially for F5. The success rates for F3 and F5 are significantly improved in EPfGA. The increasing size makes the improvement of the success rate obviously. In other case, the values of ENES are slightly improved except for F6.

Table 3. Test functions. n is the dimension. Each VTR is for the five dimensional case. $g_i(\mathbf{x})$ is used for F3 and F5.

function	domain	VTR
F1: Sphere model $\quad f(\mathbf{x}) = \sum_{i=1}^{n} (x_i - 1)^2$	$-5 \leq x_i \leq 5$	1.0×10^{-6}
F2: Griewank's Function $\quad f(\mathbf{x}) = \frac{1}{d}\sum_{i=1}^{n}(x_i - 100)^2 - \prod_{i=1}^{n}\cos\left(\frac{x_i - 100}{\sqrt{i}}\right) + 1$ $\quad d = 4000$	$-600 \leq x_i \leq 600$	1.0×10^{-4}
F3: Shekel's foxholes $\quad f(\mathbf{x}) = -\sum_{i=1}^{m} 1/[g_i(\mathbf{x}) + c_i]$ $\quad m = 30$	$0 \leq x_i \leq 10$	-9
F4: Michalewicz' Function $\quad f(\mathbf{x}) = -\sum_{i=1}^{n}\sin(x_i)\sin^{2m}\left(\frac{i x_i^2}{\pi}\right)$ $\quad m = 10$	$0 \leq x_i \leq \pi$	-4.687
F5: Generalized Langerman's Function $\quad f(\mathbf{x}) = -\sum_{i=1}^{m} c_i \exp\left[-\frac{1}{\pi}g_i(\mathbf{x})\right] \times \cos\left[\pi g_i(\mathbf{x})\right]$ $\quad m = 5$	$0 \leq x_i \leq 10$	-1.4
$\quad g_i(\mathbf{x}) = \sum_{j=1}^{n}(x_j - a_{ij})^2$		
F6: Rastrigin's Function $\quad f(\mathbf{x}) = 10n + \sum_{i=1}^{n}\left[x_i^2 - 10\cos(2\pi x_i)\right]$	$-5.12 \leq x_i \leq 5.12$	1.0×10^{-2}
F7: Rosenbrock's Function $\quad f(\mathbf{x}) = (1 - x_1)^2 + 100\sum_{i=2}^{n-1}\left(x_i^2 - x_{i+1}\right)^2$	$-2.048 \leq x_i \leq 2.048$	1.0×10^{-4}
F8: Goldberg and Richardson $\quad f(\mathbf{x}) = 1 - \prod_{i=1}^{n}\sin^a\left(b_1 x_i + b_2\right)\exp\left[-c_1(x_i - c_2)^2\right]$ $\quad a = 30, b_1 = 5.1\pi, b_2 = 0.5$ $\quad c_1 = 4\log(2)/0.64, c_2 = 0.0667$	$0 \leq x_i \leq 1$	1.0×10^{-4}
F9: Schwefel's Function $\quad f(\mathbf{x}) = -\sum_{i=1}^{n} x_i \sin\left(\sqrt{|x_i|}\right)$	$-512 \leq x_i \leq 512$	-2000

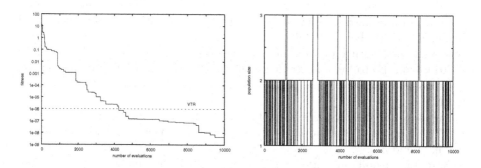

Fig. 3. Fitness and population size for F1

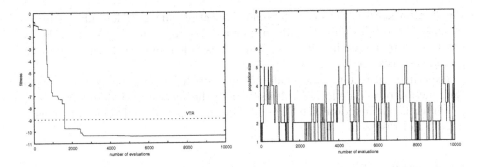

Fig. 4. Fitness and population size for F3

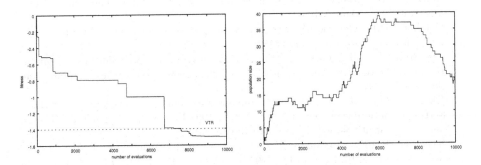

Fig. 5. Fitness and population size for F5

Table 4. Results of experiments in PfGA and EPfGA for 300 independent trials on 5 dimensional functions. The number of evaluations is 10000 per one trial. 'Succ' is the success rate, 'ENES' is the expected number of evaluation per success, and 'Size' is the average population size.

Func	PfGA			EPfGA		
	Succ [%]	ENES	Size	Succ [%]	ENES	Size
F1	100	4643.86	1.88	99.67	4435.63	2.02
F2	0.33	9097.00	1.86	0.33	8434.00	2.06
F3	6.67	3459.90	1.79	12.67	3613.39	3.45
F4	42.33	4436.37	1.85	27.00	3731.94	2.21
F5	24.33	4310.30	1.72	60.33	3975.92	23.51
F6	94.67	5573.73	1.85	67.33	6340.67	3.28
F7	4.33	5787.85	1.85	3.67	5334.64	2.25
F8	0.67	9345.00	1.58	0.67	7822.00	2.15
F9	89.67	3462.78	1.84	88.00	3207.74	5.19

6   Conclusion and Discussion

This paper proposes an EPfGA, and shows the effect of the population size. The extension have been done by mutation rate coding with additional gene to obtain the neighborhood radius and by extended selection rule with not only the fitness but also the neighborhood radius of parent. Based on the PfGA, the rule was determined by increasing diversity if search is failed and decreasing diversity if search is successful.

Experimental results show that the effect of the population size appears for the function of middle level of difficulty, and the success rate increases.

For a difficult function, like F2 and F6 in this paper, the selection rule remains a matter of research.

References

1. Holland, J.H.: Adaptation in Natural and Artificial Systems. The University of Michigan Press, Ann Arbor (1975)
2. Goldberg, D.E.: Genetic Algorithm in Search, Optimization, and Machine Learning. Addison-Wesley, Reading (1989)
3. Hinterding, R., Michalewicz, Z., Eiben, A.E.: Adaptation in Evolutionary Computation: A Survey. In: Proc. of the IEEE Int. Conf. on Evolutionary Computation, pp. 65–69 (1997)
4. Kizu, S., Sawai, H., Endo, T.: Parameter-free Genetic Algorithm: GA without Setting Genetic Parameters. In: Proc. of the 1997 Int. Symp. on Nonlinear Theory and its Applications, vol. 2/2, pp. 1273–1276 (1997)
5. Sawai, H., Kizu, S.: Parameter-free Genetic Algorithm Inspired by Disparity Theory of Evolution. In: Proc. of the 1997 Int. Conf. on Parallel Problem Solving from Nature, pp. 702–711 (1998)
6. Harik, G.R., Lobo, F.G.: A parameter-less genetic algorithm. In: Proc. of the Genetic and Evolutionary Computation Conference, pp. 258–265 (1999)
7. Lobo, F.G., Goldberg, D.E.: The parameter-less genetic algorithm in practice. Information Sciences–Informatics and Computer Science 167(1-4), 217–232 (2004)
8. Kizu, S., Sawai, H., Adachi, S.: Parameter-free Genetic Algorithm (PfGA) Using Adaptive Search with Variable-Size Local Population and Its Extension to Parallel Distributed Processing. IEICE Transactions on Information and Systems J82-D-2(3), 512–521 (1999)
9. Adachi, S., Sawai, H.: Effects of migration methods in parallel distributed parameter-free genetic algorithm. Electronics and Communications in Japan (Part II: Electronics) 85, 71–80 (2002)
10. Adachi, S., Sawai, H.: Evolutionary Computation Inspired by Gene Duplication: Application to Functional Optimization. Transactions of Information Processing Society of Japan 42(11), 2663–2671 (2001)
11. Sawai, H., Adachi, S.: A Comparative Study of Gene-Duplicated GAs Based on PfGA and SSGA. In: Proceedings of the Genetic and Evolutionary Computation Conference, pp. 74–81 (2000)

Temperature Effects on Olive Fruit Fly Infestation in the FlySim Cellular Automata Model

Vincenzo Bruno[1], Valerio Baldacchini[2], and Salvatore Di Gregorio[3]

[1] University of Calabria, Department of Physics, Arcavacata,
87036 Rende (CS), Italy
vbruno@fis.unical.it
[2] University of Calabria, Department of Ecology, Arcavacata,
87036 Rende (CS), Italy
valerio@baldacchini.it
[3] University of Calabria, Department of Mathematics, Arcavacata,
87036 Rende (CS), Italy
dig@unical.it

Abstract. FlySim is a Cellular Automata model developed for simulating infestation of olive fruit flies (*Bactrocera Oleae*) on olive (*Olea europaea*) groves. The flies move into the groves looking for mature olives where eggs are spawn. This serious agricultural problem is mainly tackled by using chemical agents at the first signs of the infestation, but organic productions with no or few chemicals are strongly requested by the market. Oil made with infested olives is poor in quality, nor olives are suitable for selling in stores. The FlySim model simulates the diffusion of flies looking for mature olives and the growing of flies due to atmospheric conditions. Foreseeing an infestation is the best way to prevent it and to reduce the need of chemicals in agriculture. In this work we investigated the effects of temperature on olive fruit flies and resulting infestation during late spring and summer.

1 Introduction

This work continues development and testing of a complex model describing how olive fruit flies (*Bactrocera Oleae*) infest olive groves (mainly *Olea europaea*, but also *Olea verrucosa*, *Olea chrysophylla*, and *Olea cuspidata*) [1], in order to obtain a Decision Support System (DSS) useful to farmers in the Mediterranean area to reduce the plague gravity [2]. In a previous paper a first, very simple, software implementa-tion of the model has been reported, focusing on the diffusion of adult flies attracted by mature olive fruits [3].

FlySim model is the evolution of a forecasting model, developed by the researchers of the *Dipartimento di Agrobiotecnologie* (*ENEA*, Roma, Italy) [4], and of a research project on traceability in agriculture, called AgroTrace, developed at the University of Calabria.

Local temperature curves are important in the life cycle of the olive fruit fly and on seasonal development of olive trees. A generation (egg, larva, pupa, adult fly) can be completed in 30 days in optimum weather conditions. Before the drupe maturation,

F. Peper et al. (Eds.): IWNC 2009, PICT 2, pp. 125–132, 2010.

females enter a state of reproductive diapause in which few or no eggs are produced. Flies may disperse to new locations during this period. When quite mature olive fruits (receptive drupes) appear, flies are attracted by drupes and females begin to produce eggs abundantly and to deposit them in the receptive drupes.

This complex system has been modelled and simulated by Cellular Automata (CA), because it evolves mainly on the basis of local interactions of its constituent parts. Nonetheless, CA modelling was yet applied to a similar problem concerning infestation caused by outbreaks of mountain pine beetle [5].

CA are a paradigm of parallel computing; they involve a regular division of the space in cells, each one characterised by a state s, which represents the actual condi-tions of the cell. The state changes according to a transition function τ that depends on the states of neighbouring cells and of the cell itself; the transition function and the neighbour-hood pattern are invariant in time and space. At time $t = 0$, cells are in states that describe initial conditions; the CA evolves changing the state of all the cells simultane-ously at discrete times, according to its transition function.

The proposed CA model FlySim is mixed, deterministic-probabilistic; it is related to olive fruit fly behaviour [4,1,6] and is based on an empirical method [7], which may be applied to some macroscopic phenomena in order to produce a proper CA model. It is based on an extension of the classical CA definition for permitting a straight correspon-dence between the system with its evolution in the physical space/time and the model with the simulations in the cellular space/time.

The main goal of this study is to isolate from the general model the self consis-tent part of variation of flies number depending on temperature in the warm season. The effect of optimal thermal conditions on fruit flies development is qualitatively well known by farmers and some studies have been reported in literature [6]. In this work we implement this portion of the model in order to simulate simple conditions on a typ-ical environment with olive groves scattered on few Km^2, and with a reasonable initial number of flies and olive drupes per cell. Those data came from direct observa-tions and discussions with many farmers, to whom results have been reported for a qualita-tive validation. The implemented simulator should describe earlier infestation in hotter regions and show possible source of infestation.

2 Description of the Model

The FlySim model is a two-dimensional CA with square cells and Moore neighbour-hood, described by **FlySim** = $\langle R, X, S, P, \tau, \Gamma \rangle$, where:

- $R = \{(x, y)|x, y \in N, 0 \le x \le l_x, 0 \le y \le l_y\}$ is a rectangular region of square cells, individuated by integer co-ordinates; N is the set of natural numbers; each cell corresponds to a portion of territory, where the phenomenon evolves;
- $X = \langle (0,0), (0,1), (0,-1), (1,0), (-1,0), (1,1), (-1,-1), (1,-1), (-1,1) \rangle$ is the neighbourhood relation; the co-ordinates of the cells in the neighbourhood of a cell c are obtained adding the c co-ordinates; they are the cell itself (called the central cell) with index 0 and the eight surrounding cells with indices $1, 2, \cdots, 8$.
- S is the finite set of states $S = S_{territory} \times S_{drupe} \times S_{fly}$, where $S_{territory}$ regards the substates associated to the terrain and climate features of the cell (e.g. temperature),

S_{drupe} regards the substates that accounts for the drupes conditions in the cell (e.g. maturation degree) and S_{fly} regards the substates that accounts for the general conditions of olive fruit flies in the cell (e.g. stage of development). The complete specifications of such substates are reported in the Table 1, 2 and 3.

- P is the finite set of global parameters of the CA, which affect the transition function. All the parameters are always global in space/time except when it is specified differently. The size of the cell edge and the time corresponding to a FlySim step are the fundamental parameters to match FlySim to the real phenomenon. The hour+date specification is a space global parameter. The determination of the maximum number of drupes in a cell is dependent by a density parameter for each cultivar. The determination of the receptive drupes is very complex, it must account for climatic history in order to determine the ratio (receptivity parameter) of the not receptive drupes, which reach a maturity degree to become receptive for the flies after the time interval p_{time}. The receptivity parameters are space global. Progress (step rate) for each maturity stage is described by a couple of temperature connected logistic curves (the former is increasing, the latter is decreasing); therefore, six parameters are necessary for each stage, except for the adult stage. The relaxa-tion rate and attractor parameters are needed for the fly diffusion algorithm. The parameters are reported in the Table 4.

- $\tau : S^9 \to S$ is the transition function; it must account for the average daily temperature, thermal accumulation (important for the fly growth), number of receptive drupes (depending on the drupe growth), number of deposited eggs, growth of the olive fruit flies for each maturity stage, except the adult stage, death rate and diffusion (outflows) of the adult olive fruit fly from a cell to the other cells of the neighbourhood. Corresponding elementary processes are specified in the next section.

- $\Gamma = \{\gamma_1, \gamma_2, \gamma_{r1}, \cdots, \gamma_m\}$ accounts for the external influences: $\gamma_1 : p_{cl} \to p_{cl}$ updates hourly p_{cl}; $\gamma_2 : N \to S_T$ updates hourly the substate temperature from a value average of around weather stations; $\gamma_{ri} : N \times S_{avT} \times p_{rj} \to p_{rj}$ update the value of the receptivity parameters for each cultivar. N is referred to the FlySim steps. The Γ functions are applied in order at each step before the τ function.

3 Elementary Processes and Transition Function

In this section we describe the elementary processes of FlySim and the transition function at each step. The new values of substate or parameter for the next step are indicated with the name of the substate or parameter, followed by the prime sign. Neighbourhood indexes are placed in square brackets. The portion of the model implemented in this work is spanning on 60 days between late spring and beginning of summer, when temperature increases and flies grow consequently.

3.1 Temperature

In the time frame of our simulation we modelled the temperature with a function of time and space. Remembering that $T(t)$ is the temperature of a cell at the step t, $T(t, x) =$

$f(t)g(x)$ is the temperature function of the CA at the step t and coordinate x, where: $f(t) = T_0 + \alpha t + A \sin(\omega t)$ and $g(x) = (\sin(4\pi x/60 - \pi/2) + 1)/20$.

The function $f(t)$ is the time modulation with a daily period and linear shift, while the spatial modulation $g(x)$ simulates regions hotter and colder due to different altitude, like hills. The parameters used are described in Table 5.

At each step the new temperature T' is computed for each cell of the CA and the corresponding thermal accumulation is updated following the formula: $cT' = \sum_{1 \le i \le 24m}(T(i) - T_s)/24m$ where $T_s = 10°C$ is the threshold temperature of flies development and $m = 4$ is the number of days.

3.2 Growth and Death Rate of Olive Fruit Flies

The maximum number of drupe in a cell can be considered constant during the time of our simulation, while the number of receptive drupes increases as drupes mature. When a egg is spawn in a drupe this accounted as damaged and is not receptive anymore. The number of total receptive drupes ($rD = \sum_{1 \le j \le n} recD_j$) at each step is computed by the following formula: $rD' = rD + \sum_{1 \le j \le n}(\max D_j - recD_j - damD_j) \cdot rj$. The damaged drupes are previously receptive drupes, where olive fruit flies lay eggs (usually no more than one egg for drupe). In this implementation of the simulator the development of olive drupes is not taken into account because they are still in a early stage of maturation and can be approximated as constant over time.

In all stages the insects are stressed by temperature too low or too high, slowing down their physiological activities. The olive fruit flies, as many other insects, has a strict correlation between temperature and development, but is not a peak of temperature that influence the development, also an optimal temperature is needed by the insect to let it grow. This is accounted as thermal accumulation, i.e. the normalized sum of temperature exceeding a threshold.

The growth of flies is a complex phenomenon that takes into account the growth at each stage (egg, larva$_1$ larva$_2$, larva$_3$, pupa and spawning rate) described by two logistic functions, one increasing, for $T < T_{switch}$, and one decreasing, for $T > T_{switch}$, connected at $T = T_{switch}$. For each stage we have six constants, with a total of 36 parameters. As example, the spawning rate is computed by $sr_1/(1 + sr_2 \cdot \exp(sr_3 \cdot T))$ and $sr_4/(1 + sr_5 \cdot \exp(-sr_6 \cdot T))$.

In order to simplify the model in the spring/summer period and the calculation needed, we consider only the increase of number of adults as a function of thermal accumulation. We take into account three threshold values of cT, $S_1 < S_2 < S_3$; below S_1 and above S_3 there is no fly development. The number of adults flies increases linearly between S_1 and S_2 and decreases between S_2 and S_3 following the formulas

$$a' = a(1 + mR(cT - S_1)/(S_2 - S_1)); \quad S_1 < cT < S_2$$
$$a' = a(1 + mR(1 - (cT - S_2)/(S_3 - S_2))); \quad S_2 < cT < S_3$$

where mR is the maximum growth rate per hour. The values used in the simulation are reported in Table 6.

In order to take into account the death of flies due to ageing factor, we considered a constant death rate per step (hour) $dR = 0.04\%$.

3.3 Diffusion of Adult Olive Fruit Flies

An attraction weight [10] $aw(aw \leq 0)$ is attributed to each cell; it is related to its olive coverage $1 - C_0$, number of its receptive drupes $rD = \sum_{1 \leq j \leq n} recD_j$, its water quantity $w : aw = a_1 \cdot (1 - C_0) \cdot \exp b_1 + a_2 \cdot rD \cdot \exp b_2 + a_3 \cdot w \cdot \exp b_3$.

Fly diffusion is a local interaction and is modelled according to a minimisation algorithm [7], which determines the movement of flies from the central cell to the other cells of the neighbourhood so that the sum of the differences of the "indicators" $(1 - aw) \cdot a$ among the cells is minimised. It involves that cells with more attraction weight (more convenient conditions) "capture" more flies and that flies diffuse little by little towards cells with higher attraction weight.

More precisely, two quantities are identified in the central cell: a "mobile" part (the number of adult flies $a[0]$), which can originate migration flows $(maf[k], 1 \leq k \leq 8)$ toward the other 8 cells of the neighbourhood and a "not mobile" part $aw[0]$; $a[k]$ and $aw[k], 1 \leq k \leq 8$ are the corresponding quantities for the neighbourhood other cells; $ind[k] = a[k] + aw[k], 0 \leq k \leq 8$ is defined.

The migration flows alter the situation: $ind^*[0] = a[0] - \sum_{1 \leq k \leq 8} maf[k] + aw[0]$, while $ind^*[k] = a[k] + maf[k] + aw[k], 1 \leq k \leq 8$. The minimisation algorithm application deter-mines $maf[k], 1 \leq k \leq 8$, such that the quantity $\sum_{0 \leq k_1 < k2 \leq 8} |ind^*[k1] - ind^*[k2]|$ is minimum.

Number of flies in a cell is obtained trivially by a balancing equation, adding migration inflows and subtracting migration outflows to the number of flies in the cell at the previous step.

4 Results

FlySim is a model still under development, with an heterogeneous group of researchers involved in it. Physicists, biologist and computer scientists are working in conjunction with farmers and their agronomists to define the whole model, which can be split in many parts: pupation of flies in the soil during winter, first flies generation in early spring, maturation of olive drupes in summer/autumn, attraction of female fly pheromone, eggs deposition, and so on. Quantitative informations are difficult to obtain, e.g. is unknown if the pheromone (fly attractor) produced by the mature drupes can overcome significantly the limits of the cell. Also the number of adult flies per cell is difficult to estimate, farmers can say if infestation is high, medium or low, but the only exact value is the percentage of olives with eggs at a certain time. In such scenario we focused simulations on the effect of temperature on flies development and diffusion in late spring-early summer when flies start spawning eggs. We had a lot of input and feedback from farmers to define roughly the relevant parameters.

The CA space is 60×60 square cells of 100m each, covering an area of about 10Km^2. The simulation spans over 60 days (60×24 steps) in late spring/early summer. The olive production span from 5000Kg/Ha to 10000Kg/Ha corresponding to 1-2 M olives/Ha. In the period under study attractive olive drupes are around 10%.

The temperature is modulated with a period of 24h, with a mean value starting at 10°C and ending at 25 °C after 60 days, with amplitude of 10 °C. It is also spatially modulated along x, as shown in figure 1, having two vertical regions with temperature slightly higher than the others. In that regions flies growth starts earlier in the simulation.

The olive trees distribution reproduce a realistic pattern of cultivated areas in a rural landscape. There are seven areas with one cultivar of olives, a mean value of 90000 drupes per cell with a random correction factor of 10% (figure 2). The initial flies distribution is uniform, with a mean value of 1450 flies per cell and also a ran-dom correction factor of 10%.

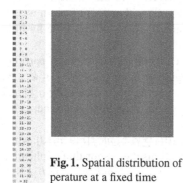

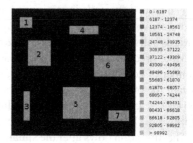

Fig. 1. Spatial distribution of temperature at a fixed time

Fig. 2. pattern of olive drupes distribution, areas are numbered for reference in the text

In figure 3 the flies distribution is shown during the simulation. In the first image diffusion of olive fruit flies was observed to produce an important "shield" effect: flies migrate toward more attractive areas but, when many contiguous attractive cells form a dense cluster, flies attack the more external olive trees and penetrate very slowly inside the cluster.

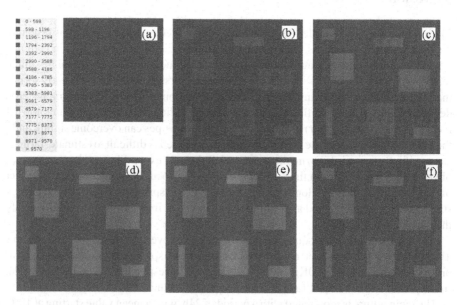

Fig. 3. The figure represents the evolution of an instance of FlySym at the end of the steps 0 (a), 72 (b), 432 (c), 648 (d), 760 (e) and 1440 (f)

This property is exploited by farmers, which surround clusters of olive trees with precocious olive trees, in order to prevent in time the flies' diffusion (at olive fruit maturing) inside the cluster by preparing opportune actions.

After 96 steps (4 days) $cT > S_1$ in the two hottest regions, then flies start to increase there and become source of infestation. The areas 2 and 6 are infested first, while area 5, situated in the center of the lower temperature region is non affected. Flies migrate slowly toward regions with lower flies density and toward cells with olive drupes. After step 384 (16 days) $cT > S_1$ in the coldest regions, while $cT > S_3$ in the hottest ones. The source of infestation is then inverted, area 5 is fully infested while area 4 partially, only in the part "up hill".

5 Conclusions

An accurate analysis of this and other simulations together with agronomists working in the field confirms that the simulations results are qualitatively plausible and compatible with observations. The model application with extensions to temperature effects on flies is very encouraging. The next step is a quantitative validation of this partial model with opportune field data covering few years. Such data are in a early stage of collection in cooperation with farmers and agronomists.

Acknowledgements

The authors are grateful to Dr Pierre Pommois for his contribution to the first model implementation, to prof. V. Carbone and to Prof. Beatrice Bitonti for the useful suggestions for the development of the model. This research has been partially funded by the Italian Ministry of Agriculture, project "Contratto di Filiera Campoverde".

References

1. Dominici, M., Pucci, C., Montanari, G.E.: Dacus Oleae (Gmel.) Ovipositing in Olive Drupes (Diptera, Tephrytidae). J. Appl. Entomol. 101, 111–120 (1986)
2. Neuenschwander, P., Michelakis, S.: The Infestation of Dacus Oleae (Gmel.) (Diptera, Tephritidae) at Harvest Time and Its Influence on Yield and Quality of Olive Oil in Crete. Z. Ang. Entomol. 86, 420–433 (1978)
3. Pommois, P., Brunetti, P., Bruno, V., Mazzei, A., Baldacchini, V., Gregorio, S.D.: FlySim: A Cellular Automata Model of Bactrocera Oleae (Olive Fruit Fly) Infestation and First Simulations. In: El Yacoubi, S., Chopard, B., Bandini, S. (eds.) ACRI 2006. LNCS, vol. 4173, pp. 311–320. Springer, Heidelberg (2006)
4. Baldacchini, V., Gazziano, S.: Environmental monitoring and forecasting. In: Olive Pest Management, final report of the project ECLAIR 209(CEE AGRE 0013-C), Development of Environmentally Safe Pest Control System for European Olives, ch. 8 (1998)
5. Bone, C., Dragicevic, S., Roberts, A.: A Fuzzy-Constrained Cellular Automata Model of Forest Insect Infestations. Ecological Modelling 192, 107–125 (2006)
6. Fletcher, B.S., Kapatos, E.T.: The Influence of Temperature, Diet and Olive Fruits on the Maturation Rates of Female Olive Flies at Different Times of the Year. Entomol. Exp. Appl. 33, 244–252 (1983)
7. Gregorio, S.D., Serra, R.: An Empirical Method for Modelling and Simulating Some Complex Macroscopic Phenomena by Cellular Automata. FGCS 16, 259–271 (1999)

Appendix 1 – Tables

Table 1. List of the substates names (to be used as variables) and their meaning for $S_{territory}$

$$S_{territory} = S_T \times S_{avT} \times S_{T1} \times S_{T2} \times \cdots \times S_{Tm} \times S_{cT} \times S_w \times S_{C0} \times S_{C1} \times \cdots \times S_{Cm}$$

$T(t)$	actual temperature at the step t
avT	average temperature in the day
$T_i, \quad 1 \leq i \leq m$	average temperatures of the last m days
cT	thermal accumulation for last m days
$C_j \quad 0 \leq j \leq n$	ratio of the n cultivars in the cell (0 for no trees in the cell)

Table 2. List of the names (used as variables) of the substates and their meaning for S_{drupe}

$$S_{drupe} = S_{maxD1} \times \cdots \times S_{maxDn} \times S_{recD1} \times \cdots \times S_{recDn} \times S_{damD1} \times \cdots \times S_{damDn}$$

$maxDj \quad 1 \leq j \leq n$	maximum number of drupes (for the j-th cultivar)
$recDj \quad 1 \leq j \leq n$	number of receptive drupes (for the olive fruit fly)
$damDj \quad 1 \leq j \leq n$	number of damaged drupes (by the olive fruit fly)

Table 3. List of the names (used as variables) of the substates and their meaning for S_{fly}

$$S_{fly} = S_e \times S_{l1} \times S_{l2} \times S_{l3} \times S_p \times S_a \times S_{ed} \times S_{l1d} \times S_{l2d} \times S_{l3d} \times S_{pd} \times S_{ad} \times S_{mf1} \times S_{mf2} \times \cdots \times S_{mf8}$$

e, l_1, l_2, l_3, p, a	number of flies at different stages of maturity in order: egg, larval larva2, larva3, pupa, adult
$maf_k \quad 1 \leq k \leq 8$	number of adults migrating to the neighbourhood cell k

Table 4. List of the set P of parameters and their meaning

$$P = \{ p_{ce}, p_{time}, p_{cl}, p_m, p_n, p_{d1} \cdots p_{dn}, p_{r1} \cdots p_{rn}, p_{aaf}, p_{ec1} \cdots p_{ec6}, p_{l1c1} \cdots p_{l3c6}, p_{c1} \cdots p_{c6},$$
$$p_{sr1} \cdots p_{sr6}, p_{a1} \cdots p_{a3}, p_{b1} \cdots p_{b3}, p_r \}$$

$ce, time \, (= 1 \text{ hour})$	size of the cell edge, time corresponding to a FlySim step
cl	internal clock (hour and date specification)
m	number of days for historical temperatures
n	number of cultivars
$r_j \quad 1 \leq j \leq n$	receptivity parameter for each cultivar
$ec_h, l_{1ch}, l_{2ch}, l_{3ch}, p_{ch}, s_{rh}$	six constants of logistic functions for:
$1 \leq h \leq 6$	egg, larva$_1$ larva$_2$, larva$_3$, pupa and spawning rate
$a_1, a_2, a_3, b_1, b_2, b_3$	couple of constants for the three cell attractors
rel	relaxation rate for fly diffusion

Table 5. List of parameters used to calculate temperature

T_0	Initial mean temperature (10°C)
T_1	Final mean temperature after t_f days (25°C)
A	Amplitude of daily temperature (10°C)
t_f	Number of days of the simulation (60 days)
$\alpha = (T_1 - T_0)/(24 t_f)$	Coefficient of temperature shift
$\omega = 2\pi/24$	24 hours period

Table 6. Parameters of flies growth, with values used in the simulation

S_1	Low cT threshold of fly development	19.95 °C/h
S_2	cT of maximum development	30.1 °C/h
S_3	High cT threshold of fly development	33.3 °C/h
mR	Maximum growth rate per hour	0.6%

Computing by Observing Changes

Matteo Cavaliere[1] and Peter Leupold[2,*]

[1] CoSBi
Trento, Italy
cavaliere@cosbi.eu
[2] Department of Mathematics, Faculty of Science
Kyoto Sangyo University, Department of Mathematics
Kyoto 603-8555, Japan
leupold@cc.kyoto-su.ac.jp

Abstract. Computing by Observing is a paradigm for the implementation of models of Natural Computing. It was inspired by the setup of experiments in biochemistry. One central feature is an observer that translates the evolution of an underlying observed system into sequences over a finite alphabet. We take a step toward more realistic observers by allowing them to notice only an occurring change in the observed system rather than to read the system's entire configuration. Compared to previous implementations of the Computing by Observing paradigm, this decreases the computational power; but with relatively simple systems we still obtain the language class generated by matrix grammars.

1 Computing by Observing

The paradigm of *Computing by Observing* was originally introduced under the name *Evolution and Observation* based on the following reflections [4,5]. Nearly all models in the area of DNA and molecular computing follow the classical computer science paradigm of processing an input directly into an output, which is the result of the computation. Only the mechanisms of processing are different from conventional models; instead of a finite state control or a programming language it is biomolecular mechanisms that are used, or rather abstractions of such mechanisms.

However, in many experiments in biology and chemistry the setup is fundamentally different. The matter of interest is not some product of the system but rather the process, i.e., the change observed in certain, selected quantities. To cite two simple examples: the predator-prey relationship, where the interesting fact is how the change in one population effects a change in the other; a reaction with catalyst often has the same product as without, but the energy curves during the reactions are different.

The goal was to formalize this approach in an architecture for computation. The resulting paradigm is that of Computing by Observing. It consists of an

---

* This work was done, while the author was funded as a post-doctoral fellow by the Japanese Society for the Promotion of Science under grant number P07810.

F. Peper et al. (Eds.): IWNC 2009, PICT 2, pp. 133–140, 2010.

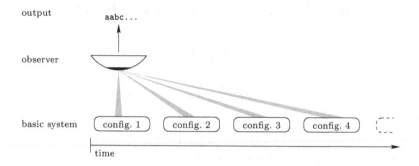

Fig. 1. The Computing by Observing architecture

underlying basic system, which evolves in discrete steps from one configuration to the next. The observer reads these configurations and transforms them into single letters; a type of classification. In this way a sequence of configurations is transformed into a simple sequence of symbols as depicted in Figure 1. This abstracts the protocol of an experiment in biology or chemistry and for us is the result of the computation.

In this combination the components are frequently more powerful than just by themselves. For example, context-free string-rewriting, as observed systems, and finite state automata as observers suffice to obtain computational completeness [5,6]. There are even context-free grammars that are universal as observed systems in the sense that one such grammar can generate every recursively enumerable language with different regular observers [2].

The relatively easy achievement of computational completeness and universal grammars suggests considering even less powerful observers. Natural candidates for such restrictions are ones that are suggested by actual limitations of physical observation of biochemical systems. Our goal here is to take another step in the direction toward more realistic systems without narrowing down the architecture too much to fit just one specific physical setup. We do this based on the observation that it is an unrealistic assumption to think that the observer could read the entire configuration in every step. Technical reasons like a lack of time and other factors will prevent this in most settings. Therefore we deal here with observers that can only read one specific measure derived from the observed system's configuration; this could model for example temperature, volume, pH-gradient etc. which are changed by the ongoing process and might be observable without any intrusion into the system itself.

When assigning some type of measure, we have two obvious choices of where to assign it to: to the single objects or to the rules. The latter could for example model the fact that energy is consumed or freed in a certain reaction. The former could model physical features of the object like mass or electrical charge. In this second case, rules like $a \rightarrow bc$ and $a \rightarrow cb$ would effect the same total change, because the objects on either side of the rules are the same in type and number. When assigning our measure to rules, we can give different values to these rules.

Different molecules formed by the same atoms can have different levels of energy, in this case bc and cb could represent two distinct configurations of this type.

This very simple example already shows that assigning a measure to objects is a special case of assigning it to rules. In the latter model we can always simulate the former by simply giving a rule $U \to V$ the value of the sum of values of objects in U minus the sum resulting from V. Therefore we will choose the variant of assigning values to rules here.

We want to mention that there are two concepts in Formal Language Theory that work in somewhat similar manner although their motivations were completely different. On the one hand, there are *control languages*. An early version restricted to left-most derivations was introduced Ginsburg and Spanier [8]. Here a grammar's rules are seen as an alphabet, and a derivation is successful only if the sequence of rules that are applied belongs to a certain language. Thus only some of the possible derivations are considered valid.

On the other hand, there are *Szilard languages* as introduced by Moriya [11]. Again the derivation rules form an alphabet, but in this case all derivations are valid. However, the Szilard language consists of the sequences of rules of a grammar rather than the terminal words that are generated. It is known that Szilard languages can be more complicated than the ones generated in the conventional manner by the respective grammars. For example, context-free grammars can have non-context-free Szilard languages. For a survey of early results on both Szilard languages and control languages we refer to the book by Salomaa [12].

2   Preliminaries

By Σ we usually denote a finite alphabet, and then Σ^* is the set of all strings over this alphabet, including the empty string λ. For a string w we denote by $|w|$ its length. $|w|_u$ denotes the number of distinct occurrences of u in w. A *prefix* of a word w is any factor starting in the first position. Language classes from the theory of generative grammars that should be familiar to the reader are regular, context-free and context-sensitive languages. For details about these we refer to the book of Salomaa [12].

The observed systems in our architecture will be string-rewriting systems. Concerning these we follow notations and terminology as exposed by Book and Otto [1]. We only recall briefly the most basic notions needed here.

Definition 1. A *string-rewriting system* R on an alphabet Σ is a subset of $\Sigma^* \times \Sigma^*$. Its elements are called *rewrite rules*, and are written either as ordered pairs (ℓ, r) or as $\ell \to r$ for $\ell, r \in \Sigma^*$. By $\mathsf{Dom}(R) := \{\ell : \exists r((\ell, r) \in R)\}$ and $\mathsf{Range}(R) := \{r : \exists \ell((\ell, r) \in R)\}$ we denote the set of all left-hand respectively right-hand sides of rules in R.

The *single-step reduction relation* induced by R is defined for any $u, v \in \Sigma^*$ as $u \Rightarrow_R v$ iff there exists an $(\ell, r) \in R$ and words $w_1, w_2 \in \Sigma^*$ such that $u = w_1 \ell w_2$ and $v = w_1 r w_2$. The *reduction relation* $\overset{*}{\Rightarrow}_R$ is the reflexive, transitive closure of $\Rightarrow_R$. If the rewriting system used is clear, we will write simply $\Rightarrow$, omitting its name. A string w is called *irreducible* with respect to R, if no rewrite rule from R can be applied to it, i.e. it does not contain any factor from $\mathsf{Dom}(R)$. The set of all such strings is denoted by $IRR(R)$.

By imposing restrictions on the set of rewriting rules, many special classes of rewriting systems can be defined. Following Hofbauer and Waldmann [9] we will call a rule (ℓ, r) *inverse context-free*, if $|r| \le 1$. A string-rewriting system will be called *inverse context-free*, if all of its rules are inverse context-free and always $|r| = 1$; otherwise the system is called *inverse context-free with deleting rules*. A string-rewriting system is called a *painter* system if for all its rules (ℓ, r) we have $|\ell| = |r| = 1$; this means the every rule just replaces one letter by another one.

3   Computing by Observing Changes

We now proceed to define the formal implementation of the ideas described in the introductory section. In contrast to earlier implementations of the Computing by Observing paradigm, here we do not need to define explicit observers. The observation is not derived from reading the observed system's configuration with a device; rather the change effected by the application of a rule in that observed system can be already directly the observation.

Definition 2. A *change-observing acceptor* is a tuple $\Omega = (\Sigma, P, F, \mathcal{O})$, where Σ is the input alphabet, P is the rule set of a string-rewriting system over an alphabet that contains Σ, F is a language that is called the *filter*, and the *Observer* $\mathcal{O}$ is a mapping from P to the alphabet of F.

So the observer assigns a letter from the alphabet of F to each rule in P. If we take a filter over an alphabet like $\{-2, -1, 0, 1, 2\}$, this can explicitly reflect the change in some quantity that the rule effects. The operation of a change-observing acceptor is as follows. The input word is processed by the string-rewriting system P. Every time a rule r is applied, $\mathcal{O}(r)$ is appended to the word called the *observation*, starting from the empty string. The input word is accepted if the string-rewriting system halts and the observation is an element of the filter F. Otherwise the input word is rejected. We illustrate the definitions and the operation of change-observing acceptors by two examples.

Example 3. Consider a test tube, in which two types of reactions can occur: one consumes (i.e. binds) energy, the other one sets energy free. Good input words for this system are ones that can be processed completely; this means they cannot consume more energy than is present in the system initially. Further, their processing should not lead to excessive setting free of energy, because this might lead to technical troubles.

We model this situation by the following change-observing acceptor: the input alphabet is $\{a, b\}$, the string rewriting system with the energy values already assigned is $\{(a \to c, 1), (b \to c, -1)\}$; thus processing the symbol a sets one unit of energy free (symbol 1 is observed when $a \to c$ is executed in the observed system), while processing of b binds one unit (symbol -1 is observed). Let k be our constant such that the energy balance should never be greater than k and never be smaller than $-k$. So our filter is the language

$$\{w : \text{for every prefix } u \text{ of } w \text{ there is } \big||u|_1 - |u|_{-1}\big| \leq k\}.$$

This filter is clearly regular. The language accepted by the system, however is

$$\{w : \big||w|_a - |w|_b\big| \leq k\}.$$

Thus in prefixes of words in the language there can be arbitrarily big imbalances between as and bs, and the resulting language is known not to be regular. So we can accept non-regular languages with very simple string-rewriting systems and regular filter. In this case the reason is that the string-rewriting system does not need to process the input string from left to right but can jump back and forth arbitrarily.

In a very straight-forward manner, this method can be extended to more than two letters, and in this way even non-context-free languages can be accepted.

Next we take a look at a string-rewriting system that is somewhat more complicated, because the rules have left sides of length greater than one. In this way, the relative position of different letters in the input string can be checked.

Example 4. The input alphabet is $\{a, b, c\}$, and we assign to the three letters the weights 1, 2, and 3 respectively. In this way the results of the observation of the rules are not assigned arbitrarily, but are computed from the weight of the letters that are involved, as the difference between the right side and left side total weight of the rules. Further the working alphabet contains a letter A of weight 6. The string-rewriting rules are the following, where we explicitly give the weight for ease of reading: $(aa \rightarrow a, -1)$, $(bb \rightarrow b, -2)$, $(cc \rightarrow c, -3)$, $(abc \rightarrow A, 0)$, $(a \rightarrow A, 5)$, $(b \rightarrow A, 4)$, $(c \rightarrow A, 3)$.

The filter applied to the observations is defined as the language $(-1-2-3)^*0$. The iteration of complete blocks $(-1 - 2 - 3)$ ensures that the same number of all three letters is deleted. The final 0 checks the structure of the original word, i.e. whether there were first as followed by bs, then cs. The other rules have the function of checking that no other letters were present; they would be rewritten before the system halts and thus cause the observation of a positive number.

So the language accepted by the observation of this inverse context-free string-rewriting system is $\{a^n b^n c^n : n > 0\}$, which is not context-free. We strongly believe that this language cannot be accepted using the simpler painter rules employed in Example 3.

These initial examples demonstrate that observation as defined above can increase the computational power of the total system compared to the power of the single components. We now proceed to characterize more exactly this computational power for specific variants of change-observing acceptors.

4 The Power of Change-Observing Acceptors

To start with, we investigate the case of the very simple painter systems, which can only replace one symbol by another.

Proposition 5. *The class of languages accepted by change-observing acceptors with painter string-rewriting systems is incomparable to the classes REG and CF.*

Proof. Example 3 has already shown that change-observing acceptors with painter string-rewriting systems can accept non-regular and even non-context-free languages. On the other hand, these acceptors cannot distinguish between, for example, the words ab and ba. The letters a and the b are always rewritten independent of their relative position to other letters; therefore any computation that is possible rewriting them separately in one of these words will also be possible in the other. Thus there are even finite languages that cannot be accepted by change-observing acceptors with painter string-rewriting systems.

So the class of languages accepted by change-observing acceptors with painter string-rewriting systems is somewhat orthogonal to the lower part of the Chomsky Hierarchy. This changes when we use inverse context-free string-rewriting. We obtain a nice characterization of a language class well-known from the theory of regulated rewriting. Change-observing acceptors with inverse context-free string-rewriting can be shown to be equivalent to state grammars, a construct from the theory of regulated rewriting that was introduced by Kasai [10]. The proof is rather long and technical, it will be published in a different context [3].

Theorem 6. *The class of languages accepted by change-observing acceptors with inverse context-free string-rewriting systems is equal to the class of languages generated by state grammars with context-free rules.*

It is worth mentioning that the class of languages characterized in Theorem 6 is maybe best known as the class of languages generated by matrix grammars, which are equivalent to state grammars [7]. Further, it is interesting to compare the results presented here with the corresponding ones for Monadic Transducers as observers. In that case painter systems and inverse context-free systems lead to the same power, namely all context-sensitive languages are recognized with either type [6]. Thus with the restriction of our observers, we lose computational power, especially for painter systems.

For matrix grammars it is a long-standing open problem whether deleting rules increase their generative power or not. In our case this does not change the expressive power.

Theorem 7. *The class of languages accepted by change-observing acceptors with inverse context-free string-rewriting systems is equal to the class of languages accepted by change-observing acceptors with inverse context-free string-rewriting systems with deleting rules.*

Proof. We only need to show that inverse context-free string-rewriting systems with deleting rules can be simulated by systems without deleting rules in our context. For a given change-observing acceptor $\Omega = (\Sigma, P, F, \mathcal{O})$ with an inverse context-free string-rewriting system P we first construct a derived string-rewriting system R that does not have deleting rules. For this we introduce a new symbol ϵ, which will be generated instead of the empty string in deleting rules. So for every rule $u \to \lambda \in P$ we add $u \to \epsilon$ to R. Their image under $\mathcal{O}$ shall be that of the original rule.

Now for rules with left sides longer than one, ϵ might occur between the symbols. Therefore we add for such a rule $u \to x \in P$ all rules $u' \to x$ where u' is obtained by inserting ϵ at the beginning, at the end, or between any two symbols of u. These are only finitely many, to be more exact $2^{|u|+1}$. Also here, their image under $\mathcal{O}$ shall be that of the original rule. It remains to solve the problem of several adjacent ϵ. For this we add the rule $\epsilon^2 \to \epsilon$ and it will be mapped by $\mathcal{O}$ to a new letter y in the observation alphabet.

It remains to define the filter. It can be very similar to the original one, because rule applications in P and R are in one-to-one correspondence in the following way: let h be a morphism mapping every letter to itself except for ϵ, which is deleted. Then if $w_1 \to_P^* w_2 \to_P w_3$ there is a word w_2' such that $h(w_2') = w_2$ and $w_1 \to_R^* w_2' \to_R w_3'$ with $h(w_3') = w_3$, and the last rule applied under R is derived from the last rule applied under P. Derivations might be longer under R if the rule $\epsilon^2 \to \epsilon$ needs to be applied. This might occur at any point. Therefore our filter is the original F with arbitrarily many y inserted between the first and last letter of every word. This insertion preserves regularity.

Of course, this does not contribute anything toward a solution of the matrix grammar problem. Since our systems reduce words while grammars generate them, one would have to consider rules of the form $\lambda \to u$ to obtain an acceptor that might be equivalent to matrix grammars with deleting rules.

5   Conclusion

We have presented a proposal for more realistic yet very general observers under the Computing by Observing paradigm. In contrast to the observers used so far, they do not need to read the entire system, rather they just notice a change in some quantity. In this way they have constant running time and do not need to physically interfere with the observed system. In this they are much closer to real recording devices.

However, our systems still rely on a few assumptions that might not be easy to meet in reality. The division into discrete steps seems like one, because biochemical systems typically do not follow any clock. However, we do not require the single steps to have equal length. In the present model where only certain changes are recorded, one step can simply be the variable time between one event and the other. A problematic case is the one where this time is zero, i.e. two or more events take place in parallel. For example, this would mean an infinite number of possible observations if we observe energy being consumed or set free like in Examples 3 and 4, namely all integers would be possible observations. Many computational models work with strong parallelism, i.e. all reactions that can take place have to take place in a given step. It would be a challenge to devise an implementation of the Computing by Observing paradigm that is able to deal with a situation where any number of reactions can occur in every step.

References

1. Book, R., Otto, F.: String-Rewriting Systems. Springer, Berlin (1993)
2. Cavaliere, M., Frisco, P., Hoogeboom, H.J.: Computing by Only Observing. In: Ibarra, O.H., Dang, Z. (eds.) DLT 2006. LNCS, vol. 4036, pp. 304–314. Springer, Heidelberg (2006)
3. Cavaliere, M., Leupold, P.: Computing by Observing: Simple Systems and Simple Observers (2009) (submitted)
4. Cavaliere, M., Leupold, P.: Evolution and Observation: A New Way to Look at Membrane Systems. In: Martín-Vide, C., Mauri, G., Păun, G., Rozenberg, G., Salomaa, A. (eds.) WMC 2003. LNCS, vol. 2933, pp. 70–87. Springer, Heidelberg (2004)
5. Cavaliere, M., Leupold, P.: Evolution and Observation — A Non-Standard Way to Generate Formal Languages. Theoretical Computer Science 321, 233–248 (2004)
6. Cavaliere, M., Leupold, P.: Observation of String-Rewriting Systems. Fundamenta Informaticae 74(4), 447–462 (2006)
7. Dassow, J., Păun, G.: Regulated Rewriting in Formal Language Theory. Springer, Berlin (1989)
8. Ginsburg, S., Spanier, E.H.: Control Sets on Grammars. Mathematical Systems Theory 2(2), 159–177 (1968)
9. Hofbauer, D., Waldmann, J.: Deleting string rewriting systems preserve regularity. Theoretical Comput. Sci. 327(3), 301–317 (2004)
10. Kasai, T.: An Hierarchy Between Context-Free and Context-Sensitive Languages. Journal of Computer and Systems Sciences 4(5), 492–508 (1970)
11. Moriya, E.: Associate languages and derivational complexity of formal grammars and languages. Information and Control 22, 139–162 (1973)
12. Salomaa, A.: Formal Languages. Academic Press, New York (1973)

Robustness of the Critical Behaviour in a Discrete Stochastic Reaction-Diffusion Medium[*]

Nazim Fatès[1] and Hugues Berry[2]

[1] INRIA, MaIA team, Nancy, France
[2] INRIA, ALCHEMY team, Saclay, France

Abstract. We study the steady states of a reaction-diffusion medium modelled by a stochastic 2D cellular automaton. We consider the Greenberg-Hastings model where noise and topological irregularities of the grid are taken into account. The decrease of the probability of excitation changes qualitatively the behaviour of the system from an "active" to an "extinct" steady state. Simulations show that this change occurs near a critical threshold; it is identified as a nonequilibrium phase transition which belongs to the directed percolation universality class. We test the robustness of the phenomenon by introducing persistent defects in the topology : directed percolation behaviour is conserved. Using experimental and analytical tools, we suggest that the critical threshold varies as the inverse of the average number of neighbours per cell.

Keywords: Reaction-diffusion media, stochastic cellular automata, phase transitions.

Foreword: See `http://webloria.loria.fr/~fates/Amybia/reacdiff.html` for accessing the complete set of experiments this paper refers to. The simulations presented in this paper were made with the *FiatLux* [1] cellular automata simulator.

1 Introduction

Let us consider the Greenberg-Hastings cellular automaton (GHCA) for the simulation of travelling-wave propagation in an excitable medium [2]. This model, and its different variants, have been intensively studied as a means of capturing the qualitative properties of various natural systems. It was also suggested that reaction-diffusion media could provide a path for building new types of computing devices [3]. For instance, we have recently proposed a bio-inspired system based on GHCA, to achieve decentralized and robust gathering of mobile agents scattered on a surface or computing tasks scattered on a massively-distributed computing medium [4]. As usual with such models, GHCA has mainly been studied using an homogeneous and regular lattice. However, in the context of massively distributed computing, one also needs to consider unreliable elements

[*] This research was funded by the AMYBIA ARC grant of the INRIA.

F. Peper et al. (Eds.): IWNC 2009, PICT 2, pp. 141–148, 2010.
© Springer 2010

and defect-based noise. A first analysis showed that in this case, phase transitions could govern the behaviour of the system [4]. Our goal now is to broaden the knowledge on stochastic reaction-diffusion media by investigating how such systems behave when various types of noise are introduced.

Specifically, we focus on two types of noise or heterogeneities : (a) *temporary failures*: the excitation from one cellular automata cell may temporarily fail to propagate to a nearest-neighbour cell; (b) *definitive failures*: we consider two types of definitive failures; (1) a cell never gets excited; this is equivalent to adding "holes" in the lattice; (2) a cell permanently fails to have influence on one of its nearest-neighbours; this is equivalent to deleting links in the lattice. We show that variation of the probability to receive an excitation triggers a non-equilibrium phase transition between an "active" and an "extinct" macroscopic state and we study how the critical threshold varies when definitive failures are introduced in the model.

2   Model and Methods

2.1   The Model

We study a two-dimensional stochastic version of the GHCA, where space is modelled by a square lattice $\mathcal{L} = \{1, \ldots, L\} \times \{1, \ldots, L\}$ and each cell is denoted by its coordinates $c = (c_x, c_y) \in \mathcal{L}$. The set of possible states for each cell is $\{0, \ldots, M\}$: state 0 is the *neutral* state, state M is the *excited* state and states 1 to $M-1$ are the *refractory states*. Each cell c is associated with a set of cells $\mathcal{N}_c$, its *neighbourhood*, which captures the locality of the influences between cells (see below). We denote by σ_c^t the state of a cell c at time t. Let E_c^t be the set of excited cells in $\mathcal{N}_c$ at time t: $E_c^t = \{c' \in \mathcal{N}_c \,|\, \sigma_{c'}^t = M\}$. The cells of the GHCA are updated according to:

$$\sigma_c^{t+1} = \begin{cases} M & \text{if } \sigma_c^t = 0 \text{ and } \operatorname{card}\{E_c^t\} > 0; \text{ with probability } p_T, \\ \sigma_c^t - 1 & \text{if } \sigma_c^t \in \{1, \ldots, M\}, \\ 0 & \text{otherwise} . \end{cases} \quad (1)$$

The value of $(1 - p_T)$ represents the rate of temporary excitation failures in the dynamics (i.e. noise).

In this paper, we explore the effects of different neighbourhoods on this system: (a) 4-nearest-neighbours (NESW adjacent cells); (b) 8-nearest-neighbours (adjacent + diagonal cells); (c) 6-nearest-neighbours (adjacent cells on a hexagonal grid) and (d) circular neighbourhoods [5]. We present results for periodic (toroidal) boundary conditions but we checked that bounded conditions (cells on the boundaries have less neighbours) give identical results. We use two methods to model definitive failures in the system: (1) We choose some cells at random before starting the simulation and definitely remove these cells from the lattice (i.e. $\sigma_c^t = 0$, $\forall t$ for these cells). Missing cells, or *holes*, are selected with probability HR (hole rate). (2) We first construct the regular neighbourhood of each cell, then we remove each link to a neighbour with probability MLR (missing link rate).

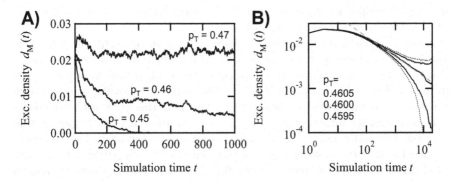

Fig. 1. Evolution of the density of excited cells $d_M(t)$. (A) Single run with three values of the transmission rate: sub-critical $p_T = 0.45$, critical $p_T = 0.46$, supra-critical $p_T = 0.47$. (B) Log-log plot of the average of 30 runs for, from bottom to top (full lines) : $p_T = 0.4595, 0.4600$ and 0.4605. The dotted lines indicate $-$ and $+$ 1 s.d. for $p_T = 0.4595$ and 0.4605, respectively. The gray dashed line indicates a decay compatible with the behavior expected for 2D directed percolation : $d_M(t) \propto t^{-0.451}$. $L = 400$ (Left) or 800 (right); von Neumann topology and $M = 4$.

2.2 Effect of Noise

To evaluate the effects of local temporary failures (noise) on the *macroscopic* behaviour of the model, we monitored the evolution with simulation time of a macroscopic quantity, the overall density of excited cells $d_M(t)$ (i.e. the total number of cells in state M at time t divided by the total number of cells). Fig. 1A shows prototypical evolutions of $d_M(t)$ for different failure levels. When excitation probability p_T is high, the system reaches a steady-state where d_M oscillates around a fixed value. In this steady state, excitation waves fluctuate in space and time, but their overall number is rather stable. This state can thus be referred to as the *active* steady-state. Now, when temporary failures are too frequent (*i.e.*, for low values of p_T), the number of excited cells rapidly vanishes. The system reaches an *extinct* state, where all cells are neutral and where the activity is definitively washed out.

Plotted in log-log coordinates (Fig. 1B), the convergence to the active steady state corresponds to an upward curvature of the density while convergence to the extinct state is accompanied by a down-ward curvature (exponential decay). As shown in this figure, a very small change in the value of p_T has huge long-term effects, as it can decide which steady state, whether extinct or active, is reached. This indicates the presence of a phase transition around a *critical threshold* p_{TC}, that separates the p_T values leading to the active state from those leading to the extinct state. Statistical physics tells us that near the critical point, the laws governing the phase transition are power laws [6]. For instance in our case, one expects $d_M(t) \sim t^{-\delta}$ at the critical threshold $p_T = p_{TC}$ which means that, in log-log plots as in Fig. 1B, the critical threshold should manifest as a straight line. It is also a well-established fact that different systems may display identical critical exponent values (e.g. δ) [6], which means these systems share identical fundamental properties. Systems sharing the same critical exponent values are said to belong to the same *universality class*.

Note that statistical physics predicts that δ is a universal quantity (i.e., its value does not depend on details of the model) for which we have precise numerical estimations. By contrast, the location of the critical threshold p_{TC} is not universal and must be evaluated by other means. In the following, we used both simulations and mean-field techniques to study how this critical threshold depends on the lattice inhomogeneity. To test for the robustness of the phase transition, our method consisted in spotting changes of concavity in log-log curves and checking that the best-fit exponent thus obtained is compatible with the expected value of δ (0.451 in our case, see [6,7,8]). Using this approach, we refined the measures on p_{TC} until the accuracy is of the order of 10^{-4}. In most cases, it was sufficient to take a grid size of $L = 800$, an observation time $T = 20000$ and to average the data over less than S = 40 runs.

3   Robustness Studies

3.1   Robustness to Variations of the Excitation Level

As a first step for the analysis of the robustness of the phase transition, we varied M and examined how p_{TC} is modified. Fig. 2A and B show the decay of the density of excited cells $d_M(t)$ during simulations carried out with $M = 4$ or 7. First of all, we observed that the critical exponent δ was in both cases compatible with directed percolation (i.e., the values determined from the simulations were statistically compatible with those for directed percolation). Beyond the data presented in this figure, we evaluated δ for M values ranging from 2 to 10 and various topologies (see our website[1]). We noticed that p_{TC} slightly increases with M (from 0.4245 to 0.4947 when M ranges from 2 to 10). At first sight, one may judge this variation moderate (less than 17%). It is an open question to determine if there is an asymptotic behaviour of the system for large values of M. From a practical point of view, an increase in M results in the apparition of longer waves, which implies than larger grids are required to measure the phase transition. However, for all the values of M we tested, the phase transition did belong to the universality class of directed percolation.

3.2   Holes and Missing Links

To evaluate the effects of permanent failures on the phase transition, we varied the hole rate HR, and the missing link rate MLR, from 0 to 0.40 by increments of 0.10. As expected, the critical threshold increased with the defect rate. More surprisingly, we observed that as the defect rates increased, the critical threshold p_{TC} became more difficult to determine. The limits of our protocol were reached for HR = 0.40 or MLR = 0.40, see Fig. 2C and D. In both cases (faulty cells or links), the limitation stemmed from the difficulty to clearly spot the changes

---

[1] See http://webloria.loria.fr/~fates/Amybia/reacdiff.html

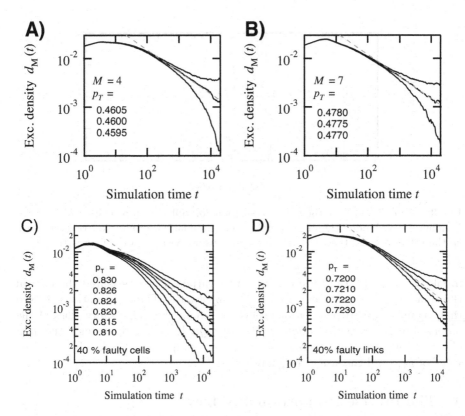

Fig. 2. (A-B) Robustness of the phase transition to the number of excitation level M. Decay of the instantaneous densities of excited cells $d_M(t)$ for $M = 4$ (A) and 7 (B). (C-D) Robustness of the phase transition with defects in the lattice. 40% of the cells (C) or of the links (D) were destroyed. The curves are obtain with $S = 15$ runs, each run starts from different initial conditions and defect configurations. Simulations conditions : $M = 4$, $L = 1200$; von Neurmann topology. Gray dashed lines as in fig.1.

of curvatures. We observed that even in large-size simulations, sub- and supra-critical curves are close to a power law. Experiments show that it is not possible to compensate this difficulty by a mere increase of the lattice size; it is an open problem to understand this "blurring effect" with more details.

However, it is impressive to observe that in all the experiments we conducted, δ remained compatible with directed percolation (other simulations are available on our web site). Hence, albeit the precision is lower, these simulations show that the phase transition and its universality class are conserved in our model, even when close to half of the cells or links are destroyed.

3.3 Changing the Topology

Fig. 3 shows determination of critical threshold p_{TC} for various topologies and $M = 4$. Clearly, we observed that the critical threshold varies greatly with the

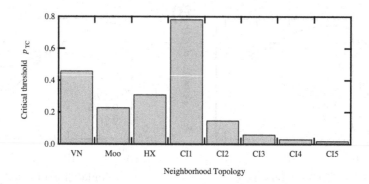

Fig. 3. Critical values of the phase transition for different topologies : VN : Von-Neuman; Moo : Moore; HX : Hexagonal; : CIx : Circular with radius x. Error estimates on the values are less than 10^{-3} in all cases, thus not visible on the figure. $M = 4$.

topology. To understand the origin of this variation, we measured the critical threshold and exponent for *more than 20 topologies* . These simulations showed that, again, directed percolation seems to remain the universality class of the observed phase transition *whatever the topology*. Interestingly enough, the data of Fig. 3 suggested that there exists an inverse proportionality law between the value of p_{TC} and the size of neighbourhood.

4   The Inverse Proportionality Law

We now present an approximation of the behaviour of the model Eq.(1) using mean-field arguments. We only give here a rapid description of this analysis, leaving a more complete account of the calculus for future presentations. Let $d_i(t)$ be the density of cells in state i at time t. Conservation principles derived from the transition rules can be expressed as a set of ordinary differential equations:

$$\begin{cases} \dot{d_0}(t) = d_1(t) - p_T d_0(t)\pi_0(t) \\ \dot{d_j}(t) = d_{j+1}(t) - d_j(t); \quad \forall j \in \{1, \dots, M-1\} \\ \dot{d_M}(t) = p_T d_0(t)\pi_0(t) - d_M(t) \end{cases} \tag{2}$$

where $\pi_i(t)$ is the conditional probability that a cell c in state i has at least one excited cell in its neighbourhood. Under several mean-field-type assumptions (e.g., spatial correlation are neglected, cell states are assumed homogeneously distributed over space), we find that the steady-states for the density of excited cells d_M^* obey:

$$d_M^* - p_T \left(1 - M d_M^*\right) \left[1 - (1 - d_M^*)^N\right] = 0. \tag{3}$$

Eq. (3) has two solutions for a given p_T: one trivial steady-state $d_{M,1}^* = 0$ $\forall p_T$ ("extinct" steady-state) and an active one, $d_{M,2}^*$ that depends on p_T. Importantly, these two steady-states intersect for some unique value of p_T: $d_{M,1}^* =$

$d_{M,2}^{*}$ for $p_T = p_T^{*}$. For $p_T < p_T^{*}$, the mean-field approach predicts that the system reaches the extinct state $(d_{M,1}^{*} = 0)$ asymptotically, while the active state $(d_{M,2}^{*})$ is attained for $p_T > p_T^{*}$. This means p_T^{*} is the mean-field approximation for p_{TC}. To evaluate p_T^{*}, a series expansion at order 1 for $d_M^{*} \to 0$ approximates Eq.(3) as $(1 - p_T N) d_M^{*} + O\left(d_M^{*\,2}\right) \approx 0$. Note that M has vanished from the order 1 term in this approximation. p_T^{*} is given by $(1 - p_T^{*} N) = 0$, so that one gets $p_{TC} \approx 1/N$. This indicates that p_{TC} should scale as the inverse of N. Moreover, p_{TC} is expected to depend on the average neighbourhood size N but not on the excitation level M.

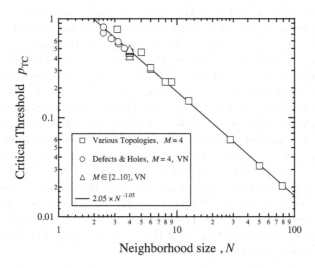

Fig. 4. Illustration of the proportionality law between p_{TC} and $1/N$. Each data point reports the value of the critical threshold for a given condition as a function of the corresponding neighbourhood size N (the average number of neighbours). The red line shows the best fit to a power law $p_{TC} = 2.05 \times N^{-1.05}$.

This prediction is tested on Fig. 4. The critical thresholds for all the simulation we have run (various topologies, various defect rates, various M values) were plotted as a function of the corresponding neighbourhood size N. Strikingly, all the points collapsed on a single line in log-log coordinates. This confirms the power law dependence of p_{TC} on N. Fitting over all the data we obtained, we get:

$$p_{TC} \approx 2/N. \tag{4}$$

Therefore, our mean-field approach nicely predicts the observed N^{-1} dependency of the critical threshold. It thus shows that the important quantity to determine the value of the critical threshold is the average number of neighbours per cell and not the detail of the topology or its homogeneity. The mean-field approach however fails to predict several quantitative features of the simulations. First of all, the value of the prefactor in the scaling law Eq. 4 was predicted to be 1, whereas the simulations rather indicate a value of 2. More intriguingly, the mean field approach predicts that p_{TC} does not depend on the number of cell states M. Yet, we have already seen above that this is not observed in our simulations: p_{TC} increases slightly with M. However, this increase is moderate enough that the observed values are still roughly in agreement with the N^{-1} scaling, as seen in Fig 4 (triangles).

5   Conclusion

The observations reported in this paper confirmed the great robustness of the directed percolation universality class. Although this statement is not new, to our knowledge, it is the first time that this robustness was confronted against such a large number of perturbations: holes in the grid, missing links, change of topology. This impressive robustness makes the phase transition in this model an interesting phenomenon for the design of novel computing devices.

Note that a *conjecture* by Janssen and Grassberger (see [6] for an overview of this topic) lists the fundamental properties that a system must display in order to belong to this universality class. Clearly, in spite of the added noise and hetero-geneity, the different systems we considered all validate these conditions, so that the amazing robustness of the phase transition in these systems unambiguously pleads in favour of the validity of this conjecture. However, as far as we know, there exists no theory to predict the value of the critical threshold. In a context of building computing devices that would use reaction-diffusion waves, deter-mining the value of the critical threshold would be of paramount importance. The inverse proportionality law we presented paves the way for obtaining generic laws (even approximate ones) to predict the position of the critical threshold in various simulation conditions.

References

1. Fatès, N.: Fiatlux CA simulator in Java, http://webloria.loria.fr/~fates/
2. Greenberg, J.M., Hassard, B.D., Hastings, S.P.: Pattern formation and periodic structures in systems modeled by reaction-diffusion equations. Bulletin of the Amer-ican Methematical Society 84(6), 1296–1327 (1978)
3. Adamatky, A., De Lacy Costello, B., Asai, T.: Reaction-Diffusion Computers. Else-vier, Amsterdam (2005)
4. Fatès, N.: Gathering agents on a lattice by coupling reaction-diffusion and chemo-taxis. Technical report, INRIA Nancy Grand-Est (2008),
   http://hal.inria.fr/inria-00132266/
5. Markus, M., Hess, B.: Isotropic cellular automaton for modelling excitable media. Nature 347, 56–58 (1990)
6. Hinrichsen, H.: Nonequilibrium Critical Phenomena and Phase Transitions into Ab-sorbing States. Advances in Physics 49, ch. 43, 847–901 (2000)
7. Ódor, G.: Universality classes in nonequilibrium systems. Reviews of modern physics 76, 663–724 (2004)
8. Fatès, N.: Asynchronism induces second order phase transitions in elementary cel-lular automata. Journal of Cellular Automata 4(1), 21–38 (2009),
   http://hal.inria.fr/inria-00138051

Quantifying the Severity of the Permutation Problem in Neuroevolution

Stefan Haflidason and Richard Neville

School of Computer Science, University of Manchester, UK
{haflidas,rneville}@cs.man.ac.uk

Abstract. In this paper we investigate the likely severity of the Permutation Problem on a standard Genetic Algorithm used for the evolutionary optimisation of Neural Networks. We present a method for calculating the expected number of permutations in an initial population given a particular representation and show that typically this number is very low. This low expectation coupled with the empirical evidence suggests that the severity of the Permutation Problem is low in general, and so not a common cause of poor performance in Neuroevolutionary algorithms.

1 Introduction

One of the defining characteristics of algorithms in the field of Neuroevolution (Evolutionary Algorithms applied to the optimisation of Neural Network topology and weights) is their use or avoidance of recombination. Due to perceived inefficiencies with recombining two neural network representations caused by the so-called *Permutation Problem*, early approaches tended to either avoid recombination [1,2,3] or employ modified operators or representations with the aim of making the recombination operator 'safe' to use [4,5,6,7,8,9,10]. In early research the Permutation Problem was considered to be a serious obstacle to the application of Evolutionary Algorithms[1] to the evolution of Neural Networks, as it was thought that the Permutation problem "may render crossover essentially impotent as a search operator, at least until the population has converged on a single convention" [11], i.e., crossover would only cease to be a deleterious operator once the population had converged around one individual.

The severity of the Permutation Problem has been questioned in the past however, with empirical results suggesting that its effects do not appear serious in practise given that an unmodified crossover operator can be seen to perform acceptably well [12,13], and because the Permutation Problem appears to occur very rarely, if at all in practise [14].

In contrast to past, purely empirical studies we present an initial analysis of the likely severity of the Permutation Problem by calculating the probability of its occurrence in the initial generation of an evolutionary run. Empirical results

---

[1] In particular Genetic Algorithms, which tended to employ crossover as their main search operator.

F. Peper et al. (Eds.): IWNC 2009, PICT 2, pp. 149–156, 2010.

from [14] suggest that the problem occurs rarely or not at all during an evolutionary run. This work aims to test the hypothesis that this rarity is due to there being very few or no permutations in the initial population. It is then hypothesised that given very few or no permutations, the typical genetic operators do not readily produce more permutations.

We begin by distinguishing between two possible types of permutation, *genotypic* permutations, where two genotype strings are permutations of each other, and *phenotypic* permutations, a subset of the genotypic permutations where whole neurons (i.e. blocks of weights) have been permuted as opposed to just single weights. The latter definition is the common interpretation of the Permutation Problem in the literature. We present an equation for the probability that a pair of individuals drawn uniformly are genotypic or phenotypic permutations of each other. This is then extended to arbitrary population sizes, allowing for the calculation of the expected number of permutations in the initial population.

Using this equation to examine even simple network spaces demonstrates the low probability of occurrence of either form of the problem in the initial generation. These results support those of the empirical investigation [14], where the number of permutations in the population is determined through explicit counting and is shown to agree approximately with the predicted value.

2   The Permutation Problem

The Permutation Problem is commonly defined as follows. Given a single layer feed-forward neural network, we can rearrange its hidden neurons in any order without affecting the function encoded by the network. In other words, the order in which we sum the contributions of the hidden neurons is unimportant in determining the network function. Given a network with N_h hidden neurons we can then form up to[2] $N_h!$ equivalent networks. Such permutations are formed in the network space and so we term them *phenotypic* permutations. For reasons of computational feasibility we mainly consider genotypic permutations in this paper. It is worth noting however that genotypic permutations have not been identified as a concern in the literature. As all phenotypic permutations are by definition also genotypic permutations their calculation gives us an upper bound on the probability that two individuals are phenotypic permutations.

An example of the Permutation Problem can be seen in Figure 1. Here, genotypes (a) and (b) are significantly dissimilar but in fact encode phenotypically-equivalent networks. This equivalence is due to the fact that the hidden neurons have been permuted; since neuron position is unimportant in this type of network model the networks then have identical behaviour and so equal chance of being recombined when selected under fitness-proportional selection.

If we apply any form of crossover operator to these genotypes we are highly likely to lose information from the shared distributed representation of these two networks due to duplication of one or more genes.

---

[2] It is important to remember to account for the duplicate permutations formed due to repeated allele values, i.e. the number of permutations is not always $N_h!$.

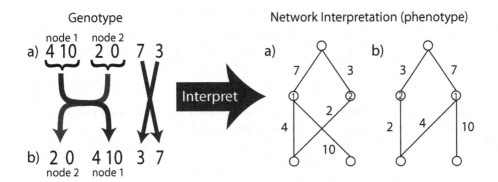

Fig. 1. An instance of the Permutation Problem: two networks composed of the same hidden neurons but in a different order. While phenotypically equivalent, the genotypes are incompatible for crossover: any recombination is likely to result in information loss. Of note however is that to form (b) from (a) the block of genes for each hidden node must be swapped, and the output weights inverted. Adapted from [15].

3   Calculating the Probability of Genotypic Permutations

In this section we outline the method for calculating the probability that two randomly-drawn individuals will be genotypic permutations of each other. We then generalise this to arbitrary population sizes.

Given an allele alphabet α and string length l we can fully enumerate all $|\alpha|^l$ possible strings, where $|\alpha|$ is the cardinality of the alphabet. This space of strings can be decomposed into groups where each group contains all strings with a particular selection of alleles, in any order. There are $\left(\!\binom{|\alpha|}{l}\!\right)$ such groups, where $\left(\!\binom{n}{k}\!\right)$ is the number of ways to select subsets of size k from n items with repetition, and is equivalent to the binomial coefficient $\binom{n+k-1}{k}$. We are interested in the probability of picking two distinct strings from the same group. For example, the group which contains strings with one copy of allele one and three copies of allele two, or the group which contains strings with two copies of allele one and two copies of allele two and so on, for all possible selections of allele values. Once we pick one string, the number of possible permutations that can be formed from it determines the size of the group and so the subset of strings which, if chosen, would form a pair which are permutations of each other.

For two allele values and genotype length l we can calculate the probability that a pair of individuals drawn at random are permutations as

$$P(2,l) = \sum_{i=0}^{l} \frac{\left(\frac{l!}{i!(l-i)!}\right)\left(\frac{l!}{i!(l-i)!}\right) - 1}{2^l} .$$

For three allele values this would be calculated as

$$P(3,l) = \sum_{i=0}^{l}\sum_{j=0}^{l-i} \frac{\left(\frac{l!}{i!j!(l-i-j)!}\right)\left(\frac{l!}{i!j!(l-i-j)!}\right) - 1}{3^l} .$$

The equation generalises to alphabets of length n as

$$P(n,l) = \sum_{i_1=0}^{l} \sum_{i_2=0}^{l-i_1} \cdots \sum_{i_{n-1}=0}^{l-(\sum^{n-2} i)} \frac{\left(\frac{l!}{\left(\prod_{j=1}^{n-1} i_j!\right)\left(l-\sum_{k=1}^{n-1} i_k\right)!}\right)\left(\frac{l!}{\left(\prod_{j=1}^{n-1} i_j!\right)\left(l-\sum_{k=1}^{n-1} i_k\right)!}\right) - 1}{n^l} .$$

(1)

To illustrate the strategy behind this arrangement we can expand the equation for the case of a binary alphabet and string length 4:

$$P(2,4) = \sum_{i=0}^{4} \frac{\left(\frac{4!}{i!(4-i)!}\right)\left(\frac{4!}{i!(4-i)!}\right) - 1}{2^4}$$

$$= \left(\frac{\left(\frac{4!}{0!4!}\right)\left(\frac{4!}{0!4!}\right) - 1}{2^4}\right) + \left(\frac{\left(\frac{4!}{1!3!}\right)\left(\frac{4!}{1!3!}\right) - 1}{2^4}\right) + $$

$$\left(\frac{\left(\frac{4!}{2!2!}\right)\left(\frac{4!}{2!2!}\right) - 1}{2^4}\right) + \left(\frac{\left(\frac{4!}{3!1!}\right)\left(\frac{4!}{3!1!}\right) - 1}{2^4}\right) + $$

$$\left(\frac{\left(\frac{4!}{4!0!}\right)\left(\frac{4!}{4!0!}\right) - 1}{2^4}\right) .$$

Here we can see that the probabilities of drawing two distinct strings from each possible group, for example the group with one copy of allele one and three copies of allele two (of which there are $\frac{4!}{1!3!} = \frac{24}{1 \cdot 6} = 4$) are considered, and the union of these disjoint events is taken which gives us the probability that we select two distinct strings from any one of the groups. Figure 2 shows the calculation of this probability for a range of genotype spaces.

If we take X to be a random variable representing the probability of encountering a pair of permutations given a particular permutation space, then the value of interest is its expectation which tells us how many permutations we can expect to see on average when drawing pairs at random. Given then the probability that a pair of individuals drawn at random are permutations of each other, we can calculate the expected number of permutations $E(X)$, which for a pair may be either zero or one:

$$E_{\text{perm}}(X) = (0)(1 - P(n,l)) + (1)(P(n,l)) = P(n,l) .$$

We then wish to calculate the expected value for a population of size p. Since the probability that one pair are permutations of each other is independent of the probability for any other distinct pair, we can achieve this by taking the probability for a single pair and multiplying by the number of possible pairings of individuals in the population. Thus for a population of size p we form the union of all $\binom{p}{2}$ events where each pair are permutations of each other, giving an expected number of permutation pairs of $\binom{p}{2}P(n,l)$. This is shown in Figure 3 where the expected number of genotypic permutations is calculated for all population sizes from 2 to 10,000 (dashed line).

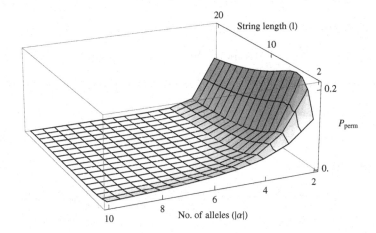

Fig. 2. The probability P_{perm} that a pair of individuals (drawn uniformly from a solution space defined by the number of alleles $|\alpha|$ and string length l) are genotypic permutations of each other

4   Calculating the Probability of Phenotypic Permutations

Calculating the probability of phenotypic permutations can be achieved using the same mechanism as for the genotypic permutations, simply by reinterpreting the parameters of Equation 1. Now, instead of number of alleles we are considering number of possible neurons, and instead of string length we have number of hidden neurons. The difficulty here is that due to the large number of possible neurons for even modest genotype spaces, calculating the probabilities is computationally infeasible with the method presented. So, for the phenotypic permutations example we consider a toy genotype space of only 5 values, $\alpha = \{-1.0, -0.5, 0, 0.5, 1.0\}$ and two hidden neurons. With such a coarse-grained genotype space we might expect that the probability of encountering individuals which are phenotypic permutations of each other would be high. As we will show however, even for this unrealistically coarse space the probability of drawing a population containing an appreciable number of permuted individuals is low.

We now present the calculation of the parameters to use in Equation 1 to calculate the expected number of phenotypic permutations in an initial population. Given an alphabet α we can count the number of unique hidden neurons ϕ in a typical feed-forward single-layer fully-connected network space as $|\alpha|^{N_i+N_o}$ where N_i and N_o are the number of inputs and outputs respectively. We can arrange these neurons into ϕ^{N_h} networks of N_h hidden neurons[3]. For our example with a network of three inputs, one output and $|\alpha| = 5$ we would then have

---

[3] However, due to the invariance of a network's function under permutation of its hidden neurons, there are only $\left(\!\!\binom{\phi}{N_h}\!\!\right)$ unique networks.

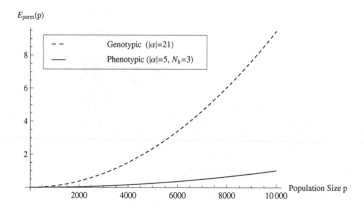

Fig. 3. Expected number of permutations $E_{\mathrm{perm}}(p)$ for a range of population sizes p, for 1) a genotype space of 21 alleles with string length 9 (genotypic permutations) and 2) for a phenotype space of a 3-input, 1-output, 3-hidden neuron neural network (phenotypic permutations) with a weight space of size 5

$\phi = |\alpha|^{N_i+N_o} = 5^4 = 625$ possible neurons[4]. We can now calculate the probability that two randomly drawn networks are phenotypic permutations of each other using Equation 1, giving $P(625, 2) = 2.6 \times 10^{-6}$. The expected number of phenotypic permutations for a range of population sizes using this representation is shown in Figure 3 (solid line).

5 Discussion

Figure 2 shows the expected number of genotypic permutations for a uniformly-drawn pair of individuals, for a range of alphabet cardinalities and string lengths. Here we can see that the expectation is high for binary alphabets but drops off rapidly for larger alphabets, which tells us that genotypic permutations are only common with binary alphabets. Even with unrealistically small alphabets, such as $\{-1.0, -0.5, 0, 0.5, 1.0\}$ which has cardinality 5, the expectation for genotypic permutations is low for most string lengths. For more realistic alphabets this expectation can be shown to decrease rapidly, suggesting that this kind of permutation is a relatively rare occurrence, at least in the initial population. Extending this to population sizes greater than two, we can see in Figure 3 that even for population sizes up to 10,000 the expected number of pairs of individuals which are permutations of each other is very low. For realistic network spaces this expectation will be considerably lower.

Compared to that of phenotypic permutations, the expected number of genotypic permutations in the initial population is higher, though it is debatable as to

---

[4] The number of unique three-input one-output networks with two hidden neurons would then be $\binom{625}{2} = 195{,}625$.

whether this is of any real concern as genotypic permutations are not frequently cited as being an obstacle in the application of Genetic Algorithms. Given that genotypic permutations may occur in any representation this would potentially have far-reaching implications, particularly for binary representations. The recombination of two individuals which are permutations of each other will result in a higher-than-average likelihood of repeated allele values in the offspring. While this may result in an increase in the production of poor offspring due to repetition of genes [9], population-based search techniques such as Evolutionary Algorithms are largely unaffected by such infrequent events in the general case.

Figure 3 shows the expected number of genotypic permutations, and an upper bound for the expected number of phenotypic permutations for the representation used in the empirical investigation [14], where a population size of 100 was used. It can be seen here that the expected number of permutations is low in both cases: $E_{\text{perm}}(100) = 9.4 \times 10^{-4}$ genotypic permutations; this is in accordance with the count of zero permutations in the actual evolutionary runs.

6   Conclusions and Future Work

It would appear that in some cases the frequency of the occurrence of the Permutation Problem may have been implicitly[5] overestimated, causing crossover to have been avoided or customised where it might have been a useful search operator in its canonical form.

The results of this paper strongly suggest that for standard GAs the Permutation Problem is not as serious as originally thought in early research into the evolution of Neural Networks. This paper has presented a method for calculating the degree of this severity in the initial population for any genotypic representation with fixed length and finite allele alphabet. This result is supported by an empirical experiment [14] on a standard benchmark problem (Non-Markovian Dual Pole Balancing), showing the low rate of occurrence of permutations in a practical setting. The equation presented (Equation 1) tells us how many permutation pairs we can expect to see in a population given a particular genotypic representation. It does not however lend insight into the probability of the appearance of permutations during an evolutionary run. The empirical results suggest that we will not on average encounter permutations during a run, however this remains to be confirmed analytically.

A limitation of this work is that due to the rapid growth of the number of multisets as a function of n and l the calculation of the probability for the Permutation Problem using Equation 1 is limited to very simple spaces such as that of the genotypic permutations. As the object of interest is phenotypic permutations, particularly for realistic network spaces, future work will be to find a more efficient method of calculation.

The presented results strongly suggest that poor performance in a given setup when using recombination is unlikely to be to due to the recombination of per-

---

[5] The authors have not seen any explicit probabilities assigned to the frequency of occurrence of the Permutation Problem in the literature.

mutations. The problem now is to better describe and qualify when and how recombination is useful in the context of Neural Networks, with the overall aim of more fully understanding the effect representation and operator choice can have so that informed decisions can be made when designing future Neuroevolutionary algorithms.

Acknowledgments

Stefan Haflidason would like to thank Arjun Chandra and Richard Allmendinger for their insightful comments and discussion of this work.

References

1. Angeline, P., Saunders, G., Pollack, J.: An evolutionary algorithm that constructs recurrent neural networks. Neural Networks (1994)
2. Yao, X.: Evolving artificial neural networks. Proceedings of the IEEE (1999)
3. Yao, X., Islam, M.: Evolving artificial neural network ensembles. Computational Intelligence Magazine (2008)
4. Belew, R., McInerney, J., Schraudolph, N.: Evolving networks: Using the genetic algorithm with connectionist learning. Tech. rept. CSE TR90-174 UCSD (1990)
5. Branke, J.: Evolutionary algorithms for neural network design and training. In: Proceedings of the 1st Nordic Workshop on Genetic Algorithms (1995)
6. Carse, B., Fogarty, T.C.: Fast evolutionary learning of minimal radial basis function neural networks using a genetic algorithm. Selected Papers from AISB Workshop on Evolutionary Computing (1996)
7. García-Pedrajas, N., Ortiz-Boyer, D., Hervás-Martínez, C.: An alternative approach for neural network evolution with a genetic algorithm: Crossover by combinatorial optimization. Neural Networks (2006)
8. Montana, D., Davis, L.: Training feedforward neural networks using genetic algorithms. In: Proceedings of the Eleventh International Joint Conference on Artificial Intelligence (1989)
9. Stanley, K.: Efficient evolution of neural networks through complexification. PhD Thesis, The University of Texas at Austin (2004)
10. Thierens, D.: Non-redundant genetic coding of neural networks. In: Proceedings of IEEE International Conference on Evolutionary Computation (1996)
11. Schaffer, J., Whitley, D., Eshelman, L.: Combinations of genetic algorithms and neural networks: a survey ofthe state of the art. Combinations of Genetic Algorithms and Neural Networks (1992)
12. Hancock, P.: Genetic algorithms and permutation problems: a comparison of recombination operators for neural net structure specification. In: Proceedings of the International Workshop on Combinations of Genetic Algorithms and Neural Networks (1992)
13. Froese, T., Spier, E.: Convergence and crossover: The permutation problem revisited. University of Sussex Cognitive Science Research Paper CSRP 596 (2008)
14. Haflidason, S., Neville, R.: On the significance of the permutation problem in neuroevolution. In: Proceedings of the 11th Annual conference on Genetic and Evolutionary Computation, GECCO (2009)
15. Yao, X., Liu, Y.: A new evolutionary system for evolving artificial neural networks. IEEE Transactions on Neural Networks (1997)

Extending the Geometrical Design
of DNA Nanostructures

Shogo Hamada[1,2] and Satoshi Murata[1]

[1] Department of Computational Intelligence and Systems Science,
Tokyo Institute of Technology, Yokohama 226-8502 Japan
[2] JSPS Research Fellow
hamada@mrt.dis.titech.ac.jp

Abstract. Structural DNA nanotechnology enables us to design and
fabricate shapes and patterns at nanoscale as a versatile platform for
nanotechnology and bio-related computing. Since the introduction of
crossover junctions, an endeavor to create nanostructures by DNA are
now flourished as self-assemblies of various 2-D and 3-D shapes. Those
achievements mainly owe to two factors: one is the geometry defined by
crossover junctions, and the other is the introduction of design approach.
The design approach itself is not dependent on any junction structure,
however the lack of choice in junctions limits the appearance of resultant
nanostructures. We found our interconnected single-duplex DNA junction extends the geometry of DNA nanostructures into a broader class
of shapes and patterns. Here we propose an abstraction method that
enables us to design variety of structures by those junctions with com-
patibility. Several demonstrations by this abstraction and possibilities
of various new shapes and patterns based on the design approach are
presented.

1 Introduction

DNA, life's genetic information carrier, is now also known as a basis for various
nanotechnologies. Structural DNA nanotechnology is one of the most promising
fields in nanotechnology for devices and platforms for bio-related computing.
The foundation of structural DNA nanotechnology is based on the invention of
DNA crossover junction structure (Fig. 1a)[1]. The structure uses an immobilized
stacked X formation of the Holliday junction found in nature, and performs
as a connection to allocate double helices side by side under appropriate ionic
concentration. The DNA double crossover molecules[2] combine two crossover
junctions in order to fix positions of two double helices in parallel. This molecule
is the first DNA-based molecule with rigidity, which pointed out that DNA can
work as a nanoscale building block for various shapes and patterns. Since then,
the crossover junction defined the way to design organization of DNA helices
with aligned formation; therefore the geometry of most DNA nanostructures is
regulated by this manner.

Two important design approach, strategies for DNA nanostructure designing,
were introduced among various shapes and patterns based on this geometry of

F. Peper et al. (Eds.): IWNC 2009, PICT 2, pp. 157–164, 2010.

crossover junctions. The structures are called "DX Tiles[3]" and 2D/3D "DNA Origami[5,6]". The former approach introduces a concept of "tiles", which uses double crossover molecule as a rigid repeating unit (a motif) for self-assembly. Both ends of each unit have single-stranded connectors called "sticky-ends" to assign appropriate connectivity using Watson–Crick complementarity between tiles; thereby it self-assembles into patterned planar structures. The latter Origami approach uses hundreds of short "staple" strands to fold a long single-stranded DNA into arbitrary shapes. Those ideas are not conceptually restricted to use crossover junctions, however, the lack of choices in junctions made the restriction: namely, the structures made by both methods are typified as parallel-aligned helices. Almost all structures at the present moment are using one of these approaches[7,8].

In this paper, we introduce an abstraction of interconnected single-duplex DNA junction, which extends the current geometry of DNA nanostructures. We show a design method of structures and the universality of this design by motif-based and origami-based approaches. Some experimental results of those designs are also demonstrated.

2   Interconnected Single-Duplex Junction (T-Junction)

2.1   Structural Description

The interconnected single-duplex junction consists of two DNA helices, one of them has a single-stranded domain in the middle of the helix and the other has at the end (Fig. 1b,c). Both single-stranded domains have complementary sequences, so they hybridize to form a T-shaped junction structure. Our previous study[9] indicates 5 and 6[bp] as preferable lengths for this connection. A branch can be placed at the major groove side of the helix. Our approximation of the phase difference in 6bp compared with the major and minor groove of

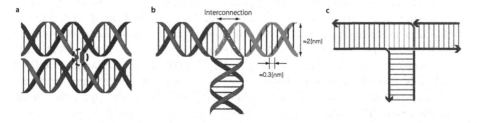

Fig. 1. DNA Junctions. **a** Helical representation of the crossover junction. Antiparallel crossover junction is shown at the black-dotted circle. **b** Helical representation of the T-junction. Each dark gray and black parts indicate the groups of double helix. The black duplex has a single-stranded region in the middle, and the dark gray has the complementary sticky-end region at the end. The light gray lines indicate the connected sticky-end region (5 bp in the figure). **c** Base-pair based diagram of the T-junction. Arrows indicate the 5'-3' direction.

double helix supports this preference. (Minor groove: 56°, Major groove: 4°.)
This result is also supported by an electrophoresis experiment (data not shown).
The most obvious difference between conventional crossover junctions and this
T-junction is the conformation of double helices. Crossover junctions basically
connect two helices side by side. On the other hand, T-junction connects two
helices at right angle. This difference in shapes provide us an extended geometry
for the structural design.

2.2 Abstraction

Most of DNA motifs were designed directly by base-pairs and actual DNA models so far, those representations are often become cumbersome especially in case
of design with several types of junctions. We found that the important points
in these designs are the relative phase angle difference and the length of helix
between junctions. In case of 2-D structures, those can be sufficiently described
by integral multiples of half-turns of helices. Our abstraction is based on a simplified representation of DNA double helices that satisfy the conditions above.
Figure 3a represents two units of the abstraction, which corresponds to a single
turn of the double helix.

A T-junction is represented as two perpendicularly connected squares on the
major groove side of DNA double helix (Fig. 3b). Note that the actual DNA's one
turn is shorter than this abstraction; however in order to represent T-junction at
the major groove, square representation is preferable. Likewise, crossover junctions are also compatible by this representation (Fig. 3c). Motifs and origami
structures can be designed using several connected blocks along with these junctions. Thanks to this abstraction, DNA nanostructures can be designed without
explicitly thinking about phase difference between helices and base-pair lengths
of double helix. T-junction based nanostructures are basically difficult to depict
the final appearance compared to crossover junction-based structures, which
can use pre-determined templates of aligned duplex in parallel. Even in those
cases, this abstraction enables us to design various nanostructures without defining relative position of helices beforehand. Also note that once we finished this
abstraction, conversion to the base-pair lengths and actual sequence can be determined semi-automatically by simple calculations (5.25[bp]/square) and with
an aid of computer algorithm.

3 Designing Shapes and Patterns

Various motifs can be designed using the abstraction. Here we show some tile-
based examples that we designed by this abstraction to show the versatility of
this technique. In the following examples we used only T-junctions, and utilized interconnected regions in the junctions as sticky-ends (Fig. 1b) between
motifs. We named these motifs as "T-motifs". Fig. 2 shows two variations of
1-D "ladder" structure. One block difference in ladder's rung makes the orientation of handrails upside down. Fig. 4 shows one of the planar lattice motifs,

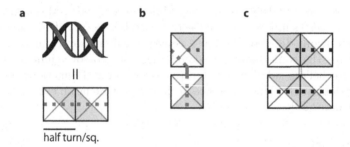

Fig. 2. T-motif "ladder" designs. **a** Ladder 15-15-15. **b** Ladder 15-20-15. Each motif is shown in opaque, and repeating units are shown in semi-transparent. Note that both structures are continuously extended in horizontal direction.

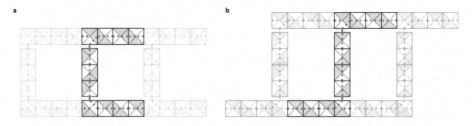

Fig. 3. Helix abstraction. **a** Two units (= a single turn of helix). One half turn of the double helix is represented as one block. Line across the square represents backbone of the DNA strand. Dark gray area represents minor groove of the helix. Dotted line, which intersects the square, distinguishes helix and represents orientation. **b** The abstraction of T-junction. **c** The abstraction of crossover junction. Plates b,c of this figure correspond to Fig. 1b,a respectively.

called "brick-wall" motif. Plate b and c describe examples of patterned lattice designs. Not just using DNA molecules, we can also bind and attach different functional molecules such as proteins and nano-particles for markers and future nanomaterials. These examples explain different sets of sticky-ends (but with only two tile-types) can create different patterns. Theoretically, we are able to expand this patterning technique into computation with logical universality called "algorithmic self-assembly". Fig. 5a describes another type of lattice structures, called "windmill". This lattice shows patterned holes of two different sizes between helices. Both "brick-wall" and "windmill" clearly represent the different geometry of lattice by a porous surface, compared to the crossover-junction based lattices with aligned helices. Another example of expanded geometry is a "wheel" structure, described in Fig. 5b and c. Making use of the single-duplex flexibility, we are able to design DNA nanostructures based on polar coordinates.

In addition, origami-method based structures can be designed by this abstraction (Fig. 6). Fig. 6a shows one-dimensional (ladder-shaped) origami structure,

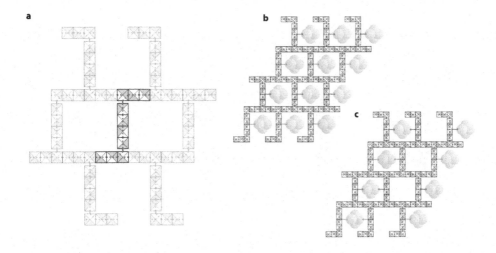

Fig. 4. Planar "brick-wall" lattice designs. **a** "Brick-wall" lattice 10-20-10. **b,c** Examples of patterned lattice by attachment of different molecules.

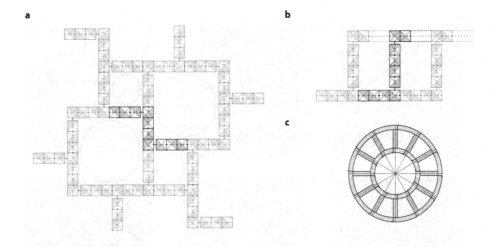

Fig. 5. Planar "windmill" lattice and "wheel" designs. **a** "Windmill" lattice 15-10-15. The dotted circles identify different size of holes between helices. **b** Abstraction and **c** base-pair based diagram of "wheel" structure. A curvature is controlled by a deletion of two units/motif at the inner side of wheel, compared with "ladder" 15-20-15.

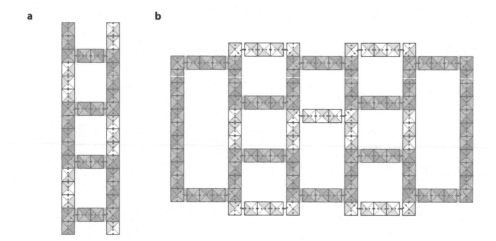

Fig. 6. Origami-method based structures. **a** Ladder. **b** Double-H structure. Units with gray overlays indicate long template DNA; Other units indicate staples. Note that original origamis are using single-stranded DNA. On the other hand, this T-junction based origami uses double strands with some interconnected regions.

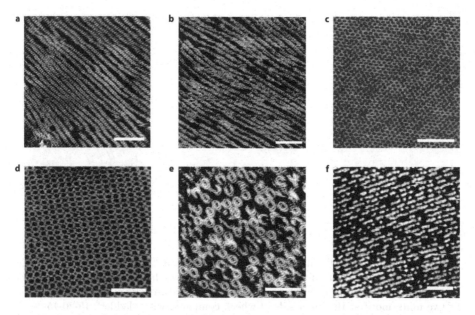

Fig. 7. AFM images. **a** Ladder 15-15-15. **b** Ladder 15-20-15. **c** Brick-wall. **d** Windmill. **e** Wheel. **f** Brick-wall lattice, patterned by streptavidin (Fig. 4c). Scale bar: 100[nm].

which can be folded into a polygonal contour by replacing specific staple duplexes. Fig. 6b describes an example of 2-D structures. We can also design fractal-like shapes by an extended template strands.

4   Experiments

Fig. 7 represents some experimental results of the design in the previous section. Although the yield of complete structures varies by types of the motifs, all shapes (a-e) are self-assembled successfully as far as we observed. Patterned brick-wall (f) seems to have lower yield of protein binding compared to the original lattice (c). This difference is due to the two-staged process of the experiment, and fractures during the observation by an AFM cantilever's scratch. The result shows the potential of this design, which several types of sticky-ends can cooperatively hybridize and form structures and patterns as designed.

5   Conclusion

We presented the abstraction method and the design of DNA nanostructures based on a combination of T-junctions. Our experimental results denote this abstraction can be used as a practical design method for various DNA nanostructures. The designs described in this paper are not the only varieties that we could make by this abstraction; it can be widely expanded into broader geometry of novel shapes and patterns. Furthermore, theoretically this method can be also applied to 3-D nanostructures. For instance, current 3-D origami method uses the only geometry of aligned helices in honeycomb pattern. We can extend this organization of helices using our methods in the same manner. Moreover, the experimental results of protein-patterned lattice implies that this design has a potential to be extended into algorithmic self-assemblies at the same level as crossover-junction based tiles[4]. This abstraction method and structures will lead us to a new application field of DNA nanostructures. We are planning to unveil the scalability of this method in size and complexity by further experiments.

Experimental Protocols

DNA sequences were designed by DNA Design software, developed in Winfree Lab at Caltech (Pasadena, CA). Oligonucleotides were synthesized and purified by Sigma-Aldrich Japan K.K. All strands were mixed in equimolar ratio with Tris-Acetic-EDTA/Mg^{2+} (TAE/Mg^{2+}, 12.5 mM) buffer. Mixed samples were gradually annealed from 95 °C to room temperature or 4 °C over 2 days in free solution or on mica. Mica-assisted self-assembly was performed under the same conditions except mica substrate was annealed together with DNA. Protein binding process is performed after annealing. Samples were then directly observed with AFM. Observations were carried out in tapping mode on a Multimode AFM with Nanoscope IIIa controller (Veeco) using NP-S cantilevers. Visit http://www.mrt.dis.titech.ac.jp/~shogo/ for details.

References

1. Seeman, N.C.: Nucleic acid junctions and lattices. J. Theor. Biol. 99(2), 237–247 (1982)
2. Fu, T.J., Seeman, N.C.: DNA double-crossover molecules. Biochemistry 32(13), 3211–3220 (1993)
3. Winfree, E., Liu, F., Wenzler, L.A., Seeman, N.C.: Design and self-assembly of two-dimensional DNA crystals. Nature 394(6693), 539–544 (1998)
4. Rothemund, P.W.K., Papadakis, N., Winfree, E.: Algorithmic self-assembly of DNA Sierpinski triangles. PLoS Biol. 2(12), e424 (2004)
5. Rothemund, P.W.K.: Folding DNA to create nanoscale shapes and patterns. Nature 440(7082), 297–302 (2006)
6. Douglas, S.M., Dietz, H., Liedl, T., Högberg, B., Graf, F., Shih, W.M.: Self-assembly of DNA into nanoscale three-dimensional shapes. Nature 459(7245), 414–418 (2009)
7. Yan, H., Park, S.H., Finkelstein, G., Reif, J.H., LaBean, T.H.: DNA-templated self-assembly of protein arrays and highly conductive nanowires. Science 301(5641), 1882–1884 (2003)
8. He, Y., Chen, Y., Liu, H., Ribbe, A.E., Mao, C.: Self-assembly of hexagonal DNA two-dimensional (2D) arrays. J. Am. Chem. Soc. 127(35), 12202–12203 (2005)
9. Hamada, S., Murata, S.: Substrate-assisted assembly of interconnected single duplex DNA nanostructures. Angew. Chem. Int. Ed. 48(37), 6820–6823 (2009)

An Optical Solution for the Subset Sum Problem[*]

Masud Hasan[1], S.M. Shabab Hossain[1],
Md. Mahmudur Rahman[1], and M. Sohel Rahman[1,2]

[1] Department of CSE, BUET, Dhaka-1000, Bangladesh
[2] Algorithm Design Group, Department of Computer Science
King's College London, Strand, London WC2R 2LS, England
{masudhasan,msrahman}@cse.buet.ac.bd,
shabab_hossain@yahoo.com,
sajib_cse_buet@yahoo.com

Abstract. Recently, a number of researchers have suggested light-based devices to solve combinatorially interesting problems. In this paper, we design a light based device to solve a generalized version of the subset sum problem which was previously handled by Oltean and Muntean [Solving the subset-sum problem with a light-based device. *Natural Computing*]. We further design a system which is capable of providing us with the solution subset of the problem in addition to the YES/NO answer to the question of whether there exists a solution or not.

1 Introduction

Optical computing is an interesting and exciting avenue for research in the area of unconventional computing. Various properties of light enables us to solve some real world problems more easily than conventional electric based counterparts. To the best of our knowledge, the idea of using light instead of electric power to perform computations dates back to 1929 when G. Tauschek obtained a patent on Optical Character Recognition (OCR) in Germany [1]. Later various researchers have used the properties of light to achieve faster solutions to problems than is possible in conventional computers. An example is the n-point discrete Fourier Transform Computation which can be performed in only unit time [6, 13]. Quite recently, an important practical step was taken by the Intel researchers who developed the first continuous wave all-silicon laser using a physical property called the Raman Effect [3, 12, 14, 15].

Very recently, a number of researchers have suggested light-based devices to solve combinatorially interesting problems. For example, a system which solves the Hamiltonian Path problem (an NP-Complete problem [5]) using light and its properties has been proposed in [8]. Similar system was devised in [11] and [10] to solve the Exact Cover Problem and the Subset Sum problem respectively, both of which are NP-Complete problems [5]. A number of interesting polynomial-time solvable (in conventional computing) problems have also been investigated from this point of view. For example, in [16], the physical concept of refraction and dispersion of light have been cleverly

---

[*] This research work is part of the B.Sc. Engg. thesis of S. M. Shabab Hossain and Md. Mahmudur Rahman.

explored to solve the sorting problem. An stable version of this sort, known as the Rainbow Sort, has been proposed in [7]. Very recently, solutions to the string matching problem has also been proposed [9] using light and optical filters for performing computations.

In this paper, we revisit the Subset Sum problem. As has been discussed above, the Subset Sum Problem is NP-Complete and has been explored before from optical computing point of view [10]. Here, we first design a system to solve a more general version of the problem than that handled in [10]. Then, we design a system which is capable of providing us with the solution subset of the problem in addition to the YES/NO answer to the question of whether there exists a solution or not. In other words, our proposed system not only solve the decision problem but also provide us with an actual solution.

2   The Device

We start this section with a formal definition of the subset sum problem.

Problem 1. **Subset Sum Problem.** We are given a set $A = \{a_1, a_2, ..., a_n\}$ of positive integers and another positive integer B. We are asked the question whether there exists a subset S of A such that the summation of the values of S equals B.

In this paper we consider a more general problem as defined below.

Problem 2. **Generalized Subset Sum Problem.** We are given a set $A=\{a_1, a_2, ..., a_n\}$ of positive integers and two other positive integers B and K. We are asked the question whether there exists a subset S of A such that the summation of the values of S equals B such that $|S| \leqslant K$.

Clearly, the Subset Sum Problem (Problem 1) is a restriction of the Generalized Subset Sum Problem (Problem 2), since the latter reduces to the former when $K = n$; therefore, the latter is also NP-Complete. We design our device to solve Problem 2 in a step by step manner. We start from the device proposed in [10] (Figure 1). The main two properties of light that are used in the device are listed below:

1. The speed of light has a limit and we can delay the ray by forcing it to pass through an optical fibre cable of a certain length.
2. The ray can be easily divided into multiple rays of smaller intensity/power. Beam splitters are used for this operation [2, 4].

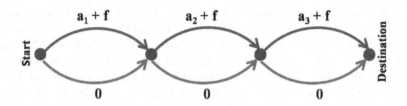

Fig. 1. First version of our device

As can be seen from the Figure 1, the device has a graph-like structure. The following two simple operations are performed by the device.

1. When passing through an arc the light ray is delayed by the amount of time assigned to the arc.
2. When passing through a node the light ray is divided into a number of rays equal to the external degree of that node.

The basic idea of the system is as follows. The system is constructed in such a way that if $B = a_1 + a_4 + a_5$, then we must be able to sense a light ray at the destination node at time $a_1 + a_4 + a_5 + (some fixed fractional Value)$. This is achieved by introducing the corresponding delays in the arcs of the designed system. For example, as can be seen from Figure 1, there are arcs representing the numbers (i.e., introducing corresponding delays) of the problem instances (plus a fixed fraction f to de discussed shortly), namely, a_1, a_2 and a_3. Clearly, the problem size of the system of Figure 1 is 3. The fractional value will help us determine the maximum number of elements in the expected subset. The two properties of the fraction f are as follows. Firstly, f is equal for all arcs, so that we can identify the number of elements participating in the solution set. And secondly, sum of all f is less then 1. The significance of these properties will be clear when we prove the correctness of this device. Now, clearly, we can easily calculate such fraction by using the formulae, $f = \frac{1}{(n+1)}$. So, for the problem instance shown in the Figure 1 with $n = 3, f = \frac{1}{(3+1)} = 0.25$ the signal passing through the arcs containing a_1, a_3 will reach the destination at time $a_1 + a_3 + 0.5$. So if we are interested in finding out whether there is a solution subset S with sum equal to B and $|S| = K$, then we wait for a signal at the moment $B + \frac{K}{(n+1)}$.

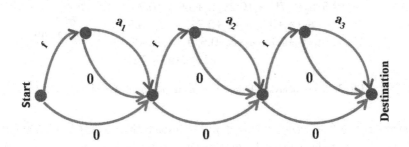

Fig. 2. Second version of our device

Clearly, the above system is not capable to answer the question whether there is a solution subset S with sum B and $|S| \leqslant K$. In the rest of this section we present a novel system to achieve our goal.

2.1 The Modified Device

We modify the device as shown in figure 2. It can be verified that in the modified device, if there is a light signal at moment $B + \frac{\beta}{(n+1)}$, then there must also be a light signal at

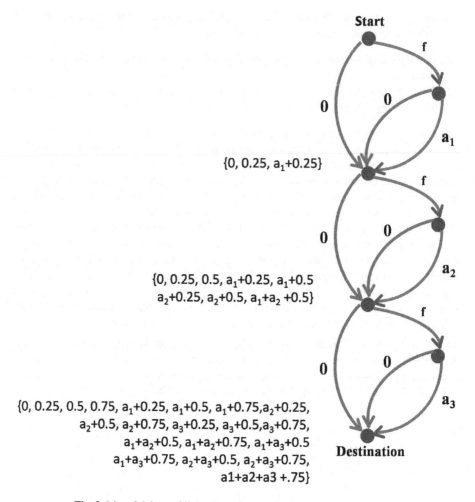

Start

f

0   0

a_1

{0, 0.25, a_1+0.25}   f

0   0

a_2

{0, 0.25, 0.5, a_1+0.25, a_1+0.5
a_2+0.25, a_2+0.5, a_1+a_2 +0.5}   f

0   0

a_3

{0, 0.25, 0.5, 0.75, a_1+0.25, a_1+0.5, a_1+0.75,a_2+0.25,
a_2+0.5, a_2+0.75, a_3+0.25, a_3+0.5,a_3+0.75,
a_1+a_2+0.5, a_1+a_2+0.75, a_1+a_3+0.5
a_1+a_3+0.75, a_2+a_3+0.5, a_2+a_3+0.75,
a1+a2+a3 +.75}   **Destination**

Fig. 3. List of delays of light signal in our proposed system for $n = 3$

moments $B + \frac{\alpha}{(n+1)}$, where $\beta < \alpha \leqslant n$. For instance if $n = 3, f = 0.25$ the device in figure 2 will produce a signal which reaches destination at time $a_1 + a_3 + 0.5$ as before. But it will also produce a signal at time $a_1 + a_3 + .75$. Now we prove that the system works correctly. In figure 3, the signals reaching each node is shown explicitly. The system ensures that if there is a ray arriving at moment $B + \frac{K}{(n+1)}$ then there is a subset S of A of sum B and $|S| \leqslant K$. The correctness of the system follows from the following theorem.

Theorem 1. *There is a signal at moment* $B + \frac{K}{(n+1)}$ *if and only if there is a solution subset S of A with sum B and $|S| \leqslant K$.*

Proof. We first prove the *if part*. Assume that, there is a solution subset S of A with sum B and $|S| \leqslant K$. We will construct a path P for the light signal to travel which

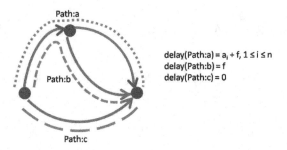

delay(Path:a) = a$_i$ + f, 1 ≤ i ≤ n
delay(Path:b) = f
delay(Path:c) = 0

Fig. 4. Unit stage of proposed n stage device

has delay equal to $B + \frac{K}{(n+1)}$. To construct the path P, we begin from the staring node and continue to next node by the $Path : a$ (see figure 4) only if $Path : a$ corresponds to element $a_i, 0 \leqslant i \leqslant n$ such that a_i is a member of solution subset S. Otherwise we continue to next node via $Path : c$. After repeating this process for n times we reach the destination. Now the ray following the path P has delay equal to $B + \frac{|S|}{(n+1)}$. Now, if $|S| = K$, we are done. Otherwise $|S| < K$ and we arbitrarily select $K - |S|$ number of $Path : c$ contained in path P and change them to $Path : b$. This process increases the delay of path P by $\frac{(K-|S|)}{(n+1)}$. So the total delay of path P is now $B + \frac{K}{(n+1)}$ as desired.

Now we prove the *only if part*. If there is a light signal reaching the destination at moment $B + \frac{K}{(n+1)}$ then there must be a subset R with sum equal to B. This is so, because, the integer part (i.e. B) of the signal delay depends only on the elements of the set A. Now, if $|R| < K$, then we are done. So assume otherwise. Then, the signal arriving at the destination at moment $B + \frac{K}{(n+1)}$ must have been delayed by $|R|$ elements and K fractions (i.e. f). But if light signal has been delayed by an element $a_i, 0 \leqslant i \leqslant n$, then, in addition to that, it has also been delayed by the corresponding fraction f (see figure 4). So if a signal is delayed by $|R|$ elements then it must be delayed by corresponding $|R|$ fractions. So $K \geqslant |R|$, which leads to a contradiction.   □

From practical point of view, there is a problem in the device of Figure 2 as follows. Even if theoretically we could have arcs of length 0, we cannot have cables of length 0 in practice. For avoiding this problem we follow the idea used in [10]. The idea is to add a constant k to the length of each alternative path of the light signal. The schematic view of the final device is depicted in figure 5.

After the above modification, it can be easily verified that, at the destination we will need to sense for a ray at the moment $B + \frac{K}{(n+1)} + n \cdot k$ instead of $B + \frac{K}{(n+1)}$, for a solution subset S with $|S| \leqslant K$.

For implementing the proposed device the following components are required:

- A light source at the start node.
- Beam-splitters for splitting a light beam into two light rays. Half silvered mirror is a good example for beam-splitter
- Optical fibers of various length.

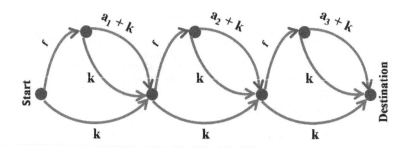

Fig. 5. Final version of our device considering implementation issues

- Light sensor at the destination node, which converts optical pulses to electric signals. Photo diode is a good example for light sensor.
- Oscilloscope, a tool for detecting a fluctuation of power generated by photo diode.

Now, we analyze our proposed device following the framework for analysis presented in [8, 10].

Precision. A problem is that we cannot measure the moment $B + \frac{K}{(n+1)} + n \cdot k$ exactly. We can do this measurement only with a given precision which depends on the tools involved in the experiments. Actually it will depend on the response time of the photodiode and the rise time of the oscilloscope. The rise-time of the best oscilloscope available on the market is in the range of picoseconds (10^{-12} seconds). This means that each signal arriving at the destination at distinct moment must maintain above mentioned time interval to be recognized correctly. This can be ensured by setting a lower bound on the length of the cable as follows. Since the speed of light is 3×10^8, the minimum cable length must be 0.0003 meter.

This value is the minimal delay that should be introduced by an arc. Also, note that, all lengths must be integer multiples of 0.0003. We cannot allow to have cables whose lengths can be written as $p \times 0.0003 + q$, where p is an integer and q is a positive real number less than 0.0003, since by combining this kind of numbers we can have a signal in the above mentioned interval and that signal may not contain valid information.

Once we have the length for the minimal delay it is quite easy to compute the length of the other cables that are used in order to induce a certain delay as follows. We assign $f = 0.0003$ meter, where f stands for the fractional value. Then we define $unit = (1 + n) \cdot f = (1 + n) \cdot 0.0003$ meter. First of all we have to multiply / divide all given numbers with such factor that the less significant digit (greater than 0) to be on the first position before the decimal place. For instance if we have the set $A = \{0.001, 4\}$ we will multiply both numbers by 1000. If we have the set $A = \{100, 2000\}$ we have to divide both numbers by 100. Now, for each $a_i \in A$, we use an arc of length $a_i \cdot unit$ where $1 \leqslant i \leqslant n$.

Hardware Complexity. Let, $f = \Delta$, where Δ is the minimum length of cable. The maximum value of a particular element $a_i \in A, 1 \leqslant i \leqslant n$, is assumed to be B, because, otherwise, a_i cannot participate to the final solution. So the maximum cable length needed for physical representation of a element a_i is, $B \cdot unit = B(1+n) \cdot \Delta = O(nB)$. So for n element set the total length of the wire is $O(n^2 \cdot B)$.

Runtime complexity. The ray corresponding to solution takes $B \cdot n \cdot \Delta + n \cdot k$ time to reach destination node. So the time complexity is $O(B \cdot n + n \cdot k)$ or $O(B \cdot n)$ assuming $B \gg k$.

Problem size. We are also interested to find the size of the instances that can be solved by our device. let, L be the maximum length of available cable. Also, assume that, $M \leqslant B$ is the maximum value of a element of set A of size n. Then, $M \cdot (1+n) \cdot \Delta = L$. Therefore, we have, $n = \frac{L}{M\Delta} - 1$. We can also calculate the time t needed to solve the problem using the equation, $t = \frac{L \cdot n + n \cdot k}{c}$, where c is the speed of light.

Power decrease. Beam splitters are used in our approach for dividing a ray in two sub rays. Hence, the intensity of the signal is decreasing. In the proposed device, there are $2n$ nodes,[1] where $|A| = n$. Now, at each node, the power of each outgoing signal is reduced to half. So for the proper detection of the light signal at the destination, the input signal must have 2^{2n} power.

3   Finding the Solution Subset

In this section, we focus on deriving a solution subset for the Generalized Subset Sum Problem. In particular, we will design a modified device which would be capable of providing us with a solution subset if the answer of the decision problem is yes. The idea is as follows. In the modified device, we assign to each arcs numbers from the given set A plus some fraction value. For instance, if the numbers $a_1 + a_4 + a_5$ generate the expected subset, then the total delay of the signal should be $a_1 + a_4 + a_5 + (some fixed fractional Value)$. The fractional value will help us determine the elements participating in the solution set S. So, as before the integer part of the delay depends on the elements of set A. And the fractional values have the following properties:

1. sum of any combination of fraction is unique.
2. sum of all fraction is less than 1.

Now, for any n, it follows from the above properties that, $f \cdot (1+1+2+4+\ldots+2^{n-1}) \leqslant 1$ So, $f \leqslant \frac{1}{2^n}$. A device based on the above mentioned delaying system is shown in figure 6. For this case, $n = 3$ so, $f = 0.125$. For each $a_i \in A$ we use an arc of length $a_i + f \cdot 2^{i-1}$ where $1 \leqslant i \leqslant n$. A solution subset S containing element a_1, a_3 will have fractional delay equal to $f + 4 \cdot f = 5f$. To complete the design, as before, we remove the 0 length arcs by adding k to the length of all arcs (Figure 7).

To answer the question whether there exists a subset S of A whose sum equals B and also to identify the elements of set S in case the answer is positive, we start scanning from time B and continue to scan at each time interval corresponding to f until we reach time $B+1$. If we receive a light signal at moment $B + (fractional Value)$ then we decode this fractional value to identify each element in the solution set ($fractional Value$

---

[1] We ignore the destination node, since no splitting needed there.

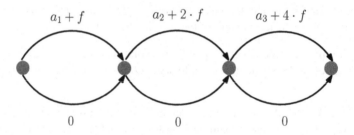

Fig. 6. Unit stage of proposed n stage device

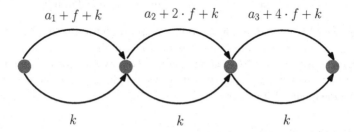

Fig. 7. Final Design of a stage of proposed n stage device

is unique for all possible solution sets S). We do this by dividing the $fractionalValue$ by f and then transforming that number to binary. For instance, in figure 6 if $B = a_1 + a_3$ then we will have a ray which will reach destination at moment $B + 5 \cdot f$. Now if we transform 5 to binary then we get 101. Since we have '1' at positions 1 and 3, we straightway can say that a_1 and a_3 are part of solution set S.

Now we analyze the proposed device. Even though the structure of the device in figure 7 is similar to the structure of the device in figure 1, the complexity of the device in figure 7 is not polynomial. The reason behind this is that, the additional fraction added to the length of the arc is of exponential order. So, the overall complexity of the device is exponential.

4   Conclusion and Future Work

In this paper we have designed a light based device to solve a generalized version of the subset sum problem which was previously handled in [10]. We further design a system which is capable of providing us with the solution subset of the problem in addition to the YES/NO answer to the question of whether there exists a solution or not.

References

1. http://en.wikipedia.org/wiki/optical_character_recognition
2. Agrawal, G.: Fibre-optic communication systems, 3rd edn. Wiley Interscience, Hoboken (2002)

3. Faist, J.: Optoelectronics: silicon shines on. Nature 433, 691–692 (2005)
4. Feitelson, D.: Optical computing: A survey for computer scientists. MIT Press, Cambridge (1988)
5. Garey, M.R., Johnson, D.S.: Computers and Intractability: A Guide to the Theory of NP-Completeness. W. H. Freeman, New York (1979)
6. Goodman, J.W.: Architectural development of optical data processing systems. Aust. J. Electr. Electron. Eng. 2, 139–149 (1982)
7. Murphy, N., Naughton, T.J., Woods, D., Henley, B., McDermott, K., Duffy, E., van der Burgt, P.J.M., Woods, N.: Implementations of a model of physical sorting. In: Adamatzky, A., Teuscher, C. (eds.) From Utopian to Genuine Unconventional Computers Workshop, pp. 79–100. Luniver Press (2006)
8. Oltean, M.: Solving the hamiltonian path problem with a light-based computer. Natural Computing 7(1), 57–70 (2008)
9. Oltean, M.: Light-based string matching. Natural Computing 8(1), 121–132 (2009)
10. Oltean, M., Muntean, O.: Solving the subset-sum problem with a light-based device. Natural Computing (to appear)
11. Oltean, M., Muntean, O.: Exact cover with light. New Generation Comput. 26(4), 329–346 (2008)
12. Paniccia, M., Koehl, S.: The silicon solution. In: IEEE Spectrum. IEEE Press, Los Alamitos
13. Reif, J., Tyagi, A.: Efficient parallel algorithms for optical computing with the discrete fourier transform primitive. Applied optics 36(29), 7327–7340 (1997)
14. Rong, H., Jones, R., Liu, A., Cohen, O., Hak, D., Fang, A., Paniccia, M.: A continuous wave raman silicon laser. Nature 433, 725–728 (2005)
15. Rong, H., Liu, A., Jones, R., Cohen, O., Hak, D., Nicolaescu, R., Fang, A., Paniccia, M.: An all-silicon raman laser. Nature 433, 292–294 (2005)
16. Schultes, D.: Rainbow sort: Sorting at the speed of light. Natural Computing 5(1), 67–82 (2006)

Design of True Random One-Time Pads in DNA XOR Cryptosystem

Miki Hirabayashi, Hiroaki Kojima, and Kazuhiro Oiwa

Kobe Advanced ICT Research Center,
National Institute of Information and Communications Technology (NICT),
588-2 Iwaoka, Iwaoka-chou, Nishi-ku, Kobe 651-2492, Japan
m_hirabayashi@office.so-net.ne.jp, kojima@nict.go.jp, oiwa@nict.go.jp

Abstract. We present a new model to realize true random one-time pad (OTP) encryption using DNA self-assembly. OTP is an unbreakable cryptosystem if the pad (random key) is truly random, never reused, and kept secret. Mathematical algorithms can generate pseudo-random numbers only. "True" random numbers can be generated from a physical process such as thermal noise. In this work, we propose a new tile-colony algorithm that can utilize the DNA hybridization process as an effective source for the random key construction, and discuss the error tolerance of this method. Our results indicate that the molecular computation using DNA motifs will provide promising OTP applications.

Keywords: DNA computing, Algorithmic self-assembly, DNA-based cryptography, One-time pad.

1 Introduction

True random numbers are essential in computational finance, numerical experiments on natural phenomena, or data encryption. We present a new algorithm using DNA-based true random key generation process for the one-time-pad (OTP) cryptosystem. OTP is known as the only theoretically unbreakable cryptosystem when the random key (pad) is truly random, never reused, and kept secret [1,2]. True random numbers are generated using a physical process such as nuclear decay or thermal noise of electronic circuits, because mathematical algorithms can provide pseudo-random numbers only. In DNA-based cryptography [3,4,5,6], we can directly use the random hybridization process of DNA motifs for the random key generation. Although several ideas of OTP using DNA strands have been presented [6,7,8], details of the random key generation utilizing hybridization events toward experimental implementation are not discussed enough. In this work, we present a new implementable multi-layer algorithm in order to use random hybridization events of DNA tiles for the random key generation.

We outline our new DNA-based OTP cryptosystem first and then discuss the error tolerance. DNA can serve not only a random number generator but also a compact medium for information storage. The DNA-based cryptosystem will offer promising applications for the information security.

F. Peper et al. (Eds.): IWNC 2009, PICT 2, pp. 174–183, 2010.

2 DNA XOR One-Time-Pad Cryptosystem with Random Hybridization

The cryptosystems using a secret random OTP are known to be absolutely unbreakable [1]. To encrypt n-bit binary input message $M = (M_1, M_2, ..., M_n)$, each bit M_i is XOR'ed with the random-key bit K_i to produce the cipher bit $C_i = M_i \oplus K_i$ for $i = 1, ..., n$. To decrypt a cipher C, bitwise XOR is performed using the sequence C in the place of M. The input message M is reproduced using self-inverse property of binary XOR: $M_i = C_i \oplus K_i$ (Fig. 1a).

Figure 1b shows the fundamental concept of DNA OTP encryption using a physical random process of DNA self-assembly for the random key construction. Experimental demonstration of logically reversible conditional DNA XOR has been presented in 2000 [9] based on previous works of DNA tilings [10,11]. Utilizing this DNA XOR algorithm, several encryption ideas have been proposed. As for the three-layer encryption model in [6], the details of the random key generation were not discussed. As for the four-layer encryption model in [8], the true random operation was introduced, but, there is the purification problem of key probes for experimental verification. Here we introduce new multi-layer algorithm into this four-layer model and lead a tile-colony method including a novel key purification process.

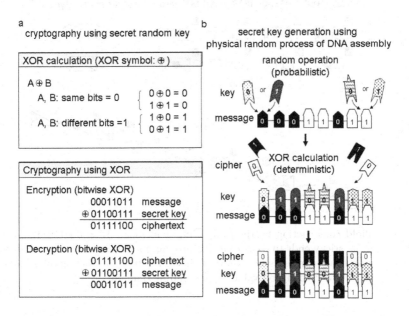

Fig. 1. Schematic drawings of OTP encryption procedures. a. Cryptography using XOR calculation. b. A DNA tile assembly model using a true random generation algorithm. Random key generation is realized by connecting of each key tile, which has a value zero or one, with probability = 0.5.

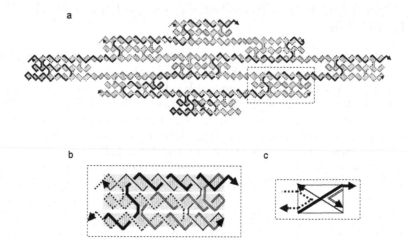

Fig. 2. TAO (triple helix, with anti-paralled crossovers, and an odd number of helical half turns between crossovers) tile. The bold black line is a reporter strand, which runs through the entire assembly.

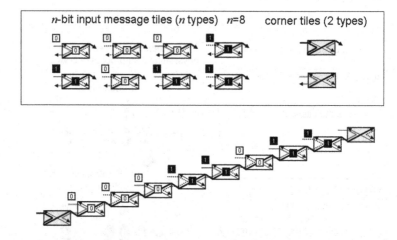

Fig. 3. Scaffold construction by the sample input message $M = 00011011$. Each tile has two specific sticky ends to connect in the programmed order and one sticky end for the encryption that represents two kinds of tile value (0 or 1). Each number indicates the tile value or the recognition value at the binding domain.

2.1 Message Encryption

A binary input message is encoded using a single tile for each bit. Figure 2 shows a TAO (triple helix, with anti-paralled crossovers, and an odd number of helical half turns between crossovers) tile [12]. This tile contains three double-helical domains such that individual oligonucleotide chains might base-pair in one helix then cross-over (strand exchange) and base-pair in another helix. It is known that the crossover points realize quite high thermal stability. A TAO tile consists of the complementary hybridization of four oligonucleotide strands shown as different line types with arrowheads on their $3'$ ends in Fig. 2. The three double-helical domains are shown with horizontal helix axes. Paired vertical lines represent crossover points where two strands exchange between two helices. Here the central helix is capped at both ends with hairpins, and the upper and lower helices serve four sticky ends. These sticky ends can encode neighbor rules for calculation utilizing algorithmic self-assembly [9,11].

In DNA cryptosystem, each tile is designed to represent each bit of binary data on a reporter strand, which runs through the entire assembly. Ligase enzyme is added to form a continuous reporter strand including the binary data of an input message, an encryption key, and a ciphertext. The reporter strand is extracted by melting apart the hydrogen bonding between strands and amplified by polymerase chain reaction (PCR) using the primers that are prepared at each end in the corner tile. Then a strand of the proper length is purified by polyacrylamide gel electrophoresis. The ciphertext and the key can be obtained separately by a cleavage site that is encoded using a restriction endonuclease or incorporated a nick using an unphosphorylated oligo between them. The purified cipher strand is sent to a destination. By applying the identical process as encryption with the encryption key, the decryption of the cipher strand is performed and the target message is extracted.

Figure 3 shows an example of the assembly of a binary input message. The numbers indicate the tile value or the recognition value at the binding domain. The first-layer tiles are designed to represent a binary input message. An n-bit binary message needs n types of input tiles with specific connecting ends in order to assemble linearly in a programmed order. The first layer provides a scaffold to create a large structure.

2.2 Random Tiling Operation

Figure 4 shows the assembly of the random operation tile utilizing the randomness of hybridization based on [8]. The random tile has both functions of XOR calculation and random number generation. The separation of processes of random number generation and key construction by this random operation tile is effective to form a multi layer, which is essential to realize an implementable algorithm. The key layer and the cipher layer are amplified through the multi-layer growth. This growth enables us to recognize one of cipher and key sets among various results with the aid of an atomic force microscope (AFM) without a preparation of random probes that are necessary for the four-layer model in [8].

Similarly to the random tiling operation in [8], we use four types of random operation tile. Each tile has one sticky end for XOR calculation, one sticky end for random number generation, and two sticky ends for connection. When the tile assembly starts, the value of random operation tile (0 or 1) is determined with probability $= 0.5$ at each slot. This equal probability is obtained by adding the same concentration of random operation tiles. The original four-layer method

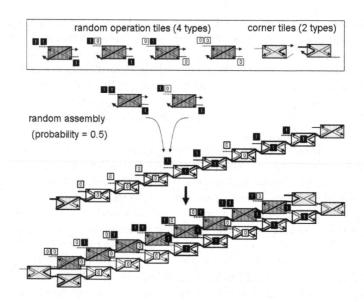

Fig. 4. Self-assembly of random operation tiles based on [8]. Each random operation tile is set to zero or one for each input value with probability $= 0.5$. The probability is obtained by adding four random operation tiles at the same concentration.

in [8] has the problem that it is difficult to extract the purified set of random keys and ciphers. It is because all possible random key assemblies are performed with same tags (primers). As mentioned above, we can not extract one target set in the reaction mixture using these prepared tags. Although we can synthesize a random probe to extract one calculation set using pseudo-random sequences generated by a mathematical algorithm, we can not utilize "true" randomness of hybridization in that case. Moreover the utilization of pseudo-random probes reduces the security of OTP encryption systems.

To solve this problem and realize implementable systems, we present a multi-layer model and introduce a tile-colony method to extract purified key and cipher strands as mentioned below.

2.3   Extraction of Cipher Strings and Key Strings

Figure 5 shows the extraction process of a cipher strand from a random operation tile based on [8]. We use four kinds of cipher tiles. The tile value is stored in the reporter strand, which runs through the entire assembly (bold black line). Each cipher bit has two specific sticky ends for calculation and two common sticky ends for connection. The calculation rule of random operation tile is recognized by the sequence at the lower right sticky end and carried by the upper left sticky end to the next layer. Figure 6 shows the extraction process of a key strand based on the carried information from a random operation layer through the cipher layer. These two extraction procedures are repeated to form a large sheet that

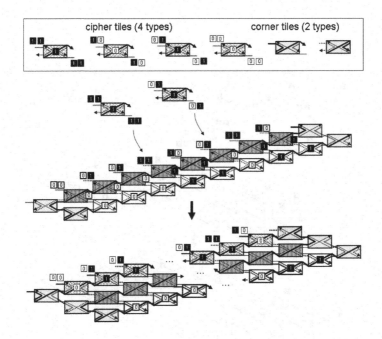

Fig. 5. Cipher extraction from random operation tiles based on [8]

enables us to observe under AFM easily. One sheet consists of a single set of key and cipher strands similarly to a DNA set of a single bacterial colony. Using the technique of a complementary DNA strand bound to an AFM cantilever [13], we will be able to pick up a target DNA sheet. This isolated set of key and cipher strands can be amplified by PCR with specific primers incorporated using corner tiles. In the calculation process, the tiles are first assembled individually from their component strands. Then all kinds of tiles are put into the reaction buffer and the self-assembly progresses automatically.

The decryption process is shown in Fig. 7. We use the three-layer model [6] for decryption. Because the decryption is performed with the purified key and the cipher strands, the multi-layer construction for purification is not necessary. Moreover our system can be used for the random key generation only. In this case, we can use a simple three-layer algorithm for encryption, because we can prepare the purified random key in advance utilizing our multi-layer system.

Key transfer is another problem. As for a public-key system using DNA, the detailed materials and methods were presented in [14]. Generally cipher strands and key strands can be hidden among dummies using steganography [15], which is a technique of hiding a message in larger data. It is difficult to extract the key and the ciphertext among dummies without knowing the correct primer pairs. Key and cipher strands are retrieved by PCR using secret primers. There are large scale of possible primer pairs. Moreover dummy strands can also have dummy meanings. Therefore, it is not easy to identify the correct message, even if we can get a significant message with a blunt force approach using random primers.

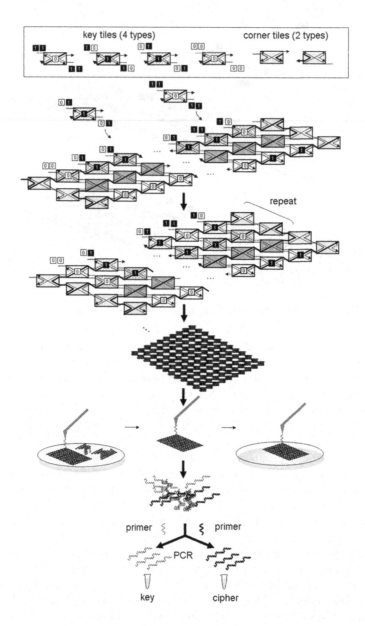

Fig. 6. Key extraction and sheet formation. Each sheet is constructed by one set of key and cipher tiles similarly to a DNA set of a single bacterial colony.

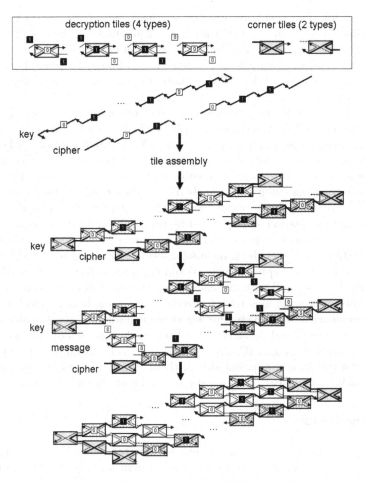

Fig. 7. Decryption by the bit-wise XOR operation

3   Analysis on Error Tolerance

We analyze the several possible errors in our system. The loss of synchronization between the input message and the key is a common critical problem in DNA cryptosystem. For example, misconstruction of a tile sheet without the first two-layer scaffold can cause this loss of synchronization. In this case the tile sheet forms an irregular shape, because the control by the first-layer scaffold is indispensable to construct the programmed structure. By selecting a sheet with appropriate size and shape, we can avoid to pick up an error sheet.

Mismatch in hybridization can cause the loss of synchronization, too. Each tile has two sticky ends for nonspecific connection to form a reporter strand, which runs through the entire assembly. The tile sheet may be constructed by these connecting ends without the involvement of other two ends. This feature allows a tile to be set at a wrong slot. We already presented the idea using the blunt ends for tile connection instead of sticky ends [7]. Thus sticky ends for calculation can play an essential role in the algorithmic self-assembly. Although the ligation efficiency is reduced by the use of blunt ends at the ligation stage, the system can obtain the significant error tolerance. It may be possible to observe the tile value directly under AFM without the ligation process, depending on the purpose.

Recently an experimental implementation of a transducer using an algorithmic self-assembly by DNA triple crossover tiles with six ends was demonstrated [16]. In this method, a PX (paranemic crossover)-JX_2 (a topoisomer of PX DNA that contains two adjacent sites where backbones juxtapose without crossing over) device was introduced to make it possible for input motifs to rotate 180° around its central dyad axis using fuel strands [17,18]. As a result, a series of n PX-JX_2 devices can provide the 2^n inputs under the control by fuel strands [16]. The input-output calculation was operated by two sticky ends of the triple crossover tile and observation of outputs was performed using a gold nanoparticle attached to each output tile under the transmission electron microscope (TEM). Particles of 5 nm were used to represent value 0 and particles of 10 nm were used to indicate value 1. Compared with our calculation system, this observation method is not appropriate for the massive information processing, because modification by gold nanoparticles costs much. Our algorithm can solve this cost problem by the multi-layer method and realize the implementable true-random-key system. As for a long input massage, PX-JX_2 devices may lead undesirable mishybridization due to the unused adjacent side of them. Our OTP-specific input tiles can reduce such errors caused by general-purpose input tiles using PX-JX_2 devices.

4   Conclusion

We have presented the novel algorithm for true random DNA OTP cryptosystem. We have detailed the computational system for experimental implementation. DNA hybridization events are expected as a useful physical process for random-number generation. Our multi-layer model makes it possible to utilize this hybridization process as a source of randomness. Our cryptosystem has significant error tolerance and will allow us to improve the quality of information security. We are currently working on experimental verification on these encryption algorithms. Through progressive platform technologies, DNA computation will provide effective applications for information processing.

References

1. Schneier, B.: Applied Cryptography: Protocols, Algorithms, and Source Code in C. John Wiley & Sons. Inc., Chichester (1996)
2. Cameron, P.J.: Notes on cryptography (2003),
   http://www.maths.qmw.ac.uk/~pjc/notes/crypt.pdf
3. Adleman, L.M., Rothemund, P.W.K., Soweis, S., Winfree, E.: On Applying Molecular Computation to the data encryption standard. Journal of Computational Biology 6, 53–63 (1999)

4. Leier, L., Richter, C., Banzhaf, W., Rauhe, H.: Cryptography with DNA binary strands. Biosystems 57, 13–22 (2000)
5. Chen, J.: A DNA-Based, Biomolecular Cryptography Design. In: 2003 IEEE International Symposium on Circuits and Systems, vol. 3, pp. 822–825 (2003)
6. Gehani, A., LaBean, T., Reif, J.: DNA-Based Cryptography. In: Jonoska, N., Păun, G., Rozenberg, G. (eds.) Aspects of Molecular Computing. LNCS, vol. 2950, pp. 167–188. Springer, Heidelberg (2003)
7. Hirabayashi, M., Kojima, H., Oiwa, K.: Effective Algorithm to Encrypt Information Based on Self-Assembly of DNA Tiles. In: Nucleic Acids Symposium Series (53), pp. 79–80 (2009)
8. Chen, Z., Xu, J.: One-Time-Pads Encryption in the Tile Assembly Model. In: Kearney, D. (ed.) Third International Conference on Bio-Inspired Computing: Theories and Applications. IEEE BICTA 2008, pp. 23–29 (2008)
9. Mao, C., LaBean, T.H., Reif, J.H., Seeman, N.C.: Logical Computation Using Algorithmic Self-Assembly of DNA Triple-Crossover Molecules. Nature 407, 493–496 (2000)
10. LaBean, T.H., Winfree, E., Reif, J.H.: Experimental Progress in Computation by Self-Assembly of DNA Tilings. In: DIMACS 5th International Meeting on DNA Based Computers, vol. 5, pp. 121–138 (1999),
    http://www.cs.duke.edu/~thl/papers/progress.DNA5.pdf
11. LaBean, T.H., Yan, H., Kopatsch, J., Liu, F., Winfree, E., Reif, H.J., Seeman, N.C.: The Construction, Analysis, Ligation and Self-Assembly of DNA Triple Crossover Complexes. J. Am. Chem. Soc. 122, 1848–1860 (2000)
12. Fu, T.-J., Seeman, N.C.: DNA Double-Crossover Molecules. Biochemistry 32, 3211–3220 (1993)
13. Kufer, S.K., Puchner, E.M., Gummp, H., Liedl, T., Gaub, H.E.: Single-Molecule Cut-and-Paste Surface Assembly. Science 319, 594–596 (2008)
14. Tanaka, K., Okamoto, A., Saito, I.: Public-Key System Using DNA as a One-Way Function for Key Distribution. Biosystems 81, 25–29 (2005)
15. Clelland, C.T., Risca, V., Bancroft, C.: Hiding Messages in DNA Microdots. Nature 399, 533–534 (1999)
16. Chakraborty, B., Jonoska, N., Seeman, N.C.: Programmable Transducer by DNA Self-Assembly. In: Geol, A., Simmel, F.C., Sosik, P. (eds.) Fourteenth International Meeting on DNA Computing, pp. 98–99 (2008)
17. Yan, H., Zhang, X., Shen, Z., Seeman, N.C.: A Robust DNA Mechanical Devices Controlled by Hybridization Topology. Nature 415, 62–65 (2002)
18. Jonoska, N., Liao, S., Seeman, N.C.: Transducers with Programmable Input by DNA Self-Assembly. In: Jonoska, N., Păun, G., Rozenberg, G. (eds.) Aspects of Molecular Computing. LNCS, vol. 2950, pp. 219–240. Springer, Heidelberg (2003)

On Designing Gliders in Three-Dimensional Larger than Life Cellular Automata

Katsunobu Imai[1], Yasuaki Masamori[2], Chuzo Iwamoto[1], and Kenichi Morita[1]

[1] Graduate School of Engineering, Hiroshima University,
Higashi-Hiroshima, 739-8527, Japan
[2] Fuso Dentsu Co.,Ltd., Japan
`imai@iec.hiroshima-u.ac.jp`

Abstract. Larger than Life (LtL) is a class of cellular automata and it can be regarded as a generalization of the game of Life. As is the case with the game of Life, a variety of glider patterns are found in LtL. A special type of glider called "bug" is known in LtL. In this paper, we extend LtL in three-dimension and show there are bugs in three-dimensional LtL with radius $3, 4, 5, 6$, and 7.

Keywords: Cellular automata, Larger than Life, three-dimension.

1   Introduction

A class of two-dimensional cellular automata, *Larger than Life*(LtL) [6], is a natural generalization of the game of Life [5] by extending the diameter of the neighborhood. So far a lot of interesting patters such as *bugs* are found in LtL [7]. Particularly a LtL rule, *Bosco*, with radius 5 neighborhood [6] has high computing efficiencies [8]. A bug is a kind of *glider* or *spaceship* patterns. These patterns are propagating in cellular space and found in the game of Life [9,10] The most famous pattern, the glider, was used to show the computational universality of the game of Life [5].

By extending the size of neighborhood, it is possible to discuss about infinite classes of Life-like cellular automata and gliders in such classes are found [7,12,1]. It is also possible to consider the "continuum" limit of LtL [7,12].

In this paper, we show another extension of LtL in three-dimension. As far as the game of Life, there are three-dimensional extensions [2,3]. In contrast to the two-dimensional case, there are many combinations of the ranges of birth and survival. In spite of the fact, Bays found several glider patterns [2,4]. In the case of radius 2, a glider pattern has already been found [11], but there is no result in the case that radius is more than 3.

Although computer simulations are effective to find gliders in the case of the game of Life, it is quite difficult to find bugs in the case of three-dimensional LtLs, because there are many candidates for their rules and they have large neighborhood sizes. So we show a methodology to design period 1 bugs and show them in the case of radius 3 to 7. So far there is no methodology to design a bug with a longer period, we also show an example of period 10 bug in the case of radius 4.

F. Peper et al. (Eds.): IWNC 2009, PICT 2, pp. 184–190, 2010.

2   Larger than Life Cellular Automata

Larger than Life (LtL) [6] $L = (r, \beta_1, \beta_2, \delta_1, \delta_2)$ is a natural generalization of the game of Life. Where r is the radius of its neighborhood, $[\beta_1, \beta_2]$ is the range of the total number of live neighborhood cells for giving birth to the center cell $(0 \rightarrow 1)$, $[\delta_1, \delta_2]$ is the range for survival of the center cell $(1 \rightarrow 1)$ The shape of its neighborhood is the square of the length $2r + 1$, thus the neighborhood includes $(2r+1)^2$ cells. Note that the game of Life can be denoted by $(1, 3, 3, 3, 4)$.

In the case of LtL, Evans [6] found a lot of special gliders called a "bug." A bug is a pattern which is translated to the place represented by a displacement vector (d_1, d_2) after a certain *period* τ by a cyclical process.

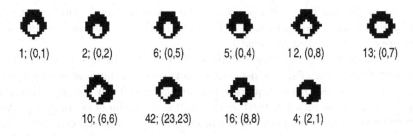

1; (0,1)   2; (0,2)   6; (0,5)   5; (0,4)   12, (0,8)   13; (0,7)

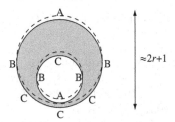

10; (6,6)   42; (23,23)   16; (8,8)   4; (2,1)

Fig. 1. Family of bugs $[\tau; (d_1, d_2)]$ supported by LtL rule $(5, 34, 45, 34, 58)$ [6]

The schematic representation of a bug with a short period is depicted in Fig. 2. The shaded part is filled with alive cells. The cells near the part A, B and C in Fig. 2 must arise, survive, and die respectively in the next time step. Because of the balance of these changes of cells, the diameter of a bug is almost determined to the size of neighborhood.

Fig. 2. Schematic representation of a bug

3   Bugs in Three-Dimensional Larger than Life

3.1   Period 1 Bugs

In this section, we design several bugs in three-dimension. Because the number of cells are very huge in three-dimension and quite difficult to find bugs in general,

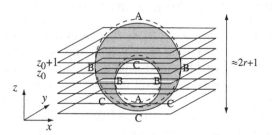

Fig. 3. Schematic representation of a three-dimensional bug

we focus on finding bugs with period 1. The schematic representation of a bug with period 1 in three-dimension is depicted in Fig. 3. The slice of bug at z_0 must appear at $z_0 + 1$ at the next time step.

Fig. 4 is a bug ($\tau = 1, d = (0,0,1)$) in three-dimensional LtL $L_3 = (3, 42, 48, 42, 52)$. Fig. 5 is its detailed structure. Each number shows the calculated value by the rules. Squared cells are alive cells. A cell with bold face number which is included in birth period will arise in the next time step. A cell with italic face number will die in the next time step. A boxed cell with normal face number will survive in the next time step. Thus for all z in Fig. 5, if the shape of the union of boxed cells with normal face number and cells with bold face number is identical to the shape of boxed cells in $z - 1$, the pattern can be a bug with period 1. Fig. 4 is designed along the track.

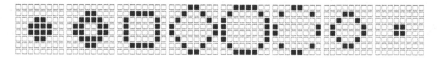

Fig. 4. A configuration of period 1 bug in $L_3 = (3, 42, 48, 42, 52)$

In the same way, we show several bugs ($\tau = 1, d = (0,0,1)$) in the case of $r = 4, \cdots, 7$. Fig. 6, 7, 8, 9 are the bugs in $L_4 = (4, 127, 156, 127, 199)$, $L_5 = (5, 192, 234, 192, 281)$, $L_6 = (6, 390, 460, 390, 638)$ and $L_7 = (7, 615, 707, 615, 1031)$ respectively.

The parameters $(\beta_1, \beta_2, \delta_1, \delta_2)$ of Life and Bosco are represented as $(0.33, 0.33, 0.33, 0.44)$ and $(0.28, 0.37, 0.28, 0.47)$ respectively by the ratio to the number of neighborhood cells. Evans's LtL usually have $0.2 \leq \delta_1 \leq \beta_1 \leq 0.27 \leq 0.3 \leq \beta_2 \leq 0.35 \leq \delta_1 \leq 0.5$ [12]. They are as follows in our case: $L_3 : (0.12, 0.14, 0.12, 0.15)$ $L_4 : (0.17, 0.21, 0.17, 0.27)$, $L_5 : (0.14, 0.18, 0.14, 0.21)$, $L_6 : (0.18, 0.21, 0.18, 0.29)$, $L_7 : (0.18, 0.21, 0.18, 0.31)$. The relative size of parameters seems to have the similar tendency to the case of two dimensional LtL, because of the limitation due to the structure of bug. In the case of L_3, its birth range cannot be changed. Although its survival range have margins

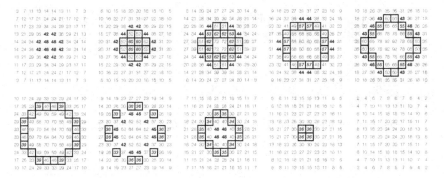

Fig. 5. The detail of period 1 bug in $L_3 = (3, 42, 48, 42, 52)$

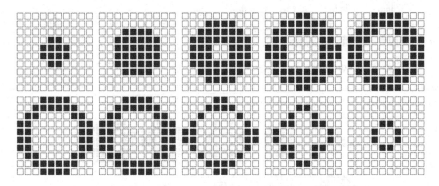

Fig. 6. A configuration of period 1 bug in $L_4 = (4, 127, 156, 127, 199)$

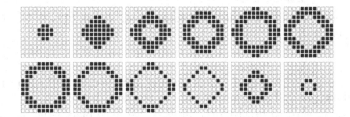

Fig. 7. A configuration of period 1 bug in $L_5 = (5, 192, 234, 192, 281)$

($\delta_1 = [40, 42], \delta_2 = [52, 54]$), it is not so large. Thus the parameters of L_3 have relatively small values and patterns apt to be annihilated.

3.2   Bugs with Longer Period

Although finding a bug with a longer period is very difficult and there is no methodology to find them so far, we found a bug with longer period by accident.

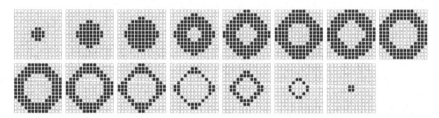

Fig. 8. A configuration of period 1 bug in $L_6 = (6, 390, 460, 390, 638)$

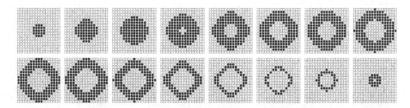

Fig. 9. A configuration of period 1 bug in $L_7 = (7, 615, 707, 615, 1031)$

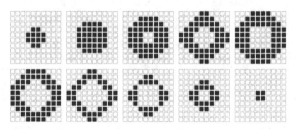

Fig. 10. A configuration of period 10, displacement vector $(0, 0, 5)$ bug in $L'_4 = (4, 141, 176, 141, 240)$

Fig. 10 is a bug with period 10 and displacement vector $(0, 0, 5)$ in LtL $L'_4 = (4, 141, 176, 141, 240)$. Thus the speed of the bug is $1/2$.

Because it is hard to describe the process, we have placed the movie file of behavior of the bug in the cite of the following URL:

http://www.iec.hiroshima-u.ac.jp/projects/ca/3dLtL/

4   Experimental Results of Collisions of Bugs

In the case of the game of Life, glider streams generated by glider guns are used to encode signals and properly designed collisions of gliders realize AND, OR, and NOT logical gates operations. There are many variations of two gliders collision. Among them vanishing and "kickback" reactions are used to embed logical circuits [5]. The kickback reaction is a collision of two gliders that a glider kicks back another glider and dissolve itself. A pattern "eater" which eliminates a glider is also employed in the framework.

So we tested front and side (90-degree) collisions of two bugs with every possible displacement in the case of L_3 by computer simulations. As the result, it turned out that almost all collisions finally eliminate all alive cells and there are only two exceptional cases: (1) forms a stable block, (2) erases a bug with no effect to another bug. The found stable block can also be usable as a eater.

Because it is hard to describe these processes here, we have placed several movie files in the cite of the following URL:

http://www.iec.hiroshima-u.ac.jp/projects/ca/3dLtL/

As we have mentioned in the previous section, the parameters of L_3 seems to be too small to support complex reactions. In the case of larger radius, reactions are quite complicated and we cannot test the all cases so far. Several examples are also found in the above cite.

5 Conclusion

We have designed several period 1 glider patterns called "bugs" in three-dimensional Larger than Life cellular automata in the case of radius 3, 4, 5, 6, and 7. We also found a period 10 bug in the case of radius 4.

Although a bug gun should be the most important building block, we can't find any bug gun so far. Take into account of the fact that no glider gun in three-dimensional Life is reported so far as well, it might be quite difficult to find a bug gun in the case of smaller radius.

In the two dimensional case, radius 5 LtL, Bosco has nice bug guns [8]. This due to the existence of a special periodic pattern, "Bosco" and a kind of stability of its bugs. So far we cannot find any pattern like Bosco. Although a bug with a longer period in a LtL with a larger radius might have a possibility to be more stable, we have no effective methodology to design such type of bugs.

Acknowledgments. The authors thank Mitsuru Matsushima of NEC Corporation. Some of simulations and inspections were performed by his simulation program. The authors also thank anonymous reviewers for their useful comments and advices.

References

1. Adachi, S., Peper, F., Lee, J., Umeo, H.: Occurrence of gliders in an infinite class of Life-like cellular automata. In: Umeo, H., Morishita, S., Nishinari, K., Komatsuzaki, T., Bandini, S. (eds.) ACRI 2008. LNCS, vol. 5191, pp. 32–41. Springer, Heidelberg (2008)
2. Bays, C.: Candidates for the game of Life in three dimensions. Complex Systems 1, 373–400 (1987)

3. Bays, C.: A new game of three-dimensional Life. Complex Systems 5, 15–18 (1991)
4. Bays, C.: Further notes on the game of three-dimensional Life. Complex Systems 8, 67–73 (1994)
5. Berlekamp, E.R., Conway, J.H., Guy, R.K.: Winning ways for your mathematical plays, vol. 2. Academic Press, London (1982)
6. Evans, K.M.: Larger than Life: Digital creatures in a family of two-dimensional cellular automata. In: Discrete Mathematics and Theoretical Computer Science Proceedings AA (DM-CCG), pp. 177–192 (2001)
7. Evans, K.M.: Larger than Life: threshold-range scaling of Life's coherent structure. Physica D 183, 43–67 (2001)
8. Evans, K.M.: Is Bosco's rule universal? In: Margenstern, M. (ed.) MCU 2004. LNCS, vol. 3354, pp. 188–199. Springer, Heidelberg (2005)
9. Gardner, M.: Mathematical games – the fantastic combinations of John Conway's new solitare game, Life. Scientific American 223, 120–123 (1970)
10. Golly game of Life, http://golly.sourceforge.net/
11. Matsushima, M., Imai, K., Iwamoto, C., Morita, K.: A Java based three-dimensional cellular automata simulator and its application to three-dimensional Larger than Life. In: Proceeding of EPSRC Workshop Cellular Automata Theory and Applications (AUTOMATA 2008), Bristol, pp. 413–415 (2008)
12. Pivato, M.: RealLife: The continuum limit of Larger than Life cellular automata. Theoretical Computer Science 372, 46–68 (2007)

Instability of Collective Flow in Two-Dimensional Optimal Velocity Model

Ryosuke Ishiwata[1,*] and Yūki Sugiyama[2]

[1] Department of Information Biology, Graduate School of Medicine and Dentistry,
Tokyo Medical and Dental University
[2] Department of Complex Systems Science, Graduate School of Information Science,
Nagoya University
sugiyama@is.nagoya-u.ac.jp

Abstract. We investigate the difference of the stability between different packings in the two-dimensional optimal velocity model, analytically. We show the phase diagram, and study the behavior of the flow in hexagonal and square arrangements by numerical simulation.

1   Introduction

Biological motion, group formation, traffic flow, and some related systems present interesting phenomena and have been studied from the physical point of view[1,2]. Collective motion and group formation are important biological problems and several models have been proposed to explain the dynamical behavior of a collective motion[3,4,5].

Recently, the two-dimensional OV model have been developed to study the pedestrian flow and biological motion, and the stability conditions are obtained by the linear analysis[6,7]. In group formation, the structure of packing is important for the stability, and its condition of the homogenous flow is dependent on such a structure. The aim of this paper is to study the stability in different packings. We consider the homogenous flow of the square and hexagonal packings. We investigate the stability condition of the homogenous flow for each packing by the linear analysis. We draw phase diagrams, and show the behavior of the flow in the characteristic phase by numerical simulation.

2   Two-Dimensional OV Model

The equation of motion of the two-dimensional OV model is given by

$$\ddot{\vec{\mathbf{x}}}_j(t) = a\left[\{\vec{\mathbf{V}}_0 + \sum_k \vec{\mathbf{F}}[\vec{\mathbf{r}}_{kj}(t)]\} - \dot{\vec{\mathbf{x}}}_j(t)\right], \tag{1}$$

$$\vec{\mathbf{F}}(\vec{\mathbf{r}}_{kj}) = f(r_{kj})(1 + \cos\theta_{kj})\vec{\mathbf{n}}_{kj}, \tag{2}$$

$$f(r_{kj}) = \alpha[\tanh\beta(r_{kj} - b) + c], \tag{3}$$

---

* Authors would like to thank R Kinukawa for their useful comments and remarks.

F. Peper et al. (Eds.): IWNC 2009, PICT 2, pp. 191–198, 2010.

where $\vec{x}_j$ is the position of the jth particle, and $\vec{r}_{kj} = \vec{x}_k - \vec{x}_j$, $r_{kj} = |\vec{r}_{kj}|$, $\vec{n}_{kj} = \vec{r}_{kj}/r_{kj}$, and θ_{kj} is defined as the angle between the vector $\vec{x}_j$ and the vector $\vec{x}_k$. a is the strength of reaction of each particle[7]. $\vec{V}_0$ is a external force. We take into account the nearest-neighbour interaction, and we define that a particle interacts with several neighbouring particles. For simplicity, each particle is supposed to move in the positive direction of the x-axis, and $\vec{V}_0 = (V_0, 0)$, $\cos\theta_{kj} = (x_k - x_j)/r_{kj}$. There exists a trivial solution as a homogenous flow,

$$(x_j, y_j) = (X_j + v_x t, Y_j + v_y t), \tag{4}$$

$$v_x = V_0 + \sum_k F_x(X_k - X_j, Y_k - Y_j), \tag{5}$$

$$v_y = \sum_k F_y(X_k - X_j, Y_k - Y_j), \tag{6}$$

where the initial position of jth particle is denoted by (X_j, Y_j), and $\vec{F} = (F_x, F_y)$.

We compare the homogenous flow for hexagonal and square packing. The sketch of each packing is shown in Fig. 1.

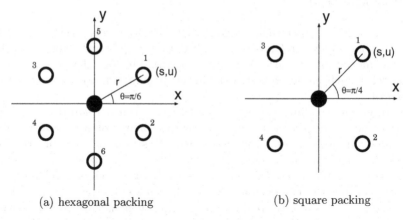

(a) hexagonal packing   (b) square packing

Fig. 1. The sketch of distribution of particles (denoted by a circle) surrounding a given particle (denoted by a filled circle): (a) hexagonal packing and (b) square packing

3   Linear Analysis

To investigate the stability in each packing, we perform the analysis of the non-linear equation (1). The linearized equation of (1) is written as

$$\ddot{x}_j = \sum_k [A_k(x_k - x_j) + B_k(y_k - y_j)] - \dot{x}_j,$$

$$\ddot{y}_j = \sum_k [C_k(x_k - x_j) + D_k(y_k - y_j)] - \dot{y}_j, \tag{7}$$

where

$$A_k = \partial_x F_x(x,y)|_{x=X_k-X_j, y=Y_k-Y_j}, \quad B_k = \partial_y F_x(x,y)|_{x=X_k-X_j, y=Y_k-Y_j},$$
$$C_k = \partial_x F_y(x,y)|_{x=X_k-X_j, y=Y_k-Y_j}, \quad D_k = \partial_y F_y(x,y)|_{x=X_k-X_j, y=Y_k-Y_j}. \quad (8)$$

Generally, the mode solutions of Eq.(7) can be written in the form

$$x_j = \epsilon_1 \exp[i\omega t + i(kX_j + mY_j)],$$
$$y_j = \epsilon_2 \exp[i\omega t + i(kX_j + mY_j)], \quad (9)$$

where (k, m) is the constant wave vector, and (ϵ_1, ϵ_2) is a polarization vector, which is not assumed to be a constant (see Sec. 3.5). From Eq.(9) and Eq.(7), we obtain

$$-\epsilon_1 \omega^2 = \epsilon_1 \bar{A} + \epsilon_2 \bar{B} - i\epsilon_1 \omega,$$
$$-\epsilon_2 \omega^2 = \epsilon_1 \bar{C} + \epsilon_2 \bar{D} - i\epsilon_2 \omega. \quad (10)$$

In hexagonal packing, $\bar{A}$, $\bar{B}$, $\bar{C}$, and $\bar{D}$ are given by

$$\bar{A} = 2[(A_1 + A_3)(\cos ks \cos mu - 1) + A_5(\cos 2mu - 1)]$$
$$+ 2i[(A_1 - A_3)\sin ks \cos mu], \quad (11)$$
$$\bar{D} = 2[(D_1 + D_3)(\cos ks \cos mu - 1) + D_5(\cos 2mu - 1)]$$
$$+ 2i[(D_1 - D_3)\sin ks \cos mu], \quad (12)$$
$$\bar{B} = 2[-(B_1 - B_3)\sin ks \sin mu] + 2i[(B_1 + B_3)\cos ks \sin mu], \quad (13)$$
$$\bar{C} = 2[-(C_1 - C_3)\sin ks \sin mu] + 2i[(C_1 + C_3)\cos ks \sin mu + C_5 \sin 2mu], \quad (14)$$

where $s = r\cos\frac{\pi}{6}$, $u = r\sin\frac{\pi}{6}$, $v = \sqrt{3}r\cos\frac{\pi}{3}$, $w = \sqrt{3}r\sin\frac{\pi}{3}$, and r is the distance among particles. In square packing, $\bar{A}$, $\bar{B}$, $\bar{C}$, and $\bar{D}$ are given by

$$\bar{A} = 2[(A_1 + A_3)(\cos ks \cos mu - 1)] + 2i[(A_1 - A_3)\sin ks \cos mu], \quad (15)$$
$$\bar{D} = 2[(D_1 + D_3)(\cos ks \cos mu - 1)] + 2i[(D_1 - D_3)\sin ks \cos mu], \quad (16)$$
$$\bar{B} = 2[-(B_1 - B_3)\sin ks \sin mu] + 2i[(B_1 + B_3)\cos ks \sin mu], \quad (17)$$
$$\bar{C} = 2[-(C_1 - C_3)\sin ks \sin mu] + 2i[(C_1 + C_3)\cos ks \sin mu], \quad (18)$$

where $s = r\cos\frac{\pi}{4}$, $u = r\sin\frac{\pi}{4}$, $v = r\cos\frac{\pi}{4}$, and $w = r\sin\frac{\pi}{4}$. The stability condition for the homogenous flow derived from mode solutions, is that the $\omega(k, m)$ does not have the negative imaginary part. In order to compare the stability of two packings, we investigate the stability condition of each mode solution: the longitudinal mode along the x-axis, the transverse mode along the x-axis, the transverse mode along the y-axis, the longitudinal mode along the y-axis, and the elliptically polarized mode.

We denote the average density by d. For an example, we set $\alpha = 0.25$, $\beta = 2.5$ and $b = 1$ in numerical calculations.

3.1   Longitudinal Mode along the x-Axis

Putting $(\epsilon_1, \epsilon_2) = (\epsilon, 0)$ and $m = 0$ into Eq.(10), equations are written as

$$-\omega^2 = \bar{A} - i\omega, \tag{19}$$
$$\bar{C} = 0, \tag{20}$$

where

$$\bar{A} = 2[(A_1 + A_3)(\cos ks - 1)] + 2i[(A_1 - A_3)\sin ks]. \tag{21}$$

The constraint $\bar{C} = 0$ is automatically satisfied in the case $m = 0$ in each packing. The stability condition is obtained as

$$a > 4\frac{(A_1 - A_3)^2}{A_1 + A_3}. \tag{22}$$

3.2   Transverse Mode along the x-Axis

Substituting $(\epsilon_1, \epsilon_2) = (0, \epsilon)$ and $m = 0$ into Eq.(10), we rewrite them as

$$\bar{B} = 0, \tag{23}$$
$$-\omega^2 = \bar{D} - i\omega, \tag{24}$$

where

$$\bar{D} = 2[(D_1 + D_3)(\cos ks - 1)] + 2i[(D_1 - D_3)\sin ks]. \tag{25}$$

The constraint $\bar{B} = 0$ is automatically satisfied in the case $m = 0$. The stability condition is

$$a > 4\frac{(D_1 - D_3)^2}{D_1 + D_3}. \tag{26}$$

We show typical critical curves of each packing.(Fig. 2)

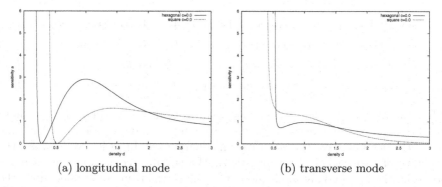

(a) longitudinal mode   (b) transverse mode

Fig. 2. Solid curve represents the critical curve of the hexagonal packing, and dotted curve represents the square packing, for the case c=0.0. (a) longitudinal mode, and (b) transverse mode.

3.3   Transverse Mode along the y-Axis

In this case $(\epsilon_1, \epsilon_2) = (\epsilon, 0)$ and $k = 0$, Eq.(10) becomes

$$-\omega^2 = \bar{A} - i\omega, \tag{27}$$
$$\bar{C} = 0. \tag{28}$$

In the hexagonal packing parameters $\bar{A}$ and $\bar{C}$ are

$$\bar{A} = 2[(A_1 + A_3)(\cos mu - 1) + A_5(\cos 2mu - 1)], \tag{29}$$
$$\bar{C} = 2i \sin mu[(C_1 + C_3) + 2C_5 \cos mu]. \tag{30}$$

In the square packing,

$$\bar{A} = 2(A_1 + A_3)(\cos mu - 1), \tag{31}$$
$$\bar{C} = 2i(C_1 + C_3) \sin mu. \tag{32}$$

We investigate the constraint to obtain the stability condition in each packing: In hexagonal packing, there are two solutions of the constraint (28), (30).

1. $\sin mu = 0$.
   $(k, m) = 0$ is a trivial stable mode. The unstable mode exists for $mu = \pi$. From (27), (29), the stability condition is

$$A_1 + A_3 > 0. \tag{33}$$

2. $(C_1 + C_3) + 2C_5 \cos mu = 0$
   The stability condition is given by following two inequalities,

$$\left| -\frac{C_1 + C_3}{2C_5} \right| \leq 1,$$
$$\bar{A} = \frac{-1}{C_5}[2C_5(A_1 + 2A_5 + A_3) + (A_1 + A_3)(C_1 + C_3) - (C_1 + C_3)^2] < 0. \tag{34}$$

In square packing the solution of constraint (28) is only $\sin mu = 0$, and the stability condition is

$$A_1 + A_3 > 0. \tag{35}$$

3.4   Longitudinal Mode along the y-Axis

Substituting $(\epsilon_1, \epsilon_2) = (0, \epsilon)$ and $k = 0$, Eq.(10) is

$$\bar{B} = 0, \tag{36}$$
$$-\omega^2 = \bar{D} - i\omega. \tag{37}$$

In each packing, the constraint $\bar{B} = 0$ leads $mu = \pi$, and the stability condition is

$$D_1 + D_3 > 0. \tag{38}$$

3.5   Elliptically Polarized Mode

Putting $(\epsilon_1, \epsilon_2) = (1, \epsilon(k, m))$, Eq.(10) becomes

$$\bar{B}\epsilon^2 + (\bar{A} - \bar{D})\epsilon - \bar{C} = 0, \tag{39}$$

$$\omega^2 - i\omega + \bar{A} + \epsilon\bar{B} = 0. \tag{40}$$

There are two solutions $\epsilon(k, m, r)$. And, there are two solutions $w(k, m, r)$ for each $\epsilon(k, m, r)$. However, their solutions are too complicated to obtain the stability condition directly. Instead, we obtain the condition by the long-wavelength approximation, $k, m \sim$ small.

In hexagonal packing

$$\omega = (A_1 - A_3 + D_1 - D_3)ks \pm [(A_1 - A_3 - D_1 + D_3)^2(ks)^2$$
$$+ 4(B_1 + B_3)(C_1 + 2C_5 + C_3)(mu)^2]^{1/2}. \tag{41}$$

Only in the case of $(B_1 + B_3)(C_1 + C_3 + 2C_5) < 0$, ω can have a negative imaginary part. The stability condition is

$$(B_1 + B_3)(C_1 + C_3 + 2C_5) > 0. \tag{42}$$

In square packing

$$\omega = (A_1 - A_3 + D_1 - D_3)ks \pm [(A_1 - A_3 - D_1 + D_3)^2(ks)^2$$
$$+ 4(B_1 + B_3)(C_1 + C_3)(mu)^2]^{1/2}. \tag{43}$$

ω can have a negative imaginary part only in the case of $(B_1 + B_3)(C_1 + C_3) < 0$. The stability condition is given as

$$(B_1 + B_3)(C_1 + C_3) > 0. \tag{44}$$

We summarize stability conditions for the variations of c (Table. 1): transverse and transverse modes along y-axis, and elliptically polarized mode

Table 1. Solutions of stability conditions (transvers and longitudinal modes along the y-axis, and elliptically polarized mode) of hexagonal and square packings in the case of c=-1.0, 0.0, and 1.0.

	mode	condition	c=-1.0	c=0	c=1.0
Hexagonal	Transverse mode	$A_1 + A_3 > 0$	$d > 0.30$	$d > 0.25$	$d > 0$
	Transverse mode	Eq.(34)	$d > 0.77$	$d > 0.46$	$d > 0$
	Longitudinal mode	$D_1 + D_3 > 0$	$d > 0.96$	$d > 0.53$	$d > 0$
	Elliptically polarized mode	Eq.(42)	$d > 0.76$	$1.67 > d > 0.46$	$1.07 > d > 0.15$
Square	Transverse mode	$A_1 + A_3 > 0$	$d > 0.62$	$d > 0.39$	$d > 0$
	Longitudinal mode	$D_1 + D_3 > 0$	$d > 0.62$	$d > 0.39$	$d > 0$
	Elliptically polarized mode	Eq.(44)	$d > 0$	$1.93 > d$	$1.24 > d > 0.17$

4   Phase Diagrams

Phase diagrams are shown in Fig. 3. The linear stablity of the solutions for transverse, longitudinal, and elliptically polarized modes are different between the hexagonal and square packings. The stable region of hexagonal packing is smaller than the stable region of square packing. The number of critical lines in square packing is less than in hexagonal packing and this difference is due to a lack of stability condition (34).

The behavior of the flow is examined by numerical simulations. The comparison for different packing is shown in Fig. 4(a) and (b) for the case, $c = 0.0$. For the same paramter, $d = 0.76$, $a = 1.2$, the flow is characterized by the emergence of the longitudinal wave in hexagonal packing(Fig. 4(a)), and the transverse wave emerges in square packing(Fig. 4(b)). These two behaviors consist in the result of the linear analysis.

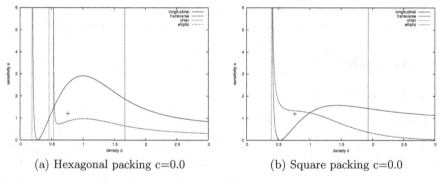

(a) Hexagonal packing c=0.0   (b) Square packing c=0.0

Fig. 3. Phase diagrams for the cases c=0.0 in each packing. Solid and dashed curves are critical curves for longitudinal and transverse modes respectively, along the x-axis. Dotted lines are critical lines for modes along the y-axis. Dashed dotted lines are critical lines for elliptically polarized modes. (a) The flow in hexagonal packing, (b) the flow in square packing. "+" represents the typical value of the parameter in the characteristic phase used for the numerical simulation.

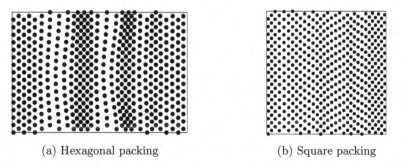

(a) Hexagonal packing   (b) Square packing

Fig. 4. Snapshots of the flow for the case c=0.0, corresponding to the typical value of the paramter in Fig 3

5 Summary

Using the linear stability analysis, we investigate stability conditions of the homogenous flow for different packings. From these results, we show the phase diagram for each packing. We justify the result of the linear analysis by the emergence of the excitation of unstable modes by the numerical simulation.

References

1. Schadschneider, A., Poeschel, T., Kuehne, R., Schreckenberg, M., Wolf, D.E.: Traffic and Granular Flow 2005 (2007)
2. Waldau, N., Gattermann, P., Knoflacher, H., Schreckenberg, M.: Pedestrian and Evacuation Dynamics 2005 (2007)
3. Sannomiya, N., Matuda, K.: IEEE Trans. Syst. Man Cybern. 14, 157 (1984)
4. Reynolds, C.W.: Comput. Graph (ACM) 21, 25 (1987)
5. Niwa, H.S.: J. Theor. Biol. 171, 123 (1994)
6. Nakayama, A., Hasebe, K., Sugiyama, Y.: Phys. Rev. E 71, 036121 (2005)
7. Nakayama, A., Hasebe, K., Sugiyama, Y.: Phys. Rev. E 77, 016105 (2008)

A Appendix

A.1 Hexagonal

The parameters $A_1, A_2, ..., A_6, B_1, B_2, ..., D_6$ are

$$\left\{ \begin{matrix} A_1 \\ A_3 \end{matrix} \right\} = f'(\frac{3}{4} \pm \frac{3\sqrt{3}}{8}) + \frac{f}{r}(\frac{1}{4} \pm \frac{\sqrt{3}}{4}), \left\{ \begin{matrix} B_1 \\ B_3 \end{matrix} \right\} = f'(\frac{3}{8} \pm \frac{\sqrt{3}}{4}) + \frac{f}{r}(-\frac{3}{4} \mp \frac{\sqrt{3}}{4}),$$

$$\left\{ \begin{matrix} C_1 \\ C_3 \end{matrix} \right\} = f'(\frac{3}{8} \pm \frac{\sqrt{3}}{4}) + \frac{f}{r}(-\frac{1}{4} \mp \frac{\sqrt{3}}{4}), \left\{ \begin{matrix} D_1 \\ D_3 \end{matrix} \right\} = f'(\frac{1}{4} \pm \frac{\sqrt{3}}{8}) + \frac{f}{r}(\frac{3}{4} \pm \frac{\sqrt{3}}{4}),$$

$$A_5 = C_5 = \frac{f}{r}, D_5 = f',$$

$$A_1 = A_2, A_3 = A_4, A_5 = A_6, B_1 = -B_2, B_3 = -B_4 = 0, B_5 = -B_6,$$
$$C_1 = -C_2, C_3 = -C_4, C_5 = -C_6, D_1 = D_2, D_3 = D_4, D_5 = D_6.$$

A.2 Square

The parameters $A_1, A_2, A_3, A_4, B_1, B_2, ..., D_4$ are

$$\left\{ \begin{matrix} A_1 \\ A_3 \end{matrix} \right\} = f'(\frac{1}{2} \pm \frac{1}{2\sqrt{2}}) + \frac{f}{r}(\frac{1}{2} \pm \frac{1}{\sqrt{2}}), \left\{ \begin{matrix} B_1 \\ B_3 \end{matrix} \right\} = f'(\pm\frac{1}{2} + \frac{1}{2\sqrt{2}}) + \frac{f}{r}(\mp\frac{1}{2} - \frac{1}{\sqrt{2}}),$$

$$\left\{ \begin{matrix} C_1 \\ C_3 \end{matrix} \right\} = f'(\pm\frac{1}{2} + \frac{1}{2\sqrt{2}}) \mp \frac{f}{r}\frac{1}{2}, \left\{ \begin{matrix} D_1 \\ D_3 \end{matrix} \right\} = f'(\frac{1}{2} \pm \frac{1}{2\sqrt{2}}) + \frac{f}{r}\frac{1}{2},$$

$$A_1 = A_2, A_3 = A_4, B_1 = -B_2, B_3 = -B_4,$$
$$C_1 = -C_2, C_3 = -C_4, D_1 = D_2, D_3 = D_4.$$

A New Differential Evolution for Multiobjective Optimization by Uniform Design and Minimum Reduce Hypervolume

Siwei Jiang and Zhihua Cai*

School of Computer Science, China University of Geosciences
Wuhan 430074, China
amosonic@gmail.com, zhcai@cug.edu.cn

Abstract. Differential evolution is a powerful and robust method to solve the Multi-Objective Problems in MOEAs. To enhance the differential evolution for MOPs, we focus on two aspects: the population initialization and acceptance rule. In this paper, we present a new differential evolution called DEMO_{DV}^{UD}, it mainly include: (1) the first population is constructed by statistical method: *Uniform Design*, which can get more evenly distributed solutions than random design, (2) a new acceptance rule is firstly presented as *Minimum Reduce Hypervolume*. Acceptance rule is a metric to decide which solution should be cut off when the archive is full to the setting size. *Crowding Distance* is frequently used to estimate the length of cuboid enclosing the solution, while *Minimum Reduce Hypervolume* is used to estimate the volume of cuboid. The new algorithm designs a fitness function *Distance/Volume* that balance the *CD* and *MRV*, which maintains the spread and hypervolume along the Pareto-front. Experiment on different multi-Objective problems include ZDTx and DTLZx by jMetal 2.0, the results show that the new algorithm gets higher hypervolume, faster convergence, better distributed solutions and needs less numbers of fitness function evolutions than NSGA-II, SPEA2 and GDE3.

Keywords: Multi-Objective Optimization, Differential Evolution, Uniform Design, Crowding Distance, Minimum Reduce Hypervolume, Genetic Distance, Inverted Genetic Distance, Spread, Hypervolume.

1 Introduction

To solve the multiple conflicting problems, Multi-Objective Evolutionary Algorithms(MOEAs) are demonstrated as useful and powerful tools to deal with complex problems such as: discontinuous, non-convex, multi-modal, and non-differentiable[1]. The main challenge for MOEAs is to obtain Pareto-optimal solutions near to true Pareto- front in terms of convergence, diversity and limited evolution times.

* The Project was supported by the Research Foundation for Outstanding Young Teachers, China University of Geosciences(Wuhan)(No:CUGQNL0911).

The Non-dominated Sorting Genetic Algorithm-II (NSGA-II) is proposed by Deb *et al*[2], which adopts a fast non-dominated sorting approach with $O(mN^2)$ computational complexity(where m is the number of objectives and N is the population size), NSGA-II combines the parent and child populations and selects the best N solutions with respect to the *Ranking* and *Crowding Distance* metrics.

The Strength Pareto Evolutionary Algorithm (SPEA2) is proposed by Zitzler *et al*[3], which incorporates a fine-grained fitness assignment strategy, a density estimation technique, and an enhanced archive truncation method. SPEA2 uses the truncation operator based on nearest *Neighbor Density Estimation* metric.

Differential evolution(DE) is a new evolutionary algorithm, which has some advantages: simple structure, ease to use, speed and robustness[4]. The developed version of Generalized Differential Evolution(GDE3) is proposed by Kukkonen *et al*[5], which is suited for global optimization with an arbitrary number of objectives and constraints. Similar to NSGA-II, GDE3 uses *Crowding Distance* metric as its acceptance rule, it can get better distributed solutions by its powerful search ability.

For long terms, the issue of population initialization has been ignored in evolution algorithms, researchers mainly focus on regeneration method and acceptance rule. *Orthogonal design* and *Uniform Design* belong to a sophisticated branch of statistics[8], they are more powerful than the *random design*.

Leung and Wang incorporated orthogonal design and quantum technique in genetic algorithm[9], which is more robust than the classical GAs for numerical optimization problems. Zeng *et al* adopts the orthogonal design method to solve the MOPs[10], it found a precise Pareto-optimal solutions for an engineer problem which has unknown before. Cai *el al* solves the MOPs by orthogonal population and ε-dominated[11], it gets better results in terms of convergence, diversity and time consume. Leung applies the uniform design to generate a good initial population and designs a new crossover operator for searching the Pareto-optimal solutions[12], which can find the Pareto-optimal solutions scattered uniformly over the Pareto frontier.

Interesting in the population initialization and acceptance rule, we propose a new differential evolution called $DEMO_{DV}^{UD}$, it construct by three aspects:

1. The first population is constructed by *Uniform Design*. It can get well scattered solutions in feasible searching space.
2. The regeneration method of differential evolution is adopted with DE/best/ 1/bin strategy. It can faster the convergence of the algorithm for MOPs.
3. Acceptance rule *Minimum Reduce Hypervolume* is firstly proposed, when the archive is full, it cuts off the solution which leads to the minimum decrease for the archive's hypervolume. Then a new fitness *Distance/Volume* is designed for DE, which maintains the spread and hypervolume along the Pareto-front.

Test the bi-objective of ZDT family and tri-objective of DTLZ family problems on jMetal 2.0[13], the results show that the new algorithm gets higher hypervolume, faster convergence, better distributed solutions and needs less numbers of fitness function evolutions than NSGA-II, SPEA2 and GDE3.

2   Uniform Design

Uniform Design belongs to a sophisticated branch of statistics, experimental methods construct the candidate solutions more evenly in feasible searching space and include more information than random design[8,9,10,11].

We define the uniform array as $U_R(C)$, where Q is the level, it's primer, R, C represent the row and column of uniform array, they must satisfy to:

$$\begin{cases} R = Q > n \\ \quad C = n \end{cases} \tag{1}$$

Where n is the number of variables. When select a proper parameters of Q, σ form table 1, uniform array can be created by Equation 2

$$U_{i,j} = (i * \sigma^{j-1} \bmod Q) + 1 \tag{2}$$

Table 1. Values of the parameter σ for different number of factors and levels

number of levels of per factor(Q)	number of factors(n)	σ
5	2-4	2
7	2-6	3
11	2-10	7
13	2	5
	3	4
	4-12	6
17	2-16	10
19	2-3	8
	4-18	14
23	2,13-14,20-22	7
	8-12	15
	3-7,15-19	17
29	2	12
	3	9
	4-7	16
	8-12,16-24	8
	13-15	14
	25-28	18
31	2,5-12,20-30	12
	3-4,13-19	22

3   Minimum Reduce Hypervolume

For Multi-Objective Optimization Problems, MOEAs often finds the optimal set including many candidate solutions which no-dominate each other. When the optimal set is full to the setting size(usually $archiveSize = 100$), we need to set an acceptance rule to decide which one should be remained or cut off from archive. It directly influences the quality of finally optimal set in convergence and spread metric.

Some acceptance rules have been presented: NSGA-II adopts *Ranking* and *Crowding Distance* metric, SPEA2 uses nearest *Neighbor Density Estimation* metric, ϵ-MOEA proposes ϵ-*dominance* concept with less time consume[6,7].

Inspired by the *Crowding Distance* metric, in this paper, we firstly present a new acceptance rule called *Minimum Reduce Hypervolume*. Hypervolume is a quality indicator proposed by Zitzler *et al*, it is adopted in jMetal 2.0[13]. Hypervolume calculates the volume covered by members of a non-dominated set of solutions (the region enclosed into the discontinuous line respect the reference point W in the figure 1 is $ADBECW$ in dashed line).

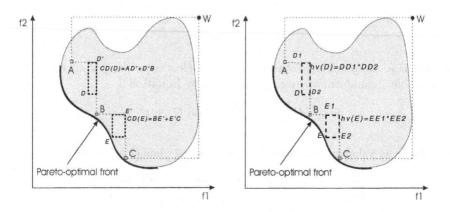

Fig. 1. The comparison of *Crowding Distance* and *Minimum Reduce Hypervolume* is described in the left and right figure

If two solutions D and E no-dominate each other and one should be cut off, NSGA-II chooses the solution D remained in archive if $CD(D) > CD(E)$, it maintains the spread along the Pareto-front.

$$\begin{cases} CD(D) = AD' + D'B \\ CD(E) = BE' + E'C \end{cases} \tag{3}$$

If one solution is deleted, it will lead to a hypervolume decrease, because the higher hypervolume means the better quality of optimal set, we will delete the solution which reduces the hypervolume minimum. *Minimum Reduce Hypervolume* chooses the solution E remained in archive if $hv(D) < hv(E)$(then the hypervolume is $ABECW$ rather than $ADBCW$), it maintains the hyervolume along the Pareto-front.

$$\begin{cases} hv(D) = DD1 * DD2 \\ hv(E) = EE1 * EE2 \end{cases} \tag{4}$$

4 Differential Evolution for MOPs by Uniform Design and Minimum Reduce Hypervolume

In this paper, we design a new hybrid acceptance rule. *Crowding distance* maintains the spread and expand solutions to feasible searching place, and *Minimum*

Reduce Hypervolume maintains the hypevolume and forces solutions near to the Pareto-front, so we combine the two properties, a new fitness assignment for solution s is designed as follows(called *Distance/Volume* fitness):

$$\begin{cases} DV(s) = CD(s) + scale * hv(s) \\ scale = \dfrac{\sum_{i=1}^{n-1} CD(s_i)}{\sum_{i=1}^{n-1} hv(s_i)} \end{cases} \tag{5}$$

The factor *scale* is designed to equal the influence of crowding distance and sub-hypervolume, but if it is too large, we set $scale = 1$ when $scale > 1000$.

To enhance the differential evolution for MOPs, we focus on two aspects: population initialization and acceptance rule. The new differential evolution based on *Uniform Design* and *Distance/Volume* fitness function is called $DEMO_{DV}^{UD}$. Population size and archive size is setting to 100, the first population is constructed by uniform design($Q = 31$) and random design($100 - Q = 69$); when archive is full, we discard the solution which has small fitness by equation 5.

Algorithm 1. Main procedure of the proposed $DEMO_{DV}^{UD}$

Construct first population with 31 uniform solutions($Q = 31$) by equation 2
Add 69 random solutions to the first population
Add the population to archive
while $eval < Max\_eval$ **do**
    for $i = 1$ to NP **do**
        Random choose two solutions from archive
        select the solution with larger DV fitness as DE base solution *best*
        Produce child c with DE/best/1/bin scheme
        Evaluate the child c and $eval++$
        if the child c dominates the one of the parent population P_t^r **then**
            $P_t^r = c$
        else if c is non-dominated by all the parent population **then**
            Rand replace one parent with probability 0.5
        else
            Discard c
        end if
        Add child c to archive.
        if $archiveSize > 100$ **then**
            Remove the candidate solution with minimum DV(s) by equation 5
        end if
    end for
end while

---

5   Experiment Results

Experiment is based on jMetal 2.0[13], which is a Java-based framework aimed at facilitating the development of metaheuristics for solving MOPs, it provides large block reusing code and fair comparison for different MOEAs.

The test problems are choose from ZDTx, DTLZx problem family including five bi-objective problems: ZDT1, ZDT2, ZDT3, ZDT4, ZDT6 and six tri-objective problems: DTLZ1, DTLZ2, DTLZ3, DTLZ4, DTLZ5, DTLZ6. Each algorithm independent runs for 50 times and maximum evolution times is 25,000.

The performance metrics can be classified into five categories:

Hypervolume. This quality indicator calculates the volume (in the objective space) covered by members of a nondominated set of solutions with a reference point. The Hypervolume (HV) is calculated by:

$$HV = volume(\bigcup_{i=1}^{|Q|} v_i) \tag{6}$$

Generational Distance. The metric is to measure how far the elements are in the set of non-dominated vectors found from those in the true Pareto-optimal set. It is defined as:

$$GD = \sqrt{\frac{\sum_{i=1}^{|n|} d_i^2}{n}} \tag{7}$$

Inverted Generational Distance. The metric is to measure how far the elements are in the true Pareto-optimal set from those in the set of non-dominated vectors found. It is defined as:

$$IGD = \sqrt{\frac{\sum_{i=1}^{|N|} d_i^2}{N}} \tag{8}$$

Spread. The Spread indicator is a diversity metric that measures the extent of spread achieved among the obtained solutions. This metric is defined as:

$$\Delta = \frac{d_f + d_l + \sum_{i=1}^{n-1} |d_i - \overline{d}|}{d_f + d_l + (n-1)\overline{d}} \tag{9}$$

Evolution Counts. This metric is to measure how many numbers of fitness function evolutions($NFFEs$) is needed, when the hypervolume of non-dominated set is larger than the 98% of hypervolume of the true Pareto-front set(Success Rate means the percentage of get the $HV_{paretoSet} \geq 0.98 * HV_{truePF}$ in 50 independent run, NA means cannot get 100% success).

$$Evolution\ counts = \{NFFEs | HV_{paretoSet} \geq 0.98 * HV_{truePF}\} \tag{10}$$

The higher Hypervolume and lower GD, IGD, Spread, Evolution Counts mean the algorithm is better. The result are compared with mean and variance, the better result between GDE3 and DEMO$_{DV}^{UD}$ is bold font.

From table 2, in term of Hypervolume, the w/t/l(win/tie/lost) value between DEMO$_{DV}^{UD}$ and GDE3 is 10/0/1, the new algorithm is only worse in problem

Table 2. The hypervolume metric of mean and variance for NSGA-II, SPEA2, GDE3 and DEMO$_{DV}^{UD}$ in 50 independent run

probs	NSGA-II		SPEA2		GDE3		DEMO$_{DV}^{UD}$	
	mean	std	mean	std	mean	std	mean	std
Hypervolume								
ZDT1	0.659343	0.000295	0.659885	0.000370	0.661925	0.000022	**0.662084**	0.000013
ZDT2	0.326077	0.000256	0.326136	0.001615	0.328631	0.000024	**0.328796**	0.000018
ZDT3	0.514758	0.000472	0.514010	0.000676	0.515936	0.000042	**0.516024**	0.000025
ZDT4	0.654138	0.005719	0.644951	0.020486	0.661999	0.000024	**0.662086**	0.000016
ZDT6	0.388331	0.001691	0.378258	0.002821	0.401343	0.000017	**0.401482**	0.000019
DTLZ1	0.647919	0.186030	0.749499	0.082703	0.757721	0.095956	**0.770371**	0.003150
DTLZ2	0.375092	0.005691	0.404556	0.001746	0.388390	0.003685	**0.388688**	0.003002
DTLZ3	0.000000	0.000000	0.000000	0.000000	0.322214	0.140052	**0.389571**	0.004504
DTLZ4	0.372841	0.005571	0.332877	0.115285	**0.384835**	0.003979	0.384668	0.004476
DTLZ5	0.092796	0.000184	0.093194	0.000139	0.094023	0.000032	**0.094092**	0.000028
DTLZ6	0.000000	0.000000	0.000000	0.000000	0.094901	0.000026	**0.094968**	0.000034

DTLZ4; the $w/t/l$ value between DEMO$_{DV}^{UD}$ and NSGA-II is 11/0/0, NSGA-II gets zero hypervolme in DTLZ3 and DTLZ6; the $w/t/l$ value between DEMO$_{DV}^{UD}$ and SPEA2 is 10/0/1, the new algorithm is only worse in problem DTLZ2, and SPEA2 gets zero hypervolme in DTLZ3 and DTLZ6.

From table 3, in term of Genetic Distance, the $w/t/l$(win/tie/lost) value between DEMO$_{DV}^{UD}$ and GDE3 is 9/0/2, the new algorithm is only worse in problem ZDT6, DTLZ5; the $w/t/l$ value between DEMO$_{DV}^{UD}$ and NSGA-II is 11/0/0; the $w/t/l$ value between DEMO$_{DV}^{UD}$ and SPEA2 is 11/0/0.

From table 4, in term of Inverted Genetic Distance, the $w/t/l$(win/tie/lost) value between DEMO$_{DV}^{UD}$ and GDE3 is 8/0/3, the new algorithm is only worse in problem ZDT2, DTLZ2, DTLZ4; the $w/t/l$ value between DEMO$_{DV}^{UD}$ and NSGA-II is 10/0/1, the new algorithm is only worse in problem DTLZ4; the $w/t/l$ value between DEMO$_{DV}^{UD}$ and SPEA2 is 10/0/1, the new algorithm is only worse in problem DTLZ2.

From table 5, in term of Spread, the $w/t/l$(win/tie/lost) value between DEMO$_{DV}^{UD}$ and GDE3 is 9/0/2, the new algorithm is only worse in problem ZDT3, DTLZ1; the $w/t/l$ value between DEMO$_{DV}^{UD}$ and NSGA-II is 11/0/0; the

Table 3. The Generation Distance metric of mean and variance for NSGA-II, SPEA2, GDE3 and DEMO$_{DV}^{UD}$ in 50 independent run

probs	NSGA-II		SPEA2		GDE3		DEMO$_{DV}^{UD}$	
	mean	std	mean	std	mean	std	mean	std
Generation Distance								
ZDT1	2.199E-4	3.623E-5	2.241E-4	3.127E-5	9.930E-5	3.210E-5	**7.252E-5**	1.874E-5
ZDT2	1.806E-4	6.895E-5	1.773E-4	3.921E-5	4.607E-5	2.202E-6	**4.588E-5**	2.258E-6
ZDT3	2.116E-4	1.322E-5	2.299E-4	1.534E-5	1.738E-4	1.243E-5	**1.700E-4**	1.165E-5
ZDT4	4.888E-4	2.358E-4	9.792E-4	1.840E-3	9.099E-5	3.139E-5	**8.127E-5**	2.494E-5
ZDT6	1.033E-3	1.108E-4	1.799E-3	2.997E-4	**5.277E-4**	1.564E-5	5.293E-4	1.617E-5
DTLZ1	1.890E-1	5.973E-1	5.181E-1	1.021E0	2.004E-3	9.746E-3	**5.976E-4**	2.763E-5
DTLZ2	1.428E-3	3.207E-4	1.363E-3	3.058E-4	6.463E-4	2.579E-5	**6.414E-4**	2.177E-5
DTLZ3	1.073E0	7.782E-1	2.353E0	1.478E0	2.196E-2	7.415E-2	**1.076E-3**	4.797E-5
DTLZ4	5.098E-3	2.786E-4	4.495E-3	1.291E-3	4.829E-3	1.693E-4	**4.737E-3**	2.294E-4
DTLZ5	3.707E-4	6.887E-5	3.770E-4	4.705E-5	**2.520E-4**	2.311E-5	2.592E-4	2.051E-5
DTLZ6	1.759E-1	2.111E-2	1.626E-1	1.424E-2	5.681E-4	2.373E-5	**5.655E-4**	2.648E-5

Table 4. The Inverted Generation Distance metric of mean and variance for NSGA-II, SPEA2, GDE3 and DEMO$_{DV}^{UD}$ in 50 independent run

probs	NSGA-II		SPEA2		GDE3		DEMO$_{DV}^{UD}$	
	mean	std	mean	std	mean	std	mean	std
	Inverted Generation Distance							
ZDT1	1.893E-4	8.636E-6	1.521E-4	4.140E-6	1.398E-4	1.144E-6	**1.351E-4**	1.100E-6
ZDT2	1.928E-4	1.012E-5	1.843E-4	2.047E-4	**1.457E-4**	1.733E-6	1.467E-4	2.583E-6
ZDT3	3.519E-4	6.766E-4	4.323E-4	9.520E-4	2.007E-4	3.445E-6	**1.924E-4**	5.210E-6
ZDT4	3.099E-4	5.045E-4	1.338E-3	1.346E-3	1.378E-4	1.088E-6	**1.354E-4**	1.314E-6
ZDT6	3.533E-4	5.251E-5	6.832E-4	9.836E-5	1.170E-4	4.315E-6	**1.088E-4**	3.969E-6
DTLZ1	1.409E-3	1.533E-3	6.567E-4	5.810E-4	6.345E-4	7.597E-4	**5.302E-4**	1.973E-5
DTLZ2	7.747E-4	3.662E-5	5.906E-4	1.104E-5	**6.802E-4**	2.263E-5	6.843E-4	2.227E-5
DTLZ3	9.306E-2	5.289E-2	7.400E-2	3.446E-2	4.244E-3	1.150E-2	**1.089E-3**	4.322E-5
DTLZ4	1.196E-3	1.134E-4	2.317E-3	2.592E-3	**1.211E-3**	9.7101E-5	1.225E-3	1.132E-4
DTLZ5	2.006E-5	1.063E-6	1.581E-5	4.951E-7	1.445E-5	2.630E-7	**1.387E-5**	3.131E-7
DTLZ6	7.254E-3	7.102E-4	6.726E-3	4.950E-4	3.518E-5	9.993E-7	**3.390E-5**	9.281E-7

Table 5. The Spread metric of mean and variance for NSGA-II, SPEA2, GDE3 and DEMO$_{DV}^{UD}$ in 50 independent run

probs	NSGA-II		SPEA2		GDE3		DEMO$_{DV}^{UD}$	
	mean	std	mean	std	mean	std	mean	std
	Spread							
ZDT1	0.380309	0.028535	0.150063	0.014035	0.146912	0.012623	**0.123448**	0.014250
ZDT2	0.379900	0.028633	0.157648	0.027377	0.135998	0.013377	**0.121396**	0.012272
ZDT3	0.747304	0.014352	0.710772	0.005718	**0.710740**	0.004612	0.716190	0.011558
ZDT4	0.388983	0.040338	0.322715	0.148944	0.139899	0.013194	**0.115616**	0.013786
ZDT6	0.358147	0.029767	0.233547	0.036104	0.127141	0.010248	**0.085469**	0.013509
DTLZ1	0.925933	0.236171	0.904105	0.316227	**0.733342**	0.043591	0.745113	0.039776
DTLZ2	0.705338	0.053090	0.530350	0.031301	0.642637	0.027290	**0.635581**	0.045237
DTLZ3	1.052288	0.128959	1.258939	0.139970	0.639076	0.040248	**0.632920**	0.045891
DTLZ4	0.672005	0.045341	0.488522	0.182093	0.644641	0.037459	**0.643950**	0.132165
DTLZ5	0.457166	0.050954	0.238807	0.037331	0.161843	0.012136	**0.110917**	0.015109
DTLZ6	0.805454	0.051306	0.578990	0.041086	0.164002	0.012606	**0.111253**	0.015057

$w/t/l$ value between DEMO$_{DV}^{UD}$ and SPEA2 is 8/0/3, the new algorithm is only worse in problem ZDT3,DTLZ2, DTLZ4.

From table 6, in term of Evolution Counts, the $w/t/l$(win/tie/lost) value between DEMO$_{DV}^{UD}$ and GDE3 is 7/4/0, in problem DTLZ1, DTLZ2, DTLZ3 and DTLZ4, both the new algorithm and GDE3 gets 0% Success Rate; the $w/t/l$ value between DEMO$_{DV}^{UD}$ and NSGA-II is 7/4/0; the $w/t/l$ value between DEMO$_{DV}^{UD}$ and SPEA2 is 7/4/0. NSGA-II and SPEA2 can not get 100% success Rate in problem ZDT2, they get 0% success Rate in problem ZDT6 and all tri-objective problems.

The comparison for four algorithms in five categories : Hypervolume, Generation Distance, Inverted Genetic Distance, Spread and Evolution Count/Success Rate, it shows that the new algorithm is more efficient to solve the MOPs. Now, we summarize the highlight as follows:

1. population initialization is constructed by *Uniform Design*, which can get more evenly scatter solutions in feasible space than *Random Design*, it will provide good guide information for next offspring and speed the convergence.

Table 6. The Evolution times metric of mean and variance for NSGA-II, SPEA2, GDE3 and DEMO$_{DV}^{UD}$ in 50 independent run. Success Rate means the percentage of get the $HV_{paretoSet} \geq 0.98 * HV_{truePF}$ in 50 independent run, NA means cannot get 100% success.

probs	NSGA-II		SPEA2		GDE3		DEMO$_{DV}^{UD}$	
	mean	std	mean	std	mean	std	mean	std
Evaluation Counts: $HV_{paretoSet} \geq 0.98 * HV_{truePF}$ or Success Rate								
ZDT1	14170.0	670.6	15998.0	912.7	9590.0	350.0	**3110.6**	450.7
ZDT2	NA(74%)	NA(74%)	NA(62%)	NA(62%)	11254.0	426.2	**4764.7**	734.9
ZDT3	12980.0	764.5	15374.0	810.9	10446.0	398.1	**4110.6**	1079.7
ZDT4	NA(80%)	NA(80%)	NA(46%)	NA(46%)	16306.0	751.6	**5136.3**	3024.5
ZDT6	NA(0%)	NA(0%)	NA(0%)	NA(0%)	4724.0	443.0	**2639.6**	478.3
DTLZ1	NA(0%)	NA(0%)	NA(0%)	NA(0%)	NA(0%)	NA(0%)	NA(0%)	NA(0%)
DTLZ2	NA(0%)	NA(0%)	NA(0%)	NA(0%)	NA(0%)	NA(0%)	NA(0%)	NA(0%)
DTLZ3	NA(0%)	NA(0%)	NA(0%)	NA(0%)	NA(0%)	NA(0%)	NA(0%)	NA(0%)
DTLZ4	NA(0%)	NA(0%)	NA(0%)	NA(0%)	NA(0%)	NA(0%)	NA(0%)	NA(0%)
DTLZ5	NA(0%)	NA(0%)	NA(0%)	NA(0%)	9088.0	429.3	**3910.0**	612.5
DTLZ6	NA(0%)	NA(0%)	NA(0%)	NA(0%)	4042.0	202.1	**2326.3**	336.0

2. Acceptance rule is firstly presented as *Distance/Volume* fitness, which expands solutions to feasible searching place and forces solutions near to the Pareto-front.

3. DEMO$_{DV}^{UD}$ can get higher hypervolume, faster convergence and well distributed solutions than NSGA-II, SPEA2 and GDE3.

4. DEMO$_{DV}^{UD}$ need less $NFFES$ to get 98% hypervolume of true Pareto-front than NSGA-II, SPEA2 and GDE3.

6   Conclusion and Further Research

In this paper, a new differential evolution for MOPS called DEMO$_{DV}^{UD}$ is presented. We adopt the statistical experimental method *Uniform Design* to construct the first population; we design a new fitness assignment as *Distance/Volume*. Experiment on ZDTx and DTLZx problems by jMetal 2.0, the results show that the new algorithm gets higher hypervolume, faster convergence, better distributed solutions and needs less number of fitness function numbers than NSGA-II, SPEA2 and GDE3.

The new acceptance rule *Minimum Reduce Hypervolume* is powerful to maintain the hypervolume and force the solution near to the Pareto-front, in the future, we can use this property to optimize ϵ-MOEA and $pa\epsilon$-MOEA.

References

1. Coello, C.A.C.: Evolutionary multi-objective optimization: A historical view of the Field. IEEE Computational Intelligence Magazine 1(1), 28–36 (2006)
2. Deb, K., Pratap, A., Agarwal, S., Meyarivan, T.: A fast and elitist multiobjective genetic algorithm: NSGA - II. IEEE Transactions on Evolutionary Computation 6(2), 182–197 (2002)

3. Zitzler, E., Laumanns, M., Thiele, L.: SPEA2: Improving the strength Pareto evolutionary algorithm, Technical Report 103, Computer Engineering and Networks Laboratory (2001)
4. Storn, R., Price, K.: Differential evolution CA simple and efficient heuristic for global optimization over continuous spaces. Journal of Global Optimization 11, 341–359 (1997)
5. Kukkonen, S., Lampinen, J.: GDE3: The third evolution step of generalized differential evolution. Proceedings of the 2005 IEEE Congress on Evolutionary Computation (1), 443–450 (2005)
6. Deb, K., Mohan, M., Mishra, S.: Evaluating the epsilon-domination based multiobjective evolutionary algorithm for a quick computation of Paretooptimal solutions. Evolutionary Computation 13(4), 501–525 (2005)
7. Laumanns, M., Thiele, L., Deb, K., Zitzler, E.: Combining convergence and diversity in evolutionary multi-objective optimization. Evolutionary Computation 10(3), 263–282 (2002)
8. Fang, K.T., Ma, C.X.: Orthogonal and uniform design. Science Press, Beijing (2001) (in Chinese)
9. Leung, Y.W., Wang, Y.: An orthogonal genetic algorithm with quantization for global numerical optimization. IEEE Transactions on Evolutionary Computation 5(1), 41–53 (2001)
10. Zeng, S.Y., Kang, L.S., Ding, L.X.: An orthogonal multiobjective evolutionary algorithm for multi-objective optimization problems with constraints. Evolutionary Computation 12, 77–98 (2004)
11. Cai, Z.H., Gong, W.Y., Huang, Y.Q.: A novel differential evolution algorithm based on epsilon-domination and orthogonal design method for multiobjective optimization. In: Obayashi, S., Deb, K., Poloni, C., Hiroyasu, T., Murata, T. (eds.) EMO 2007. LNCS, vol. 4403, pp. 286–301. Springer, Heidelberg (2007)
12. Leung, Y.-W., Wang, Y.: Multiobjective programming using uniform design and genetic algorithm. IEEE Transactions on Systems, Man, and Cybernetics, Part C 30(3), 293 (2000)
13. Durillo, J.J., Nebro, A.J., Luna, F., Dorronsoro, B., Alba, E.: jMetal: A Java Framework for Developing Multi-Objective Optimization Metaheuristics, Departamento de Lenguajes y Ciencias de la Computación, University of Málaga, E.T.S.I. Informática, Campus de Teatinos, ITI-2006-10, December (2006), http://jmetal.sourceforge.net

Noise Effects on Chaos in Chaotic Neuron Model

Naofumi Katada and Haruhiko Nishimura

Graduate School of Applied Informatics, University of Hyogo,
Harborland Center BLDG, 1–3–3 Higashikawasaki-cho,
Chuo-ku, Kobe, Hyogo, 650–0044 Japan
nkatada@hyogo-c.ed.jp, haru@ai.u-hyogo.ac.jp

Abstract. In nonlinear dynamical systems, noise can lead to qualitative changes in their behaviors, not quantitative changes. We study the effects of noise on the nonlinear chaotic neuron in the state having chaos. We show that noise induces the systems producing more order and the coherence of their noise-induced order becomes maximal for a certain noise amplitude. This effect is indicated to be a transition from chaotic behavior to ordered(non-chaotic) behavior by evaluating the Lyapunov exponent with noise.

Keywords: Chaos, neuron, noise, induced order.

1   Introduction

When considering systems in general, attention is usually paid to the existence of noise and its effects. In principle, it is believed that noise disrupts the system response, and considerable efforts have been made to suppress the emergence of noise. Indeed, depending on the strength of the noise, it introduces proportional quantitative changes in the overall behavior of linear systems, and it brings anomalies in the system response. However, since noise induces qualitative rather than quantitative changes in non-linear systems, there arises the possibility of observing phenomena that never occur in linear systems.

In fact, it can be demonstrated that the existence of noise plays positive roles in non-linear systems, including detection of weak signals below the threshold level, system synchronization phenomena, and stabilization of coupled oscillator systems, etc. [1,2,3]. In the particular case where the system is chaotic, interesting phenomena have been reported such as the transition from periodic to chaotic state due to the presence of noise, or the disappearance of chaos followed by the so-called Noise-Induced Order (NIO) [4,5,6]. NIO was discovered in a study of a model of Belousov-Zhabotinsky (BZ) reaction (an oxidoreduction process in chemistry) [4], and has later been employed in studies of electric circuits, laser beam experiments and models of chemical reactions. Furthermore, studies oriented toward biological phenomena have been conducted, such as the influence of noise on the Hodgkin-Huxley equation for nerve membrane excitation and the FizHugh-Nagumo model of nerve pulses [7] .

The present research examines in detail the responsiveness of chaotic neurons, which are non-linear dynamical systems in a state of chaos, and explores

F. Peper et al. (Eds.): IWNC 2009, PICT 2, pp. 209–217, 2010.

the presence of Noise-Induced Order (NIO) and its properties by virtue of computer experiments. By focusing on the Lyapunov exponent as the measure for evaluation of chaotic behavior, the relation between NIO and chaos is explored.

2   Chaotic Neuron Model

Chaos is observed in many cases where data obtained from real brain nerve systems is subjected to temporal analysis [8,9,10]. Furthermore, several steps have been confirmed to exist between the electric response of a single neuron cell isolated from a neural network and the overall behavior of the brain as a collection of neural circuit networks in the cases of ion channels, the giant nerve axons of Peronia verruculata and Loligo bleekeri, the behavior of self-stimulated hippocampal pyramidal neurons in rats, olfactory bulb EEGs of rabbits, EEGs of humans in different conditions, etc. As a collective term, the chaos observed in such neural systems is called neural chaos. In addition, the existence of chaos at the level of single neuron also finds qualitative support in the theoretical rationalization of the Hodgkin-Huxley equation which is constructed on the basis of physiological knowledge about nerve axons [11]. This means that actual neurons are oversimplified in neuron models following the conventional McCulloch-Pitts form, and for that reason factors contributing to the emergence of chaos are disregarded.

In this situation, the chaotic neuron model (hereafter, chaotic neurons) has been proposed as a relatively simple model capable of reflecting the responsiveness of actual neurons while taking chaos into consideration. Chaotic neurons also regard the refractory effect (a property where the threshold raises after the neuron is fired and it becomes more difficult to fire it again), as well as analogue input-output characteristics and time decay effects.

For this chaotic neuron [12,13,14], the relation between input y and output X is obtained by

$$X(t+1) = f\big(y(t+1)\big) \qquad (1)$$

$$y(t+1) = -\alpha \sum_{d=0}^{t} k_r^d X(t-d) - \theta \qquad (2)$$

Here, k_r : Decay constant for the refractoriness of the neuron itself $(0 \leq k_r < 1)$, α : Scaling parameter of the refractoriness $(\alpha \geq 0)$, θ : Threshold of the neuron. Furthermore, in the present paper, $f(y) = tanh(y/2\varepsilon)$, i.e. the neuron output X, is formulated by $-1 \leq X \leq 1$. Considering the exponential behavior (k^d) of the time-history effect, y in equation (2) follows the dynamics below (see Appendix).

$$y(t+1) = k_r y(t) - \alpha X(t) + a \qquad (3)$$

Here, for the sake of simplicity, $-\theta(1 - k_r) \equiv a$.

3   Noise-Induced Order in Chaotic Neuron Model

Taking the parameters $k_r = 0.7$ and $\alpha = 1.0$ and the steepness parameter of the sigmoid function $\varepsilon = 0.001$, the bifurcation diagram of $y(t)$ corresponding to a is shown in Fig. 1. Non-periodical regions are abundant as different bands in a. We will continue the analysis by taking the band including $a = 0.75$ (the place indicated by an arrow in the figure).

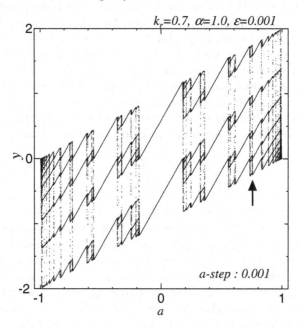

Fig. 1. Bifurcation diagram of the chaotic neuron as a function of the parameter a

For $a = 0.75$, the Lyapunov exponent λ [15] corresponding to the time series of the internal state y of the neuron is 0.26, and chaotic regions (bands) exist in its vicinity. In order to understand what influence noise exerts on the non-periodic time series behavior of the neuron state, a noise term F is introduced into the evolution equation for the neuron in the following way.

$$y(t + 1) = k_r y(t) - \alpha X(t) + a \qquad (4)$$
$$X(t + 1) = f(y(t + 1) + F(t + 1)) \qquad (5)$$

First, we consider the case of Gaussian white noise. In this case, F satisfies the following relation.

$$\langle F(t) \rangle = 0, \qquad \langle F(t)F(t') \rangle = D^2 \delta_{t,t'} \qquad (6)$$

Here, $\langle \quad \rangle$ indicates time average, and $\delta_{t,t'}$ is the Kronecker delta (1 for $t = t'$, 0 for $t \neq t'$). D is the noise intensity parameter, and represents the standard deviation for the normal distribution of the noise event probability

$$p(F) = \frac{1}{\sqrt{2\pi D^2}} \, exp\left(-\frac{F^2}{2D^2}\right) \tag{7}$$

The left side of Fig. 2 shows the behavior of the neuron for each of the cases $D = 0.0001, 0.001, 0.01, 0.1, 1.0$. In the case of $D = 0.0001$, the neuron displays irregular low peaks at different times, quite similar to the case where the neuron is not influenced by noise $(D = 0)$. For $D = 0.001$ through $D = 0.01$, the noise intensity grows, the emergence of low peaks diminishes, and the peaks of the fluctuations approach uniform behavior. Undulations appear in the case of $D = 0.1$ where the noise intensity is further increased, and non-uniformities arise over long periods of time. Up to $D = 1.0$, the disorder increases and different values emerge, which suggests that the behavior is dominated by the noise term F. In this way it is understood that the behavior of the neuron displays rather severe changes depending on the intensity of the noise.

In order to evaluate quantitatively the differences between the above instances of time-series behavior, we introduce a temporal autocorrelation

$$C(\tau) = \frac{\langle \tilde{X}(t)\tilde{X}(t+\tau)\rangle}{\langle \tilde{X}^2\rangle} \tag{8}$$

Here, $\tilde{X}(t) = X(t) - \langle X(t)\rangle$ and $\langle \ \ \rangle$ indicates time average.

The right side of Fig. 2 shows the temporal autocorrelation graphs corresponding to the time series on the left side. Considering the duration of the magnitude of C on the time interval τ, the correlation becomes stronger from $D = 0.0001$ through 0.01 as the intensity of the noise grows, in other words, the orderliness in the behavior of the neuron increases. Furthermore, the correlation is lost for $D = 0.1$ through $D = 1.0$. This fact suggests that the presence of an appropriate amount of noise increases the orderliness in the behavior of the neuron.

Furthermore, in order to evaluate quantitatively the graphs corresponding to the temporal autocorrelations in Fig. 2, we define a proper correlation index

$$C_p = \sum_{\tau=0}^{M} C^2(\tau) \tag{9}$$

This is large as long as the values of C with respect to τ are consistently large. In fact, taking $M = 299$ and applying it to the five cases in Fig. 2 where $D = 0.0001$, $0.001, 0.01, 0.1$ and 1.0, C_p takes the values 3.0, 3.1, 4.2, 3.1 and 2.1 respectively, with the largest value 4.2 occurring in case where $D = 0.01$ and the correlation (i.e. the orderliness) is high. Consequently, the graph in Fig. 3 shows the results from C_p from performing a finer breakdown of the values of noise intensity D. The value of the proper correlation index C_p increases following the increase of the noise intensity D and again starts decreasing once D has exceeded a certain value. In other words, this confirms that the presence of an appropriate amount of noise (in this case, white noise with intensity $D = 0.02$) can induce a large increase in the orderliness in the behavior of the neuron. Figure 4 shows the localized feature of the orbit in the noisy case($D = 0.02$) different from that in

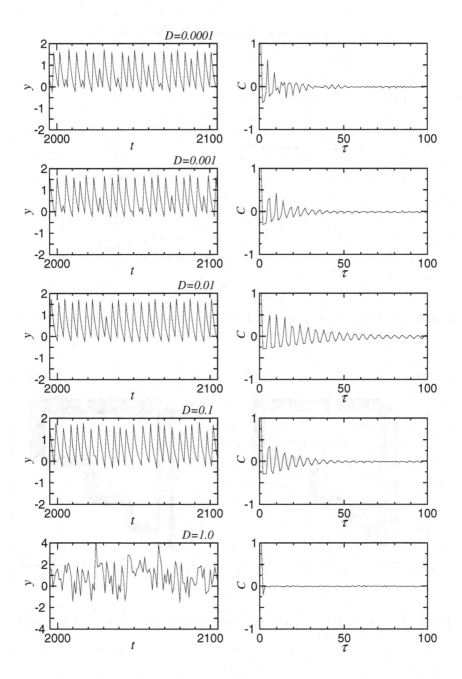

Fig. 2. Time evolution of the chaotic neuron and its autocorrelation function for $a =$ 0.75 with different noise intensities. From top to bottom D increases from 0.0001 to 1.0.

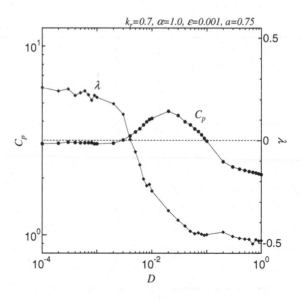

Fig. 3. Correlation index C_p and the Lyapunov exponent λ as functions of the noise intensity parameter D for the chaotic neuron

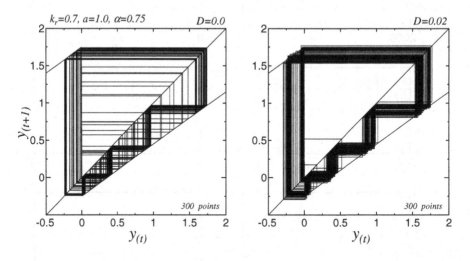

Fig. 4. Orbits in the return map of the chaotic neuron. Noiseless case($D = 0$) and the case with noise($D = 0.02$).

the noiseless case($D = 0$). This change in the orbits can be recognized in the invariant density as presented in Fig. 5.

In addition to the dependency between C_p and D, we have plotted together the respective values of the Lyapunov exponent for each value of D into Fig. 3.

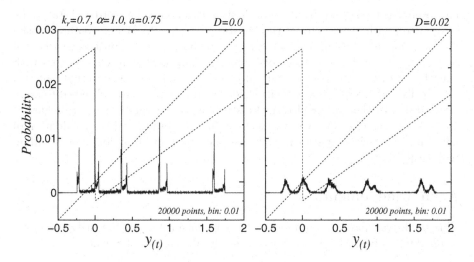

Fig. 5. Invariant density of the return map of the chaotic neuron. Noiseless case($D = 0$) and the case with noise($D = 0.02$).

Comparing the two sides, the sign of the Lyapunov exponent λ changes from positive to negative when the noise intensity is in the vicinity of 4×10^{-3}, where, in agreement with this, the correlation index C_p starts increasing. Furthermore, λ is negative at the peak of C_p. This suggests that the order induced by the application of noise is an process of shifting from a chaotic state to an ordered state. The results for uniform pseudorandom number noise and Gaussian white noise are almost the same.

It should be noted that the Lyapunov exponent calculated with the presence of noise has a different meaning from the exponent originally defined for deterministic systems [16,17]. Strictly speaking, since the Lyapunov exponent is an index of the stability of reference trajectories, its value is diminished when the concept of a trajectory is destroyed in the presence of noise. That is, the application of noise fragments trajectories which follow the original deterministic equations, even though the degree might be low. However, since we can reproduce intrinsically unreproducible time series of the noise patterns by administering random number series in a computer experiment, we can set the reference trajectory to include the noise, thus enabling λ to be formally calculated.

4   Summary

In the present research, we have conducted a computer experiment whereby we explore in detail the influence of noise on the behavior of a single chaotic neuron.As a result, we have confirmed the emergence of Noise-Induced Order (NIO) in the chaotic neuron in the presence of an appropriate amount of noise. By considering the evaluation of the Lyapunov exponents, those cases can be interpreted as transitions from a chaotic state to a non-chaotic ordered state. The comparison with the results of the logistic map and the Belousov-Zhabotinsky reaction models is now under consideration in all its aspects. Future research will be directed toward the analysis of the results obtained from experiments with neural networks composed of mutually coupled neurons. The activity of real nerve cells and groups of nerve cells inside the brain is usually subjected to noise attributable to thermal or electrical disturbances. Neural systems are regarded as models of biological nerve systems including the brain, and the elucidation of noise-induced characteristics in this case may be important for understanding biological functions.

Acknowledgements

This work was partially supported by a Grant-in-Aid for Seientific Research (c)(20500211) from the Japan Society for the Promotion of Science.

References

1. Gammaitoni, L., Hänggi, P., Jung, P., Marchesoni, F.: Stochastic Resonance. Rev. Mod. Phys. 70, 223–287 (1998)
2. Pikovsky, A., Rosenblum, M., Kurths, J.: Synchronization: A Universal Concept in Nonlinear Science. Cambridge University Press, Cambridge (2001)
3. Anishchenko, V.S., Astakhov, V.V., Neiman, A.B., Vadivasova, T.E., Schimansky-Geier, L.: Nonlinear Dynamics of Chaotic and Stochastic System. Springer, Heidelberg (2002)
4. Matsumoto, K., Tsuda, I.: Noise-induced order. J. Stat. Phys. 31, 87–106 (1983)
5. Crutchfield, J.P., Farmer, J.D., Huberman, B.A.: Fluctuations and Simple Chaotic Dynamics. Phys. Rep. 92, 46–82 (1982)
6. Kaneko, K., Tsuda, I.: Complex Systems: Chaos and Beyond - A Constructive Approach with Applications in Life Sciences. Springer, Heidelberg (2000)
7. Pikovsky, A.S., Kurths, J.: Coherence Resonance in a Noise-Driven Excitable System. Physical Review Letters 78, 775–778 (1997)
8. Freeman, W.J.: Neurodynamics: an exploration in mesoscopic brain dynamics. Springer, Heidelberg (2000)
9. Arbib, M.A. (ed.): The Handbook of Brain Theory and Neural Networks, 2nd edn. Bradford Books (2002)
10. Korn, H., Faure, P.: Is there chaos in the brain? II. experimental evidence and related models. C.R. Biologies 326, 787–840 (2003)
11. Takabe, T., Aihara, K., Matsumoto, G.: Response characteristics of the Hodgkin-Huxley equations to pulse-train stimulation. IEICE Trans. Commun. (A), J71-A(3), 744–750 (1988) (in Japanese)
12. Aihara, K.: Chaotic Neural Networks. In: Kawakami, H. (ed.) Bifurcation Phenomena in Nonlinear Systems and Theory of Dynamical Systems, pp. 143–161. World Scientific, Singapore (1990)
13. Aihara, K., Takabe, T., Toyoda, M.: Chaotic Neural Networks. Physics Letters A 144, 333–340 (1990)
14. Nishimura, H., Aihara, K.: Coaction of neural networks and chaos. Journal of The Society of Instrument and Control Engineers 39(3), 162–168 (2000) (in Japanese)

15. Parker, T.S., Chua, L.O.: Practical Numerical Algorithms for Chaotic Systems. Springer, Heidelberg (1989)
16. Paladin, G., Serva, M., Vuipiani, A.: Complexity in Dynamical Systems with Noise. Physical Review Letters 74, 66–69 (1995)
17. Witt, A., Neiman, A., Kurths, J.: Characterizing the Dynamics of Stochastic Bistable Systems by Measures of Complexity. Phys. Rev. E55, 5050–5059 (1997)

Appendix

This appendix shows the derivation of equation (3) from equation (2).

$$
\begin{aligned}
y(t+1) &= -\alpha \sum_{d=0}^{t} k_r^d X(t-d) - \theta \\
&= -\alpha \sum_{d=1}^{t} k_r^d X(t-d) - \alpha X(t) - \theta \\
&= k_r \left\{ -\alpha \sum_{d=1}^{t} k_r^{d-1} X(t-d) \right\} - \alpha X(t) - \theta \\
&= k_r \left\{ -\alpha \sum_{d-1=0}^{t-1} k_r^{d-1} X(t-1-(d-1)) \right\} - \alpha X(t) - \theta \\
&= k_r (y(t) + \theta) - \alpha X(t) - \theta \\
&= k_r y(t) - \alpha X(t) - \theta(1 - k_r)
\end{aligned}
$$

Application of Improved Grammatical Evolution to Santa Fe Trail Problems

Takuya Kuroda, Hiroto Iwasawa, Tewodros Awgichew, and Eisuke Kita

Graduate School of Information Science, Nagoya University, Nagoya 464-8601, Japan

Abstract. Grammatical Evolution (GE) is one of the evolutionary algorithms, which can deal with the rules with tree structure by one-dimensional chromosome. Syntax rules are defined in Backus Naur Form (BNF) to translate binary number (genotype) to function or program (phenotype). In this study, three algorithms are introduced for improving the convergence speed. First, an original GE are compared with Genetic Programming (GP) in the function identification problem. Next, the improved GE algorithms are applied to Santa Fe Trail problem. The results show that the improved schemes are effective for improving the convergence speed.

Keywords: Grammatical Evolution, Backus Naur Form, Santa Fe Trail.

1 Introduction

Evolutionary algorithms are techniques implementing mechanisms inspired by biological evolution such as reproduction, mutation, recombination, natural selection and survival of the fittest. They are classified into Genetic Algorithms (GA)[1,2], Evolutionary Programming (EP)[3], Genetic Programming (GP)[4,5] and so on.

In the GA, a population of chromosomes of candidate solutions to an optimization problem evolves toward better solutions by using the selection, crossover and mutation operations. Genetic Programming (GP) is an evolutionary algorithm-based methodology to find computer programs that perform a user-defined task. GP evolves computer programs represented in memory as tree structures. Simulated Annealing Programming (SAP) is also designed to find the programs, which comes from Simulated Annealing (SA).

Banzhaf has presented the GP in which programs are represented in one-dimensional chromosome[6]. A one-dimensional chromosome is translated to program according to the template defined in advance. In this case, the programs are often invalid in the syntax. On the contrary, Whigham has presented grammatically-based genetic programming[7]. Like as Banzhaf, the template is used for translating chromosome to program. Since the chromosome is defined in tree structure, like GP, many complicated genetic operators are necessary. Ryan et. al. have presented Grammatical Evolution (GE)[8,9,10], in which the difficulty of the Banzhaf algorithm is improved by the syntax definition of grammatically-based genetic programming.

The aim of this study is to introduce three schemes for improving the convergence performance of GE. First, the original GE and GP are compared in function identification problem. After that, three schemes are discussed in the Santa Fe trail problem,

F. Peper et al. (Eds.): IWNC 2009, PICT 2, pp. 218–225, 2010.

whose object is to find programs to control the artificial ants collecting foods in the region.

The remaining of this paper is as follows. The algorithm of the original GE is shown in section 2 and the results in the function identification problem are discussed in section 3. Three schemes for improving original algorithm are explained in section 4. The schemes are compared with the original GE in Santa Fe Trail Problem in section 5. The results are summarized again in section 6.

2 Original Grammatical Evolution

The algorithm of original Grammatical Evolution (GE) is as follows.

1. Define a syntax in BNF, which translates genotype (binary number) to phenotype (function or program).
2. Generate randomly initial individuals to construct an initial population.
3. Translate chromosome to function according to the BNF syntax.
4. Estimate fitness of chromosome.
5. Use genetic operators to generate new individuals.
6. Terminate the process if the criterion is satisfied.
7. Go to step 3.

The translation from genotype to phenotype is as follows.

1. Translate a binary number to a decimal number for every n-bits.
2. Define a leftmost decimal as β, a leftmost nonterminal symbol as α, and the number of candidate rules for α as n_α.
3. Calculate the remainder $\gamma = \beta \% n_\alpha$.
4. Select γ-th rule of the candidate rules for α.
5. If nonterminal symbols exist, go to step 2.

In the genetic programming (GP), the programs often grow rapidly in size over time (bloat). For overcoming the difficulty, the maximum size of the programs is restricted in advance. The similar idea is applied to the GE. The maximum size of the programs is restricted to $L_{\max}$.

3 Function Identification Problem

3.1 Problem Settings

The function identification problem is defined as

Find a function $\bar{f}$

when discrete data $\{(x_1, y_1), (x_2, y_2), \cdots, (x_n, y_n)\}$ is given.

where the parameter n denotes the total number of the discrete data. When an exact function f is given, the discrete data are referred to as $y_i = f(x_i)$.

We will consider that the exact function f is given as

$$f(x) = x^4 + x^3 + x^2 + x. \tag{1}$$

The discrete data are generated by estimating equation (1) at $x = -1, -0.9, -0.8, \cdots, 0.9, 1$.

Table 1. BNF Syntax for Function Identification Problem

(A)	`<expr> ::=`	`<expr><expr><op>`	(A0)
	`|`	`<x>`	(A1)
(B)	`<op> ::=`	`+`	(B0)
	`|`	`−`	(B1)
	`|`	`*`	(B2)
	`|`	`/`	(B3)
(C)	`<x> ::=`	`x`	(C0)

3.2 Syntax and Parameters

The fitness is defined as the mean least square error of f and $\bar{f}$ as

$$fitness = \sqrt{\frac{1}{21}\sum_{i=1}^{21}[f(x_i) - \bar{f}(x_i)]^2}. \tag{2}$$

where f and $\bar{f}$ denote the exact function and the function predicted in GE, respectively.

A GE syntax in BNF is shown in Table 1. The start symbol is `<expr>`.

The parameters for GE and GP are shown in Table 2 and 3, respectively. Tournament selection, one-elitist strategy and one-point crossover are employed for both GE and GP. The mutation operator is applied for GE alone.

The fitness values are estimated as the value averaged over 50 runs.

3.3 Result

The convergence history of the best fitness value is shown in Fig.1. The abscissa and the ordinate denote the number of generation and fitness value, respectively. The

Table 2. GE Parameters

Generation	1000
Population size	100
Chromosome	100
Tournament size	5
Crossover rate	0.5
Mutation rate	0.1
Translation bit-size	4bit
Maximum size	$L_{max} = 100$

Table 3. GP Parameters

Generation	1000
Population size	100
Crossover rate	0.9

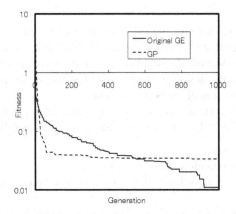

Fig. 1. Result of Function Identification Problem

convergence speed of GE is slower than that of GP. Finally, GE can find a better so-lution than GP.

4   Improved Schemes of Grammatical Evolution

4.1   Difficulties of Original GE

Rule Selection. We will consider a leftmost decimal as β, a leftmost nonterminal sym-bol as α, and the number of candidate rules for α as n_α. Since a rule is selected by the remainder $\gamma = \beta \% n_\alpha$, the rule selection is very sensitive to β. Even when the value of β alters by only one, the selected rule is changed. This may disturb the development of the better scheme in the population. The following scheme 1 is designed for overcoming this difficulty.

Selection Probability of Rules. The original GE selects rules according to the re-mainder and therefore, the selection probability for all candidate rules is identical. For example, in Table 1, the rule <expr> is translated to <expr><expr><op> (A0) or <x> (A1). The selection probability of the rule (A0) is identical to that of the rule (A1). However, biased selection probability may be better in some problems for improving the convergence speed.

The rules are classified into the recursive (non-terminal) and terminal rules. For ex-ample, in Table 1, the rule (A) is recursive rule and the others are terminal rules. The iterative use of recursive rule makes the phenotype (function or program) longer and more complicated. On the other hand, the terminal rule terminates the development of the phenotype. Since the roles of the recursive and the terminal rules are different, it is appropriate that the different selection probability is taken for each rule. The following scheme 2 and 3 are designed to control the selection probability of the recursive and the terminal rules, respectively.

4.2   Improvement of GE

Scheme 1.   In the original GE, the rules are selected by the remainder. The scheme 1 adopts the special roulette selection, instead of the remainder selection. The roulette selection is popular selection algorithm in GA. In the scheme 1, the roulette selection probability for all candidates rules is identical. The use of the roulette selection can encourage the development of the better schema.

We will consider a leftmost decimal as β, a leftmost nonterminal symbol as α, and the number of candidate rules for α as n_α. The algorithm is as follows.

1. Calculate the parameter $s_\alpha = \beta/n_\alpha$.
2. Generate a uniform random number p $(0 < p \le \beta)$.
3. If $(k-1)s_\alpha < p \le ks_\alpha$, select k-th rule from the candidate rules for α $(1 \le k \le n)$.

Scheme 2.   The selection probability of the recursive rule is controlled according to the length of the generated program. The maximum length of the programs is specified in advance. If the length of the programs is shorter than the maximum length $L_{\max}$, the selection probability is increased. If not so, the probability is decreased.

The selection probability of the recursive rule i is calculated as

$$P_i^r = 1 - \frac{L}{L_{\max}} \tag{3}$$

where L and $L_{\max}$ denote the length and the maximum length of the programs.

Scheme 3.   Occurrence numbers of the terminal rules in all individuals are estimated. The selection probability P_i^N of the terminal rule i is calculated as

$$P_i^N = \frac{N_i}{\sum_{j=1}^{N^N} N_j} \tag{4}$$

where N_i and N^N denote the occurrence number of the terminal rule i and the total occurrence number of all terminal rules, respectively.

5   Santa Fe Trail Problem

5.1   Problem Setting

The object of Santa Fe trail problem is to automatically find the program to control artificial ants which efficiently collect foods in the region[4,5]. In this study, the region is 32×32-grid region in which there are 89 foods (Fig.2). The ant behavior is defined as follows.

1. Ants start from upper-left cell.
2. Ants move one cell, turn to left or turn to right at every time-steps.
3. Ants move in vertical and horizontal directions alone.
4. Ants cannot get across the walls.
5. Ants have sensor to find food.
6. Ants have life energy which is expended one unit every time step.

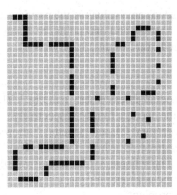

Fig. 2. Santa Fe Trail Problem

Table 4. BNF Syntax for Santa Fe Trail Problem

(A)	`<code>  ::= <code><expr>`	(A0)
	`       | <expr>`	(A1)
(B)	`<expr> ::= <if_statement>`	(B0)
	`       | <op>`	(B1)
(C)	`<if_statement> ::= if_food_ahead{<op>}else{<op>}`	(C0)
(D)	`<op> ::= right`	(D0)
	`     | left`	(D1)
	`     | forward`	(D2)

5.2   Syntax and Parameters

A syntax in BNF is shown in Table 4. The expression "`right`", "`left`" and "`forward`" denote facing to the right, facing to the left and moving forward, respectively. The expression "`if_food_ahead{<op1>}else{<op2>}`"denotes that the first argument `<op1>` is performed if a food exists in the front cell of the ant and the second argument `<op2>` is performed if a food does not exist. The start symbol is `<code>`.

The fitness is defined as follows.

$$Fitness = 89 - F_o \tag{5}$$

where the parameter F_o is the total number of foods which the ant collects.

The simulation algorithm is as follows.

1. Specify an initial ant energy as 400 and the number of collected foods as 0; $F_o = 0$.
2. Move the ant and decrement the ant energy.
3. If a food exists, take the food and $F_o = F_o + 1$.
4. If the energy is equal to 0, estimate the fitness from equation (5).
5. Go to step 2.

The parameters are specified to same values as the previous example (Table 2). Tournament selection, one-elitist strategy and one-point crossover are employed. The fitness values are the value averaged over 50 runs.

5.3 Result

The history of the best individual fitness is shown in Fig.3. The abscissa and the ordinate denote the generation and the average fitness, respectively.

The comparison of the original GE and scheme 1 shows that the use of scheme 1 improves the convergence speed. However, the use of the scheme 1+2 becomes the convergence speed worse. The scheme 3 does not affect the performance of the algorithm. Therefore, we can conclude that only the scheme 1 is effective for improving the search performance of this problem.

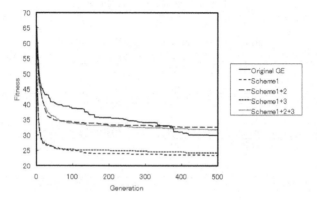

Fig. 3. Convergence History of Best Individual Fitness

6   Conclusion

This paper described some improvement of the grammatical evolution. First we explained the grammatical evolution simply and introduced three improvement schemes; scheme 1, scheme 2 and scheme 3.

In the first example, the original GE and the original GP were compared in the function identification problem. The results have shown that the convergence speed of GE was slower than that of GP and that GE could find better solution than the GP.

In the second example, three schemes were compared in Santa Fe trail problem. The results showed that the scheme 1 was very effective. The scheme 1 uses roulette selection, instead of remainder selection, in order to encourage the development of the better scheme in the chromosome. The above result indicates the effectiveness of roulette selection for the problem to be solved. On the other hand, the scheme 2 and 3 were not effective. The scheme 2 and 3 give the biased selection probability for the candidates of recursive and terminal rules, respectively. In the BNF syntax of Santa Fe trail problem, there are only two or three candidates for each rule. Therefore, the biased probability may be similar to the uniform probability. However, we have to discuss their effectiveness in more detailed.

References

1. Holland, J.H.: Adaptation in Natural and Artificial Systems, 1st edn. The University of Michigan Press, Ann Arbor (1975)
2. Goldberg, D.E.: Genetic Algorithms in Search, Optimization and Machine Learning, 1st edn. Addison Wesley, Reading (1989)
3. Fogel, D.B., Atmar, J.W.: Proc. 1st annual Conference on Evolutionary Programming. Evolutionary Programming Society (1992)
4. Koza, J.R. (ed.): Genwetic Programming II. The MIT Press, Cambridge (1994)
5. Koza, J.R., Bennett III, F.H., Andre, D., Keane, M.A. (eds.): Genewtic Programming III. Morgan Kaufmann Pub., San Francisco (1999)
6. Banzhaf, W.: Genotype-phenotype-mapping and neutral variation – A case study in genetic programming. In: Davidor, Y., Schwefel, H.P., Männer, R. (eds.) PPSN 1994. LNCS, vol. 866, pp. 322–332. Springer, Heidelberg (1994)
7. Whigham, P.A.: Grammatically-based genetic programming. In: Rosca, J.P. (ed.) Proceedings of the Workshop on Genetic Programming: From Theory to Real-World Applications, pp. 33–41 (1995)
8. Ryan, C., Collins, J.J., O'Neill, M.: Grammatical evolution: Evolving programs for an arbiturary language. In: Banzhaf, W., Poli, R., Schoenauer, M., Fogarty, T.C. (eds.) EuroGP 1998. LNCS, vol. 1391, pp. 83–95. Springer, Heidelberg (1998)
9. Ryan, C., O'Neill, M.: Grammatical Evolution: Evolutionary Automatic Programming in an Arbitrary Language. Springer, Heidelberg (2003)
10. Brabazon, A., O'Neill, M.: Biologically Inspired Algorithms for Financial Modelling. Springer, Heidelberg (2006)

Limit Theorem for a Time-Dependent Coined Quantum Walk on the Line

Takuya Machida[1] and Norio Konno[2]

[1] Department of Applied Mathematics, Faculty of Engineering,
Yokohama National University, Hodogaya, Yokohama, 240-8501, Japan
soul.street@srt.osu.seg.ynu.ac.jp,
[2] konno@ynu.ac.jp

Abstract. We study time-dependent discrete-time quantum walks on the one-dimensional lattice. We compute the limit distribution of a two-period quantum walk defined by two orthogonal matrices. For the symmetric case, the distribution is determined by one of two matrices. Moreover, limit theorems for two special cases are presented.

1 Introduction

The discrete-time quantum walk (QW) was first intensively studied by Ambainis *et al.* [1]. The QW is considered as a quantum generalization of the classical random walk. The random walker in position $x \in \mathbb{Z} = \{0, \pm 1, \pm 2, \dots\}$ at time $t (\in \{0, 1, 2, \dots\})$ moves to $x - 1$ at time $t + 1$ with probability p, or $x + 1$ with probability $q (= 1 - p)$. In contrast, the evolution of the quantum walker is defined by replacing p and q with 2×2 matrices P and Q, respectively. Note that $U = P + Q$ is a unitary matrix. A main difference between the classical walk and the QW is seen on the particle spreading. Let $\sigma(t)$ be the standard deviation of the walk at time t. That is, $\sigma(t) = \sqrt{\mathbb{E}(X_t^2) - \mathbb{E}(X_t)^2}$, where X_t is the position of the quantum walker at time t and $\mathbb{E}(Y)$ denotes the expected value of Y. Then the classical case is a diffusive behavior, $\sigma(t) \sim \sqrt{t}$, while the quantum case is ballistic, $\sigma(t) \sim t$ (see [1], for example).

In the context of quantum computation, the QW is applied to several quantum algorithms. By using the quantum algorithm, we solve a problem quadratically faster than the corresponding classical algorithm. As a well-known quantum search algorithm, Grover's algorithm was presented. The algorithm solves the following problem: in a search space of N vertices, one can find a marked vertex. The corresponding classical search requires $O(N)$ queries. However, the search needs only $O(\sqrt{N})$ queries. As well as the Grover algorithm, the QW can also search a marked vertex with a quadratic speed up, see Shenvi *et al.* [2]. It has been reported that quantum walks on regular graphs (e.g., lattice, hypercube, complete graph) give faster searching than classical walks. The Grover search algorithm can also be interpreted as a QW on complete graph. Decoherence is an important concept in quantum information processing. In fact, decoherence on QWs has been extensively investigated, see Kendon [3], for example. However,

F. Peper et al. (Eds.): IWNC 2009, PICT 2, pp. 226–235, 2010.

we should note that our results are not related to the decoherence in QWs. Physically, Oka et al. [4] pointed out that the Landau-Zener transition dynamics can be mapped to a QW and showed the localization of the wave functions.

In the present paper, we consider the QW whose dynamics is determined by a sequence of time-dependent matrices, $\{U_t : t = 0, 1, \ldots\}$. Ribeiro et al. [5] numerically showed that periodic sequence is ballistic, random sequence is diffusive, and Fibonacci sequence is sub-ballistic. Mackay et al. [6] and Ribeiro et al. [5] investigated some random sequences and reported that the probability distribution of the QW converges to a binomial distribution by averaging over many trials by numerical simulations. Konno [7] proved their results by using a path counting method. By comparing with a position-dependent QW introduced by Wójcik et al. [8], Bañuls et al. [9] discussed a dynamical localization of the corresponding time-dependent QW.

In this paper, we present the weak limit theorem for the two-period time-dependent QW whose unitary matrix U_t is an orthogonal matrix. Our approach is based on the Fourier transform method introduced by Grimmett et al. [10]. We think that it would be difficult to calculate the limit distribution for the general n-period ($n = 3, 4, \ldots$) walk. However, we find out a class of time-dependent QWs whose limit probability distributions result in that of the usual (i.e., one-period) QW. As for the position-dependent QW, a similar result can be found in Konno [11].

The present paper is organized as follows. In Sect. 2, we define the time-dependent QW. Section 3 treats the two-period time-dependent QW. By using the Fourier transform, we obtain the limit distribution. Finally, in Sect. 4, we consider two special cases of time-dependent QWs. We show that the limit distribution of the walk is the same as that of the usual one.

2 Time-Dependent QW

In this section we define the time-dependent QWs. Let $|x\rangle$ ($x \in \mathbb{Z}$) be infinite components vector which denotes the position of the walker. Here, x-th component of $|x\rangle$ is 1 and the other is 0. Let $|\psi_t(x)\rangle \in \mathbb{C}^2$ be the amplitude of the walker in position x at time t, where $\mathbb{C}$ is the set of complex numbers. The time-dependent QW at time t is expressed by

$$|\Psi_t\rangle = \sum_{x \in \mathbb{Z}} |x\rangle \otimes |\psi_t(x)\rangle. \tag{1}$$

To define the time evolution of the walker, we introduce a unitary matrix

$$U_t = \begin{bmatrix} a_t & b_t \\ c_t & d_t \end{bmatrix}, \tag{2}$$

where $a_t, b_t, c_t, d_t \in \mathbb{C}$ and $a_t b_t c_t d_t \neq 0$ ($t = 0, 1, \ldots$). Then U_t is divided into P_t and Q_t as follows:

$$P_t = \begin{bmatrix} a_t & b_t \\ 0 & 0 \end{bmatrix}, Q_t = \begin{bmatrix} 0 & 0 \\ c_t & d_t \end{bmatrix}. \tag{3}$$

The evolution is determined by

$$|\Psi_{t+1}\rangle = \sum_{x\in\mathbb{Z}} |x\rangle \otimes (P_t\,|\psi_t(x+1)\rangle + Q_t\,|\psi_t(x-1)\rangle).\tag{4}$$

Let $||\,|y\rangle\,||^2 = \langle y|y\rangle$. The probability that the quantum walker X_t is in position x at time t, $P(X_t = x)$, is defined by

$$P(X_t = x) = ||\,|\psi_t(x)\rangle\,||^2.\tag{5}$$

Moreover, the Fourier transform $|\hat{\Psi}_t(k)\rangle$ $(k \in [0, 2\pi))$ is given by

$$|\hat{\Psi}_t(k)\rangle = \sum_{x\in\mathbb{Z}} e^{-ikx}\,|\psi_t(x)\rangle,\tag{6}$$

with $i = \sqrt{-1}$. By the inverse Fourier transform, we have

$$|\psi_t(x)\rangle = \int_0^{2\pi} \frac{dk}{2\pi} e^{ikx}\,|\hat{\Psi}_t(k)\rangle.\tag{7}$$

The time evolution of $|\hat{\Psi}_t(k)\rangle$ is

$$|\hat{\Psi}_{t+1}(k)\rangle = \hat{U}_t(k)\,|\hat{\Psi}_t(k)\rangle,\tag{8}$$

where $\hat{U}_t(k) = R(k)U_t$ and $R(k) = \begin{bmatrix} e^{ik} & 0 \\ 0 & e^{-ik} \end{bmatrix}$. We should remark that $R(k)$ satisfies $R(k_1)R(k_2) = R(k_1 + k_2)$ and $R(k)^* = R(-k)$, where $*$ denotes the conjugate transposed operator. From (8), we see that

$$|\hat{\Psi}_t(k)\rangle = \hat{U}_{t-1}(k)\hat{U}_{t-2}(k)\cdots\hat{U}_0(k)\,|\hat{\Psi}_0(k)\rangle.\tag{9}$$

Note that, when $U_t = U$ for any t, the walk becomes a usual one-period walk, and $|\hat{\Psi}_t(k)\rangle = \hat{U}(k)^t\,|\hat{\Psi}_0(k)\rangle$. Then the probability distribution of the usual walk is

$$P(X_t = x) = \left|\left|\int_0^{2\pi} \frac{dk}{2\pi} e^{ikx}\hat{U}(k)^t\,|\hat{\Psi}_0(k)\rangle\right|\right|^2.\tag{10}$$

In Sect. 4, we will use this relation. In the present paper, we take the initial state as

$$|\psi_0(x)\rangle = \begin{cases} {}^T[\alpha,\,\beta]\ (x = 0) \\ {}^T[0,\,0]\ (x \neq 0) \end{cases},\tag{11}$$

where $|\alpha|^2 + |\beta|^2 = 1$ and T is the transposed operator. We should note that $|\hat{\Psi}_0(k)\rangle = |\psi_0(0)\rangle$.

3   Two-Period QW

In this section we consider the two-period QW and calculate the limit distribution. We assume that $\{U_t : t = 0, 1, \ldots\}$ is a sequence of orthogonal matrices with $U_{2s} = H_0$ and $U_{2s+1} = H_1$ $(s = 0, 1, \ldots)$, where

$$H_0 = \begin{bmatrix} a_0 & b_0 \\ c_0 & d_0 \end{bmatrix}, \ H_1 = \begin{bmatrix} a_1 & b_1 \\ c_1 & d_1 \end{bmatrix}. \tag{12}$$

Let

$$f_K(x; a) = \frac{\sqrt{1 - |a|^2}}{\pi(1 - x^2)\sqrt{|a|^2 - x^2}} I_{(-|a|,|a|)}(x), \tag{13}$$

where $I_A(x) = 1$ if $x \in A$, $I_A(x) = 0$ if $x \notin A$. Then we obtain the following main result of this paper:

Theorem 1

$$\frac{X_t}{t} \Rightarrow Z, \tag{14}$$

where $\Rightarrow$ means the weak convergence (i.e., the convergence of the distribution) and Z has the density function $f(x)$ as follows:

(i) If $\det(H_1 H_0) > 0$, then

$$f(x) = f_K(x; a_\xi) \left[1 - \left\{|\alpha|^2 - |\beta|^2 + \frac{(\alpha\overline{\beta} + \overline{\alpha}\beta) b_0}{a_0}\right\} x\right], \tag{15}$$

where $|a_\xi| = \min\{|a_0|, |a_1|\}$.

(ii) If $\det(H_1 H_0) < 0$, then

$$f(x) = f_K(x; a_0 a_1) \left[1 - \left\{|\alpha|^2 - |\beta|^2 + \frac{(\alpha\overline{\beta} + \overline{\alpha}\beta) b_0}{a_0}\right\} x\right]. \tag{16}$$

If the two-period walk with $\det(H_1 H_0) > 0$ has a symmetric distribution, then the density of Z becomes $f_K(x; a_\xi)$. That is, Z is determined by either H_0 or H_1. Figure 1 (a) shows that the limit density of the two-period QW for $a_0 = \cos(\pi/4)$ and $a_1 = \cos(\pi/6)$ is the same as that for the usual (one-period) QW for a_0, since $|a_0| < |a_1|$. Similarly, Fig. 1 (b) shows that the limit density of the two-period QW for $a_0 = \cos(\pi/4)$ and $a_1 = \cos(\pi/3)$ is equivalent to that for the usual (one-period) QW for a_1, since $|a_0| > |a_1|$.

Proof. Our approach is due to Grimmett et al. [10]. The Fourier transform becomes

$$|\hat{\Psi}_{2t}(k)\rangle = \left(\hat{H}_1(k)\hat{H}_0(k)\right)^t |\hat{\Psi}_0(k)\rangle, \tag{17}$$

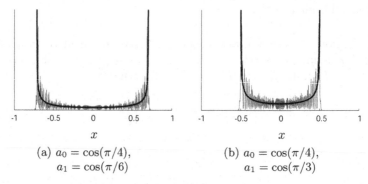

(a) $a_0 = \cos(\pi/4)$,
    $a_1 = \cos(\pi/6)$

(b) $a_0 = \cos(\pi/4)$,
    $a_1 = \cos(\pi/3)$

Fig. 1. The limit density function $f(x)$ (thick line) and the probability distribution at time $t = 500$ (thin line)

where $\hat{H}_\gamma(k) = R(k)H_\gamma$ $(\gamma = 0, 1)$. We assume

$$H_\gamma = \begin{bmatrix} \cos\theta_\gamma & \sin\theta_\gamma \\ \sin\theta_\gamma & -\cos\theta_\gamma \end{bmatrix}, \tag{18}$$

with $\theta_\gamma \neq \frac{\pi n}{2}$ $(n \in \mathbb{Z})$ and $\theta_0 \neq \theta_1$. For the other case, the argument is nearly identical to this case, so we will omit it. The two eigenvalues $\lambda_j(k)$ $(j = 0, 1)$ of $\hat{H}_1(k)\hat{H}_0(k)$ are given by

$$\lambda_j(k) = c_1 c_2 \cos 2k + (-1)^j i\sqrt{1 - (c_1 c_2 \cos 2k + s_1 s_2)^2}, \tag{19}$$

where $c_\gamma = \cos\theta_\gamma$, $s_\gamma = \sin\theta_\gamma$. The eigenvector $|v_j(k)\rangle$ corresponding to $\lambda_j(k)$ is

$$|v_j(k)\rangle = \begin{bmatrix} s_1 c_2 e^{-2ik} - c_1 s_2 \\ \left\{ c_1 c_2 \sin 2k + (-1)^j \sqrt{1 - (c_1 c_2 \cos 2k + s_1 s_2)^2} \right\} i \end{bmatrix}. \tag{20}$$

The Fourier transform $|\hat{\Psi}_0(k)\rangle$ is expressed by normalized eigenvectors $|v_j(k)\rangle$ as follows:

$$|\hat{\Psi}_0(k)\rangle = \sum_{j=0}^{1} \langle v_j(k)|\hat{\Psi}_0(k)\rangle |v_j(k)\rangle. \tag{21}$$

Therefore we have

$$|\hat{\Psi}_{2t}(k)\rangle = \left(\hat{H}_1(k)\hat{H}_0(k) \right)^t |\hat{\Psi}_0(k)\rangle$$

$$= \sum_{j=0}^{1} \lambda_j(k)^t \langle v_j(k)|\hat{\Psi}_0(k)\rangle |v_j(k)\rangle. \tag{22}$$

The r-th moment of X_{2t} is

$$E((X_{2t})^r) = \sum_{x \in \mathbb{Z}} x^r P(X_{2t} = x)$$

$$= \int_0^{2\pi} \frac{dk}{2\pi} \langle \hat{\Psi}_{2t}(k)| \left(D^r |\hat{\Psi}_{2t}(k) \rangle \right)$$

$$= \int_0^{2\pi} \sum_{j=0}^{1} (t)_r \lambda_j(k)^{-r} (D\lambda_j(k))^r \left| \langle v_j(k)|\hat{\Psi}_0(k)\rangle \right|^2$$

$$+ O(t^{r-1}), \tag{23}$$

where $D = -i(d/dk)$ and $(t)_r = t(t-1) \times \cdots \times (t-r+1)$. Let $h_j(k) = D\lambda_j(k)/2\lambda_j(k)$. Then we obtain

$$E((X_{2t}/2t)^r) \to \int_{\Omega_0} \frac{dk}{2\pi} \sum_{j=0}^{1} h_j^r(k) |\langle v_j(k)|\hat{\Psi}_0(k)\rangle|^2 \quad (t \to \infty). \tag{24}$$

Substituting $h_j(k) = x$, we have

$$\lim_{t \to \infty} E((X_{2t}/2t)^r) = \int_{-|c_\xi|}^{|c_\xi|} x^r f(x)\, dx, \tag{25}$$

where

$$f(x) = f_K(x; c_\xi) \left[1 - \left\{ |\alpha|^2 - |\beta|^2 + \frac{(\alpha\bar{\beta} + \bar{\alpha}\beta)s_1}{c_1} \right\} x \right], \tag{26}$$

and $|c_\xi| = |\cos\theta_\xi| = \min\{|\cos\theta_0|, |\cos\theta_1|\}$. Since $f(x)$ is the limit density function, the proof is complete. □

4   Special Cases in Time-Dependent QWs

In the previous section, we have obtained the limit theorem for the two-period QW determined by two orthogonal matrices. For other two-period case and general n-period ($n \geq 3$) case, we think that it would be hard to get the limit theorem in a similar fashion. Here we consider two special cases in the time-dependent QWs and give the weak limit theorems.

4.1   Case 1

Let us consider the QW whose evolution is determined by the following unitary matrix:

$$U_t = \begin{bmatrix} ae^{iw_t} & b \\ c & de^{-iw_t} \end{bmatrix}, \tag{27}$$

with $a, b, c, d \in \mathbb{C}$. Here $w_t \in \mathbb{R}$ satisfies $w_{t+1} + w_t = \kappa_1$, where $\kappa_1 \in \mathbb{R}$ and $\mathbb{R}$ is the set of real numbers. Note that κ_1 does not depend on time. In this case, w_t

can be written as $w_t = (-1)^t(w_0 - \frac{\kappa_1}{2}) + \frac{\kappa_1}{2}$. Therefore the period of the QW becomes two. We should remark that $\begin{bmatrix} a & b \\ c & d \end{bmatrix}$ $(\equiv U)$ is a unitary matrix. Then we have

Theorem 2

$$\frac{X_t}{t} \Rightarrow Z_1, \tag{28}$$

where Z_1 has the density function $f_1(x)$ as follows:

$$f_1(x) = f_K(x; a) \left\{ 1 - \left(|\alpha|^2 - |\beta|^2 + \frac{a\alpha\overline{b}\overline{\beta}e^{iw_0} + \overline{a}\overline{\alpha}b\beta e^{-iw_0}}{|a|^2} \right) x \right\}. \tag{29}$$

Proof. The essential point of this proof is that this case results in the usual walk. First we see that U_t can be rewritten as

$$U_t = \begin{bmatrix} e^{iw_t/2} & 0 \\ 0 & e^{-iw_t/2} \end{bmatrix} \begin{bmatrix} a & b \\ c & d \end{bmatrix} \begin{bmatrix} e^{iw_t/2} & 0 \\ 0 & e^{-iw_t/2} \end{bmatrix}$$

$$= R\left(\frac{w_t}{2}\right) U R\left(\frac{w_t}{2}\right). \tag{30}$$

From this, the Fourier transform $|\hat{\Psi}_t(k)\rangle$ can be computed in the following.

$$|\hat{\Psi}_t(k)\rangle = \left\{ R(k)R\left(\frac{w_{t-1}}{2}\right) U R\left(\frac{w_{t-1}}{2}\right) \right\} \left\{ R(k)R\left(\frac{w_{t-2}}{2}\right) U R\left(\frac{w_{t-2}}{2}\right) \right\}$$

$$\cdots \left\{ R(k)R\left(\frac{w_0}{2}\right) U R\left(\frac{w_0}{2}\right) \right\} |\hat{\Psi}_0(k)\rangle$$

$$= R\left(-\frac{w_t}{2}\right) \left\{ R\left(\frac{w_t}{2}\right) R(k)R\left(\frac{w_{t-1}}{2}\right) U \right\}$$

$$\times \left\{ R\left(\frac{w_{t-1}}{2}\right) R(k)R\left(\frac{w_{t-2}}{2}\right) U \right\}$$

$$\times \cdots \times \left\{ R\left(\frac{w_1}{2}\right) R(k)R\left(\frac{w_0}{2}\right) U \right\} R\left(\frac{w_0}{2}\right) |\hat{\Psi}_0(k)\rangle$$

$$= R\left(-\frac{w_t}{2}\right) \{R(k+\kappa_1)U\}^t R\left(\frac{w_0}{2}\right) |\hat{\Psi}_0(k)\rangle. \tag{31}$$

Therefore we have

$$|\psi_t(x)\rangle = \int_0^{2\pi} \frac{dk}{2\pi} e^{ikx} |\hat{\Psi}_t(k)\rangle = \int_{\kappa_1}^{2\pi+\kappa_1} \frac{dk}{2\pi} e^{i(k-\kappa_1)x} |\hat{\Psi}_t(k-\kappa_1)\rangle$$

$$= e^{-i\kappa_1 x} R\left(-\frac{w_t}{2}\right) \int_{\kappa_1}^{2\pi+\kappa_1} \frac{dk}{2\pi} e^{ikx} (R(k)U)^t |\hat{\Psi}_0^R(k)\rangle, \tag{32}$$

where $|\hat{\Psi}_0^R(k)\rangle = R\left(\frac{w_0}{2}\right)|\hat{\Psi}_0(k - \kappa_1)\rangle$. Then the probability distribution is

$$P(X_t = x)$$

$$= \left\{ e^{i\kappa_1 x} \left(\int_{\kappa_1}^{2\pi+\kappa_1} \frac{dk}{2\pi} e^{ikx} \left(R(k)U\right)^t |\hat{\Psi}_0^R(k)\rangle \right)^* R\left(\frac{w_t}{2}\right) \right\}$$

$$\times \left\{ e^{-i\kappa_1 x} R\left(-\frac{w_t}{2}\right) \left(\int_{\kappa_1}^{2\pi+\kappa_1} \frac{dk}{2\pi} e^{ikx} \left(R(k)U\right)^t |\hat{\Psi}_0^R(k)\rangle \right) \right\}$$

$$= \left\| \int_{\kappa_1}^{2\pi+\kappa_1} \frac{dk}{2\pi} e^{ikx} \hat{U}(k)^t |\hat{\Psi}_0^R(k)\rangle \right\|^2, \tag{33}$$

where $\hat{U}(k) = R(k)U$. This implies that Case 1 can be considered as the usual QW with the initial state $|\hat{\Psi}_0^R(k)\rangle = R\left(\frac{w_0}{2}\right)|\hat{\Psi}_0(k - \kappa_1)\rangle$ and the unitary matrix U. Then the initial state becomes

$$|\hat{\Psi}_0^R(k)\rangle = {}^T[e^{iw_0/2}\alpha, \, e^{-iw_0/2}\beta], \tag{34}$$

that is,

$$|\psi_0(x)\rangle = \begin{cases} {}^T[e^{iw_0/2}\alpha, \, e^{-iw_0/2}\beta] & (x = 0) \\ {}^T[0, \, 0] & (x \neq 0) \end{cases}. \tag{35}$$

Finally, by using the result in Konno [12,13], we can obtain the desired limit distribution of this case.   □

4.2   Case 2

Next we consider the QW whose dynamics is defined by the following unitary matrix:

$$U_t = \begin{bmatrix} a & be^{iw_t} \\ ce^{-iw_t} & d \end{bmatrix}. \tag{36}$$

Here $w_t \in \mathbb{R}$ satisfies $w_{t+1} = w_t + \kappa_2$, where $\kappa_2 \in \mathbb{R}$ does not depend on t. In this case, w_t can be expressed as $w_t = \kappa_2 t + w_0$. Noting $U_t = R\left(\frac{w_t}{2}\right)UR\left(-\frac{w_t}{2}\right)$, we get a similar weak limit theorem as Case 1:

Theorem 3

$$\frac{X_t}{t} \Rightarrow Z_2, \tag{37}$$

where Z_2 has the density function $f_2(x)$ as follows:

$$f_2(x) = f_K(x; a) \left\{ 1 - \left(|\alpha|^2 - |\beta|^2 + \frac{a\alpha\overline{b}\overline{\beta}e^{-iw_0} + \overline{a}\overline{\alpha}b\beta e^{iw_0}}{|a|^2} \right) x \right\}. \tag{38}$$

If $w_t = 2\pi t/n \, (n = 1, 2, \ldots)$, $\{U_t\}$ becomes an n-period sequence. In particular, when $n = 2$ and $a, b, c, d \in \mathbb{R}$, $\{U_t\}$ is a sequence of two-period orthogonal matrices. Then Theorem 3 is equivalent to Theorem 1 (i).

5 Conclusion and Discussion

In the final section, we draw the conclusion and discuss our two-period walks. The main result of this paper (Theorem 1) implies that if $\det(H_1 H_0)$ > 0 and $\min\{|a_0|, |a_1|\} = |a_0|$, then the limit distribution of the two-period walk is determined by H_0. On the other hand, if $\det(H_1 H_0) > 0$ and $\min\{|a_0|, |a_1|\} = |a_1|$, or $\det(H_1 H_0) < 0$, then the limit distribution is determined by both H_0 and H_1.

Here we discuss a physical meaning of our model. We should remark that the time-dependent two-period QW is equivalent to a position-dependent two-period QW if and only if the probability amplitude of the odd position in the initial state is zero. In quantum mechanics, the Kronig-Penney model, whose potential on a lattice is periodic, has been extensively investigated, see Kittel [14]. A derivation from the discrete-time QW to the continuous-time QW, which is related to the Schrödinger equation, can be obtained by Strauch [15]. Therefore, one of interesting future problems is to clarify a relation between our discrete-time two-period QW and the Kronig-Penney model.

Acknowledgment

This work was partially supported by the Grant-in-Aid for Scientific Research (C) of Japan Society for the Promotion of Science (Grant No. 21540118).

References

1. Ambainis, A., Bach, E., Nayak, A., Vishwanath, A., Watrous, J.: One-dimensional quantum walks. In: Proceedings of the 33rd Annual ACM Symposium on Theory of Computing, pp. 37–49 (2001)
2. Shenvi, N., Kempe, J., Whaley, K.B.: Quantum random-walk search algorithm. Phys. Rev. A 67, 052307 (2003)
3. Kendon, V.: Decoherence in quantum walks - a review. Mathematical Structures in Computer Science 17, 1169–1220 (2007)
4. Oka, T., Konno, N., Arita, R., Aoki, H.: Breakdown of an electric-field driven system: A mapping to a quantum walk. Phys. Rev. Lett. 94, 100602 (2005)
5. Ribeiro, P., Milman, P., Mosseri, R.: Aperiodic quantum random walks. Phys. Rev. Lett. 93, 190503 (2004)
6. Mackay, T.D., Bartlett, S.D., Stephenson, L.T., Sanders, B.C.: Quantum walks in higher dimensions. J. Phys. A 35, 2745–2753 (2002)
7. Konno, N.: A path integral approach for disordered quantum walks in one dimension. Fluctuation and Noise Letters 5, L529–L537 (2005)
8. Wójcik, A., Łuczak, T., Kurzyński, P., Grudka, A., Bednarska, M.: Quasiperiodic dynamics of a quantum walk on the line. Phys. Rev. Lett. 93, 180601 (2004)
9. Bañuls, M.C., Navarrete, C., Pérez, A., Roldán, E.: Quantum walk with a time-dependent coin. Phys. Rev. A 73, 062304 (2006)
10. Grimmett, G., Janson, S., Scudo, P.F.: Weak limits for quantum random walks. Phys. Rev. E 69, 026119 (2004)

11. Konno, N.: One-dimensional discrete-time quantum walks on random environments. Quantum Inf. Proc. 8, 387–399 (2009)
12. Konno, N.: Quantum random walks in one dimension. Quantum Inf. Proc. 1, 345–354 (2002)
13. Konno, N.: A new type of limit theorems for the one-dimensional quantum random walk. J. Math. Soc. Jpn. 57, 1179–1195 (2005)
14. Kittel, C.: Introduction to Solid State Physics, 8th edn. Wiley, Chichester (2005)
15. Strauch, F.W.: Connecting the discrete and continuous-time quantum walks. Phys. Rev. A 74, 030301 (2006)

Top-Predator Survivor Region Is Affected by Bottom-Prey Mortality Rate on the Monte-Carlo Simulation in Lattice Model

Minori Nagata[1] and Hiroyasu Nagata[2]

[1] Division of Mechanical Engineering and Materials Science,
Yokohama National University, Yokohama 240-8501, Japan
b0941089@ynu.ac.jp
[2] Department of Systems Engineering, Shizuoka University,
Hamamatsu 432-8561, Japan
thnagat@ipc.shizuoka.ac.jp

Abstract. This article is a continuance of [9] that study brings into focus on the n species systems food chain. Computer simulation is important today, and its results may be reflected to policy. Now, we change the mortality rate of the bottom-prey, in order to inspect the survivor region of the top-predator that is crucial for the conservation of the ecosystem. We carry out Monte-Carlo simulations on finite-size lattices composed of the species . The bottom-prey mortality rate is changed from 0 to the extinction value. Thereafter, we find the steady state densities against the species n and plot the predator survivor region. To realize the conservation of the top-predator population, substantial amount of hardship is anticipated, because the bottom-prey density gradually becomes a little.

Keywords: Bottom-prey, top-predator, survivor region, finite size lattice.

1 Introduction

In the field of ecology, the population dynamics of the species for time dependence are studied by the spatial models and the non-spatial. The former approach mainly uses either the lattice models or the partial differential equations. Lattice models are instead of cellular automaton. The lattice Lotka-Volterra-model was first introduced by [1], is usually applied in ecology. Simulations are carried out by the local interactions and the global. The mean-field theory is treated just equivalently to Lotka-Volterra equation. Our study is important for the biological conservation and the control [1-3]. The aim of the present paper is to report the survivor region of the top-predator is affected by the bottom-prey mortality change dynamically. Recently, coworkers in our laboratory studied the population dynamics. They changed the value of top-predator mortality rate in the $n \geq 4$ species systems [9].

In this paper, we deal with more general systems composed of $n \geq 6$ species to observe universal properties for the bottom-prey mortality rate. We got the

F. Peper et al. (Eds.): IWNC 2009, PICT 2, pp. 236–243, 2010.

following knowledge by this simulation; The top-predator cannot live without density of the prey. Therefore, when we protect the top-predator, we must predict the population dynamics of the prey.

2 Model and Method

2.1 Model

We deal with the square lattice systems composed of the n species. The size of the lattice is finite L, and the total number of the lattice sites is given by $L \times L$. Each lattice site is labeled by either the vacant site X_0 or the occupied site X_j by species j $(j = 1, ..., n)$. Interactions are defined by

$$X_j + X_{j-1} \longrightarrow 2X_j \quad \text{(rate } r_j), \tag{1a}$$
$$X_j \longrightarrow X_O \quad \text{(rate } m_j). \tag{1b}$$

The reaction (1a) means the reproduction process of the species j, and the parameter r_j is the reproduction rate. The reaction (1b) denotes the death process and parameter m_j is the mortality rate. In reality, the values of parameter m_j are considered as small. Therefore, in the present paper, we suppose to $m_j = 0.01 + 0.0025 \times (j - 2)$ except for the bottom-prey m_1, and r_j are constant 1.0, in order to treat very simply. Our system (1) corresponds to the contact process [4] for $n = 1$ and to the prey-predator model [5, 6] for $n = 2$. Provided that interaction occurs globally, the mean-field theory is applicable as explained. In Fig. 1, the model (1) is schematically shown.

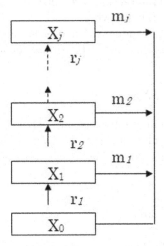

Fig. 1. The chart of the model for j-species system: the species X_j eats X_{j-1} $(j = 1, ..., 6)$. X_0 means the resource. In this chart, the top-predator is indicated by X_6.

2.2   Simulation Method

The reactions (1) are carried out two different ways: the local interactions and the global. We first describe the simulation method for the local interaction:

1) Initially, we distribute individuals X_j of species j on the square lattice L, where $j = 1, ..., n$. The initial distribution is not so important, since the system evolves into a stationary state.
2) Spatial pattern is updated in the following two steps:

   (i) we perform two-body reaction (1a): Choose one lattice site randomly, and then randomly specify one of eight neighboring sites, that is Moore neighborhood. If the pair of chosen sites are (X_j, X_{j-1}), then, the latter site is changed into X_j by the rate r_j. For any other pairs, we skip to the next step.
   (ii) we perform one-body reaction (1b). Choose one lattice point randomly; if the site is occupied by X_j, the site will become the vacant site (X_0) by the rate m_j.

3) Repeat step 2) $L \times L = 200^2)$ times, where, is the total number of lattice points. This is the Monte Carlo step [1].
4) Repeat step 3) until the system reaches a stationary state. Here we have employed periodic boundary conditions, namely replacing any index that exceeds the boundary, L, by using the modulo operator on the index; $L + 1$ is replaced by 1.

Next, we describe the method for the global interaction (mean-field simulation: MFS) in which long-ranged interactions are allowed. The step (i) in 2) for local interaction is replaced with the following:

   (i)' Choose one lattice site randomly. Choose the next site randomly. If these sites are (X_j, X_{j-1}) in order, then the latter site is changed into X_j.

3   Mean-Field Theory

If the size L of lattice system is infinitely large ($L \to \infty$), then the population dynamics for mean-field simulation (MFS) is given by the mean-field theory. For the top predator j is indicated,

$$\dot{x_j} = r_j x_j x_{j-1} - m_j x_j. \tag{2}$$

Except for the top predator are,

$$\dot{x_{j-1}} = r_{j-1} x_{j-1} x_{j-2} - r_{j-1} x_{j-1} x_{j+1} - m_{j-1} x_{j-1}. \tag{3}$$

Here the dots represent the time derivative, and x_j is the density of species j. The time is measured by the Monte Carlo step [1]. The right-hand side of the expression (3) represents the following meanings: the first term express a prey of the species $j - 2$, the second term denote the predatory by $j + 1$ and the last term means the death process. Then the total density is unity, so the density of vacant site is given by

$$x_0 = 1 - \sum_{j=1}^{n} x_j. \tag{4}$$

If the equilibrium densities are coexistence for $3 \leq n \leq 6$, the coexisting equilibrium is locally stable, but in which all densities are positive in a slight range.

4   Simulation Results

4.1   Steady-State Density

Simulations are carried out both the local interactions and the global. We describe the results of local interaction. The system evolves into the stable stationary state in every n. The figure 2 summarizes the coexistence and the top-predator survivor region. Here the maximum values of mortality rate m_1 were plotted against the species n.

In the cases of $3 \leq n$, we see, from the Fig. 2, that the existence region for top-predator becomes small, as the number of species n is on the increase. On the other hand, in the case of the global interaction, species $n = 3$ and 6 are indicated with Fig. 4. The system evolves into a stationary state. The survival region of the top-predator do not sight same appearance both interactions. The surviving region of the local interactions becomes narrow than the global. In addition, we understand similar results against the increase of n.

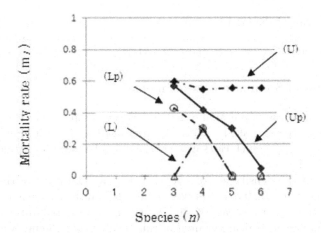

Fig. 2. The coexistence and the top-predator survivor region are summarized for the local interactions. The maximum value of m_1 plotted against n. (U): Upper existence limit, (U_p): Upper top-predator existence limit, (L): Lower existence limit, (L_p): Lower top-predator existence limit.

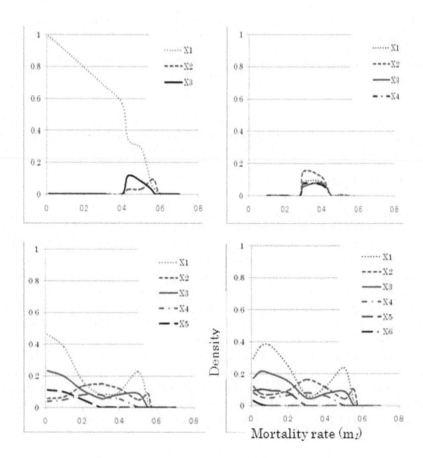

Fig. 3. The result of the local interaction, $n = 3$ to 6 species systems. The values of mortality rate m_1 are from 0 to extinction value. We repeat the same experiment 5 times with different random-seed, and take average.

4.2   Results of Simulation Experiment

Next, we indicate the population dynamics. We first report the case of the local interaction. A typical spatial pattern of experiment is shown in Fig. 5. We find universally for $n \geq 2$, which the survivor region can be observed, where the value of m_1 is existed.

Examples of the density are displayed in Fig. 3. The value of the mortality rate m_1 is changed from 0 to extinction value, while in the time passes $t = 10000$. Therein, on the way, the steady state density of the top-predator existed in the slight region. The ecological meaning of this result is important. Namely, when we conserve the top-predator, it becomes very hard the time dependence of the target species. The top-predator cannot live without the prey density.

Next, we report the results of the global interaction. In Fig. 4, indicated with $n = 3$ and 6, when m_1 increase, there are the stable state density. Therein the

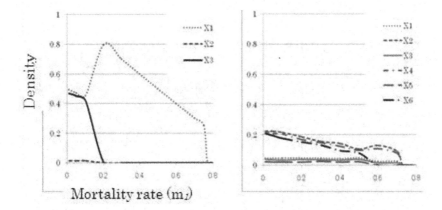

Fig. 4. Result of global interaction, $n = 3$ and 6 species systems

density of the top-predator rapidly decreased. Hence, we obtain similar results that appeared in the local interaction. After all, the top-predator cannot live without density of the prey in the same as the local interaction.

5   Discussions and Conclusion

Before, we dealt with the top-predator mortality rate changing systems of the n-species [9]. In the present paper, we newly study the bottom-prey mortality rate change systems of the n-species. By the use of the Monte Carlo simulations, are carried out on the finite size of the square lattice. Our results indicate that the survivor region comes from the conservation of the top-predator against the prey density. Since the population size of the top-predator is generally small, several conservation policies for top-predator have been performed [7, 8]. In most cases, however, the time dependence of the top-predator is not so simple;

We think about this model. As an example, in old days of Japan, there was a policy for the human management that is gIkasazu-Korosazu, in Japaneseh; the farmers were forgiven neither to dying nor to living. They had to pay the feudal lord for the taxes that were higher than half of the crop. Namely, if the predator has eaten up the prey, predator-self cannot live. On the contrary, the prey cannot survive without something to eat the food.

In this paper, the density was explained from a stochastic aspect. If we protect the bottom-prey, its population size increases. Doing so, the top-predator may become increased in the food chain.

Finally, we discuss the problem necessary of the bottom-prey whether promotes the coexistence of the multiple species or not. The same result was not obtained with the global simulation, but, our result suggests that the bottom-prey density promote the coexistence, so long as an ecosystem contains many species. At least, the predator cannot live without eating the prey.

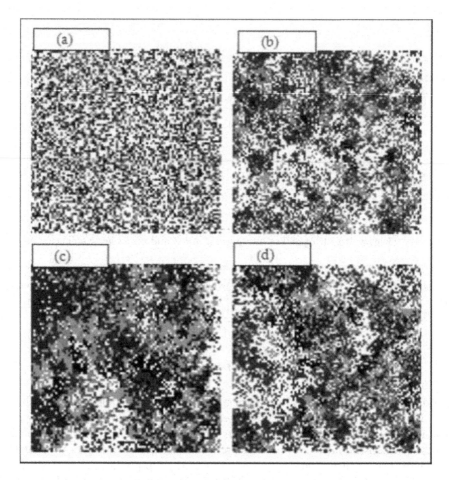

Fig. 5. A typical result of the time dependence density spatial pattern, 6 species, local interaction, $L = 200$, $m_1 = 0.04$, (a) is $t = 0$, initial random pattern, (b) $t = 100$, (c) $t = 1000$, (d) $t = 10000$. The colors represent red (X_6), black (X_5), grey (X_4), green (X_3), brown (X_2), blue (X_1) and white (X_0) respectively.

Lastly, In the n species systems, the simulations have the unexpected results by the initial condition of the parameter-sets. The initial placement is not affected it is clear. To make the matters worse, the steady state of density does not always exist, so we are start looking for the steady state conditions.

The local interaction is main purposes that are direct simulations. But whether reached the reasonable results we did not understand. Therefore we are verifying the validity with the simulation compares between the global interactions that is the mean-field approximation and the mean-field theory.

In the actual world there are varieties of hierarchy number n for the food chain. We learn many things by a model. An important is a task for the future. As you are told this question, how many stages are consisting? and so on.

References

1. Tainaka, K.: Lattice model for the Lotka-Volterra system. J. Phys. Soc. Jpn. 57, 2588–2590 (1988)
2. Paine, R.T.: Food web complexity and species diversity. The American Naturalist 100, 65–75 (1966)
3. Tilman, D., Downing, J.A.: Biodiversity and stability in grassland. Nature 367, 363–365 (1994)
4. Harris, T.E.: Contact interaction on a lattice. Ann. Prob. 2, 969–988 (1974)
5. Satulovsky, J.E., Tome, T.: Stochastic lattice gas model for a predator -prey system. Phys. Rev. E 49, 5073–5079 (1994)
6. Sutherland, B.R., Jacobs, A.E.: Self-organization and scaling in a lattice prey-predator model. Complex Systems 8, 385–405 (1994)
7. Krivan, V.: Optimal Foraging and Predator-Prey Dynamics. Theoretical Population Biology 49, 265–290 (1996)
8. McShea, W.J., Underwood, H.B., Rappole, J.H. (eds.): The science of overabundance deer ecology and population management. Smithsonian Institution Press, Washington (1997)
9. Nagata, H., Morita, S., Yoshimura, J., Nitta, T., Tainaka, K.: Perturbation experiments and fluctuation enhancement in finite size of lattice ecosystems: Uncertainty in top-predator conservation. Ecological Informatics 3, 191–201 (2008)

Simulation and Theoretical Comparison between "Zipper" and "Non-Zipper" Merging

Ryosuke Nishi[1], Hiroshi Miki[2], Akiyasu Tomoeda[3,4],
Daichi Yanagisawa[1,5], and Katsuhiro Nishinari[4,6]

[1] School of Engineering, University of Tokyo, Japan
[2] Department of Material Sciences, Interdisciplinary Graduate School of
Engineering Sciences, Kyushu University, Japan
[3] Meiji Institute for Advanced Study of Mathematical Sciences,
Meiji University, Japan
[4] Research Center for Advanced Science and Technology,
The University of Tokyo, Japan
[5] Research Fellow of the Japan Society of the Promotion of Science, Japan
[6] PRESTO, Japan Science and Technology Corporation
tt097086@mail.ecc.u-tokyo.ac.jp

Abstract. Heavy traffic congestion occurs daily at an intersection of highway traffic. For releasing this congestion, the effect of "zipper" merging is discussed in this paper. This is the merging of vehicles on two lanes alternatively, and is achieved by only the local communication of vehicles between two lanes before merging. This "zipper" merging is compared with the "non-zipper" merging, which is the merging without interactions of vehicles before merging, in terms of flux rate. This comparison is performed by using simulations with a new multiple-lane cellular automaton model which has slow-to-start effect. Numerical results show that the inverse flux rate is observed between "zipper", and "non-zipper" merging as the change of slow-to-start effect. Moreover, theoretical flux rate of the "non-zipper" merging is constructed. It is shown that these theoretical results coincide with the simulation results very well.

Keywords: Traffic flow, Zipper merging, Cellular automaton.

1 Introduction

In recent years, traffic dynamics has attracted much interest of mathematicians, and physicists, and thus it has been studied more and more diligently [1] [2]. Researchers have mainly developed the analysis of the traffic flow on one-lane road by using continuous models [3] and cellular automaton (CA) models [4][5]. Recently, the analysis of the traffic flow on multiple-lane roads with an intersection or a junction is expected to be developed for easing traffic congestion. The flow at an intersection or a junction was studied by using game theory [6], and agent based simulations [7], and the simulations of Optimal Velocity model [8], which introduced cooperation between two lanes.

F. Peper et al. (Eds.): IWNC 2009, PICT 2, pp. 244–251, 2010.

However, these previous works did not study in detail the configuration of vehicles on two lanes before an intersection or a junction, which is crucial to the efficiency of merging. Our previous work focused on the alternative configurations for realizing "zipper" merging. We proposed a simple, and bottom-up, and economical method to induce this configuration. This method is to draw a compartment line between two lanes at the merging area. This line prohibits vehicles from changing lanes, while permits them to see other vehicles on another lane. Vehicles moving along this line is expected to see other vehicles on another lane, and to adjust their own configuration to the alternative configuration. We showed for the first time that alternative configurations as a whole are achieved by only hesitative local interactions, by using simulation and mean-field theory.

As a next study, this paper is devoted to the discussion regarding the effect of "zipper" merging. We compare the flux rate at a merging section between the flow with or without this hesitative communication, by using numerical simulations. Simulations are performed by a new multiple-lane CA model named Multiple Lanes Slow-to-Start (MLSlS) model, which has slow-to-start effect [10]. Moreover, theoretical flux rate is constructed, and compared with the simulation results.

2   Modeling

We use CA models as dynamics of vehicles because they can treat the excluded-volume effect directly, and they can represent realistic fandamental diagrams, e.g, flux vs. density with meta-stable branches [11].

In MLSlS model, i-th vehicle moves one cell straight forward with probability v_i^t between time t and time $t + 1$ provided that the next cell is empty as shown in figure 1. v_i^t is given as

$$
v_i^t = \begin{cases} 0, & \Delta x_{1,i}^t = 0, \\ r, & \Delta x_{1,i}^t \geq 1 \text{ and } \Delta x_{2,i}^t = 0, \\ q, & \Delta x_{1,i}^t \geq 1 \text{ and } \Delta x_{2,i}^t = 1, \\ p, & \Delta x_{1,i}^t \geq 1 \text{ and } \Delta x_{2,i}^t \geq 2, \end{cases} \tag{1}
$$

where $\Delta x_{1,i}^t$ is defined as the distance between i-th vehicle and the proceeding one on the same lane at time t, and $\Delta x_{2,i}^t$ is defined as the distance between i-th vehicle and the proceeding one on another lane at time t. In this model, merging is expressed by the movement from two-lane road to one-lane road as shown in figure 1. When two vehicles are trying to merge simultaneously, only one of them, which is chosen at random, is permitted to move. We set these four values of v_i^t as $0 \leq r \leq q \leq p$. Under this condition, vehicles decrease their hopping probability by responding to other vehicles on another lane. This hesitative behavior induces vehicles to achieve alternative configurations.

MLSlS model uses the slow-to-start effect introduced by Benjamin et al. [10]. This effect expresses the delay of acceleration caused by the inertia of vehicles. When the following distance of i-th vehicle changes from $\Delta x_{1,i}^{t-1} = 0$ to $\Delta x_{1,i}^t \geq 1$,

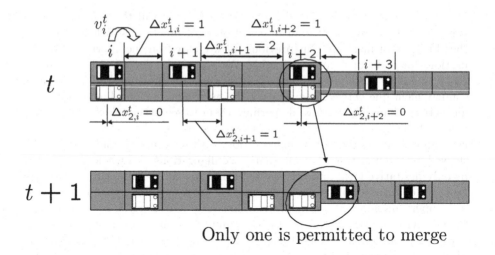

Only one is permitted to merge

Fig. 1. Moving straight forward, and merging from two-lane road to one-lane road in MLSIS model. i-th vehicle move one cell straight forward with probability v_i^t between time t and time $t+1$ provided that the next cell is empty. When two vehicles are trying to merge simultaneously, only one of them chosen at random is permitted to move. $\Delta x_{1,i}^t$, and $\Delta x_{2,i}^t$ are defined as the distance between i-th vehicle and the proceeding one on the same lane, and that on another lane at time t respectively.

hopping probability becomes sv_i^t, where s is the parameter of slow-to-start effect which is normalized as $0 \leq s \leq 1$. Small s means that the slow-to-start effect is strong, i.e., the inertia of vehicles is large.

3   Simulations and Theoretical Calculation

Our idea for easing traffic congestion at merging area, which is shown in figure 2-(a), is to draw a compartment line in merging area as shown in figure 2-(b). This line is a bottom-up method for achieving zipper merging. Each vehicle on the two lanes along the line recognizes other vehicles on another lane, and takes apart from them, and then vehicles as a whole form the alternative configurations before merging. In simulations, roads as shown in 2-(a), and (b) are represented by simple road models as shown in figure 2-(c), and (d) respectively. The road model with no line is named *Road before our measure* (*RBM*), and that with the line is named *Road after our measure* (*RAM*). The road is divided into identical cells, and the merging area is represented by a two-lane road (lane1 and lane 2) connected with a one-lane road (lane 12), which takes shape of the character "Y". This simple road model holds the bottle neck effect which is essential in merging. Note that in our simulations, two lanes are treated equally, not treated as one for main, the other for branch. Vehicles on *RBM* and *RAM* are updated in parallel. They enter in the leftmost cell of each lane with probability α as long as the cell is empty. They move straight forward without changing lanes at

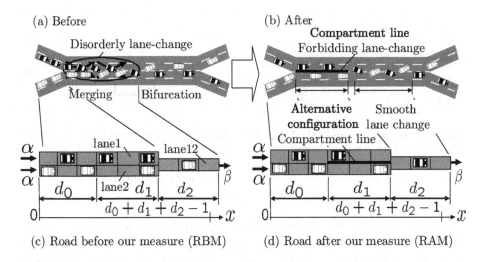

(c) Road before our measure (RBM)   (d) Road after our measure (RAM)

Fig. 2. (a): a road with a merging area without our plan. Disordred lane-change causes traffic congestions. (b): the road with our plan, which is to draw a compartment line. Vehicles along this line interact with other vehicles on the opposite lane, and achieve the alternative configurations before merging. (c), and (d): the model of road (a), and (b) respectively. Merging is modeled by two-lane road connected into one-lane road.

$0 \le x \le d_0 + d_1 - 1$, and merge from $x = d_0 + d_1 - 1$ to $x = d_0 + d_1$, and go out of the rightmost cell at $x = d_0 + d_1 + d_2 - 1$ with probability β. Vehicle on RBM at $0 \le x \le d_0 + d_1 - 1$, and vehicles on RAM at $0 \le x \le d_0 - 1$ can not recognize other vehicles on another lane, which is represented in MLSlS model by $\Delta x^t_{2,i} = \infty$. While vehicles on RAM interact at $d_0 \le x \le d_0 + d_1 - 1$ where the line is drawn.

Next, we define two important quantities in the simulations. Firstly, we quantify the degree of the alternative configurations of vehicles named *Geminity* (G) [9]. G is a function of x, and $G(k)$ denotes the degree of the alternative pattern of vehicles at $x = k$. In the simulations, $G(k)$ is calculated by counting the state of the four cells at $x = \{k, k+1\}$. There are 10 kinds of state labelled as $S(n)$ ($n = 1, 2, ...10$) as shown in figure 3. The symmetry between lane 1 and lane 2 is taken into account for eliminating the similar state. We defined $c(n)_k$ ($n = 1, 2, ...10$) as the total number of each $S(n)$ at $x = \{k, k+1\}$ through the period of each simulation, which is given as $T_1 \le t \le T_2 - 1$. When there is at least one vehicle at $x = k$, the state of the four cells at $x = \{k, k+1\}$ can be $n = 3, 5, 6, 7, 8, 9, 10$ in $S(n)$. Only $S(3)$ among them represents the perfect alternative state which is defined by figure 4. Thus, $G(k)$ is given by $c(n)_k$ as $G(k) = c(3)_k / (c(3)_k + c(5)_k + c(6)_k + c(7)_k + c(8)_k + c(9)_k + c(10)_k)$. G ranges from 0 to 1. The large value of $G(k)$ denotes that the alternative configuration of vehicles at $x = k$ is highly achieved. Secondly, we define Q as the mean flux at the rightmost cell. Q is measured through the period of each simulation which is given as $T_1 \le t \le T_2 - 1$. We compare Q with the range of $0 \le s \le 1$, and G with the

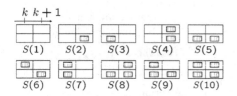

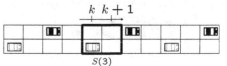

Fig. 3. 10 kinds of states in the four cells at $x = \{k, k+1\}$, which is defined as $S(n)$ ($n = 1, \ldots, 10$)

Fig. 4. Complete alternative configurations of vehicles. When at least one vehicle exists at $x = k$, this complete alternative state is given only by $S(3)$.

range of $0 \le x \le d_0 + d_1 - 2$ between RBM and RAM. The conditions are given as follows. We fix $p = 1$, and $d_0 = 400$, and $d_1 = 50$, and $d_2 = 10$, and $\beta = 1$, and $T_1 = 10000$, and $T_2 = 20000$, and $q = r$. Six kinds of condition is given as $\{q, \alpha\} = \{(1/10, 1/4), (1/10, 1/3), (1/2, 1/4), (1/2, 1/3), (9/10, 1/4), (9/10, 1/3)\}$, which are named case (a)-(f) respectively. Note that case (a), and (c), and (e) have no difference in RBM, and case (b), and (d), and (f) also have no difference in RBM. We use for simplicity Q_{RBM}^j, and Q_{RAM}^j ($j = a, b, c, d, e, f$) which denotes Q in the case (j) of RBM, and that in the case (j) of RAM respectively.

The results are shown in figure 5, and figure 6. In figure 5, Q_{RBM}^j is larger than Q_{RAM}^j ($j = a, b$). While $Q_{RBM}^c < Q_{RAM}^c$ at $0 \le s \le 0.5$, and Q_{RBM}^c is same as Q_{RAM}^c at $0.5 \le s \le 1$. $Q_{RBM}^d < Q_{RAM}^d$ at $0 \le s \le 0.7$, and $Q_{RBM}^d > Q_{RAM}^d$ at $0.7 \le s \le 1$. Q_{RBM}^j is same as Q_{RAM}^j ($j = e, f$). In figure 6, G increases to 1 in the case (a)-(d), while not increase to 1 in the case (e), and (f) in RAM. In RBM, G does not increase to 1 in all cases.

$G = 1$ observed in the case (a)-(d) in RAM denotes that the perfect "zipper" merging is achieved by only the local hesitative behavior of vehicles. Moreover, Q_{RBM}^d and Q_{RAM}^d show for the first time that the reversal flux rate between "zipper" and "non-zipper" merging occurs as the change of the slow-to-start effect. Particularly, $Q_{RBM}^d < Q_{RAM}^d$ at $0 \le s \le 0.7$ shows that decreasing hopping probability before merging makes flux rate better. This improvement of flux rate is expected to be analyzed deeply, and to be applied to real traffic.

We discuss the Q_{RBM}^j and Q_{RAM}^j ($j = a, b, c, d, e, f$) as follows. $Q_{RAM}^a < Q_{RAM}^c$, and $Q_{RAM}^b < Q_{RAM}^d$ are observed because the strength of the bottle neck in the line area $d_0 \le x \le d_0 + d_1 - 1$ of RAM becomes weaker as $q(= r)$ becomes larger. Q_{RAM}^j same as Q_{RBM}^j ($j = e, f$) is observed because "zipper" merging is not achieved in RAM. Figure 6 shows that the necessary numbers of cell to achieve $G = 1$ becomes larger as q becomes larger. In the case $q = 9/10$, the line length $d_1 = 50$ is not enough to achieve $G = 1$.

Moreover, we calculate theoretically the flux at the rightmost cell of RBM. With the slow-to-start effect, using Markov chains or queuing theory is difficult. Instead, we use an expected travel time, which gives the inverse of the flux. We

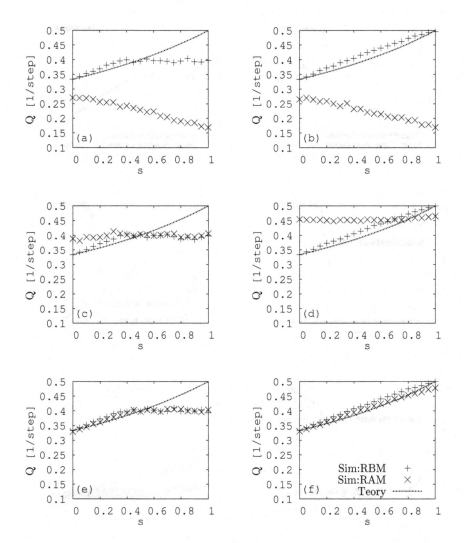

Fig. 5. Q^j_{RBM} versus s, and Q^j_{RAM} versus s ($j = a, b, c, d, e, f$) obtained by simulations, together with Q_{th} versus s obtained by theoretical calculations. The reversal of flux rate is observed between Q^d_{RBM} and Q^d_{RAM}. Q_{th} fits well to Q^j_{RBM}. Case (a)-(f) is $\{q, \alpha\} = \{(1/10, 1/4), (1/10, 1/3), (1/2, 1/4), (1/2, 1/3), (9/10, 1/4), (9/10, 1/3)\}$ respectively. Parameters are given as $p = 1$, and $d_0 = 400$, and $d_1 = 50$, and $d_2 = 10$, and $\beta = 1$, and $T_1 = 10000$, and $T_2 = 20000$, and $q = r$.

suppose the situation around the merging point, such that vehicles are occupied with $x \leq d_0 + d_1 - 1$. We define T_{th} as the expected time spent for moving from $x \leq d_0 + d_1 - 1$ to $x = d_0 + d_1$, added with that for moving from $x \leq d_0 + d_1$ to $x = d_0 + d_1 + 1$. The former time is given by $1[step] \times sp + (1 + 1/p)[step] \times (1 - sp)$. the later is given by $1/p[step]$, where we suppose that $x = d_0 + d_1 + 1$ is empty, i.e,

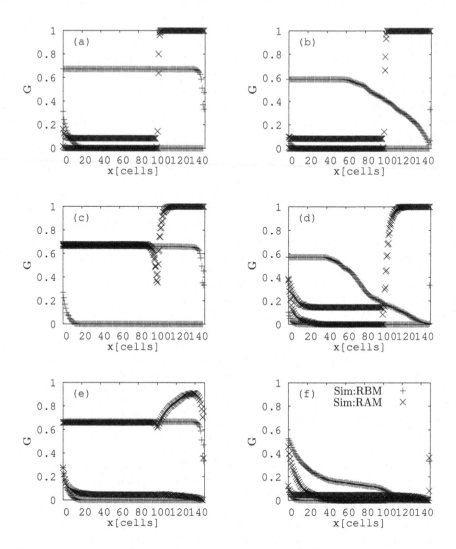

Fig. 6. G versus x obtained by simulations. Case (a)-(f), and parameters are given similarly to figure 5. the line area in RAM is $100 \leq x \leq 149$. $G = 1$ is achieved in the case of (a)-(d) in RAM, while not achieved in the other cases.

vehicles move with the probability 1 at $x = d_0 + d_1$. Theoretical flux is defined as Q_{th}, which is given approximately as

$$Q_{th} = (T_{th})^{-1} = [\{sp + (1 + 1/p)(1 - sp)\} + 1/p]^{-1} = p/\{p(1 - s) + 2\}. \quad (2)$$

We draw Q_{th} together with simulation results in figure 5. Q_{th} fits to the simulation results Q_{RBM}^{j} $(j = b, d, f)$ at $0 \leq s \leq 1$. While Q_{th} fits to the simulation results Q_{RBM}^{j} $(j = a, b, c)$ only at $0 \leq s \leq 0.5$. This discrepancy occurs because the maximum of Q^j_{RBM} is restricted by α. The maximum flux is given by $2\alpha/(1+\alpha)$ with the call loss at $x = 0$. Q^j_{RBM} becomes its maximum as long as $2\alpha/(1+\alpha) \leq Q_{th}$, i.e., $s \geq 1/2 - 1/(2\alpha) + 2/p$. In the case of $j = a, b, c$, this is $s \geq 1/2$.

4 Conclusive Discussion

We have achieved the "zipper" merging by only the local hesitative behaviors. Numerical results show for the first time that the inverse flux rate is observed between "zipper" and "non-zipper" merging as the change of slow-to-start. Moreover, theoretical flux rate of "non-zipper" merging is successfully constructed. This result is expected to be deepened to be applied for the real traffic.

References

1. Helbing, D.: Rev. Mod. Phys. 73, 1067 (2001)
2. Chowdhury, D., Santen, L., Schadschneider, A.: Phys. Rep. 329, 199 (2000)
3. Bando, M., Hasebe, K., Nakayama, A., Shibata, A., Sugiyama, Y.: Phys. Rev. E 51, 1035 (1995)
4. Evans, M.R., Rajewsky, N., Speer, E.R.: J. Stat. Phys. 95, 45 (1999)
5. Kanai, M., Nishinari, K., Tokihiro, T.: Phys. Rev. E 72, 035102(R) (2005)
6. Kita, H.: Transportation Research Part A 33(3), p. 305 (1999)
7. Hidas, P.: Transportation Research Part C 13(1), 37 (2005)
8. Davis, L.C.: Physica A 361(2), p. 606 (2006)
9. Nishi, R., Miki, H., Tomoeda, A., Nishinari, K.: Phys. Rev. E. 79, 066119 (2009)
10. Benjamin, S.C., Johnson, N.F., Hui, P.M.: J. Phys. A: Math. Gen. 29, 3119 (1996)
11. Nishinari, K., Fukui, M., Schadschneider, A.: J. Phys. A: Math. Gen. 37, 3101 (2004)

Universality of 2-State 3-Symbol Reversible Logic Elements — A Direct Simulation Method of a Rotary Element

Tsuyoshi Ogiro[1], Artiom Alhazov[1,2], Tsuyoshi Tanizawa[1], and Kenichi Morita[1]

[1] Hiroshima University, Graduate School of Engineering
Higashi-Hiroshima, 739-8527, Japan
morita@iec.hiroshima-u.ac.jp
[2] Institute of Mathematics and Computer Science, Academy of Sciences of Moldova
Academiei 5, Chişinău, MD-2028, Moldova

Abstract. A reversible logic element is a primitive from which reversible computing systems can be constructed. A rotary element is a typical 2-state 4-symbol reversible element with logical universality, and we can construct reversible Turing machines from it very simply. There are also many other reversible element with 1-bit memory. So far, it is known that all the 14 kinds of non-degenerate 2-state 3-symbol reversible elements can simulate a Fredkin gate, and hence they are universal. In this paper, we show that all these 14 elements can "directly" simulate a rotary element in a simple and systematic way.

Keywords: Reversible logic element, reversible computing, rotary element.

1 Introduction

Reversible computing [1,2,3,8] is a paradigm of computation that reflects physical reversibility, one of the fundamental microscopic laws of Nature. So far, various reversible systems such as reversible Turing machines, and reversible cellular automata have been investigated. A reversible logic element, whose function is described by a one-to-one mapping, is a primitive from which reversible computing systems can be constructed. There are two types of reversible elements: one without memory, which is usually called a reversible logic gate, and one with memory. Reversible logic gates were first studied by Petri [10]. Then Toffoli [11,12], and Fredkin and Toffoli [3] studied them in connection with physical reversibility. They showed a Toffoli gate [11,12] and a Fredkin gate [3] are both logically universal. Hence, every reversible Turing machine can be built by them.

On the other hand, Morita [6] proposed a special type of a reversible logic element with 1-bit memory called a rotary element (RE), and showed that reversible Turing machines can be constructed from it. The construction is very much simpler than to use reversible gates. Furthermore, Morita [8] showed that an RE can be easily implemented in the Billiard Ball Model, which is a reversible

F. Peper et al. (Eds.): IWNC 2009, PICT 2, pp. 252–259, 2010.

physical model of computation proposed by Fredkin and Toffoli [3]. An RE is a specific 2-state 4-symbol (i.e., it has 4 input lines and 4 output lines) reversible logic element, and thus there are also many other elements of such a type. Morita et al. [7] classified all the 2-state k-symbol reversible elements for $k = 2, 3, 4$, and showed that there exist 4 ($k = 2$), 14 ($k = 3$), and 55 ($k = 4$) essentially different and non-degenerate ones. They also investigated whether there are universal reversible elements that are simpler than an RE. Later, Ogiro et al. [9] proved that all the 14 kinds of 2-state 3-symbol elements are universal by showing that a Fredkin gate can be simulated by a circuit composed of each of them. From the fact that any RSM is simulated by a circuit constructed only of Fredkin gates [5], and the fact that an RE is an RSM, we can see an RE can be also implemented by each of the 2-state 3-symbol elements.

In this paper, we show that each of the non-degenerate 2-state 3-symbol reversible elements can "directly" simulate an RE. Of course, as stated above, it is possible to construct a circuit that simulates an RE via a Fredkin gate. However, if we take such an indirect method, a lot of logic elements are needed as a whole. Moreover, many signals in the circuit should be synchronized at each of the gates. Here, we give a simple and systematic construction method of reversible circuits that directly simulate an RE such that only one signal exists in the circuit. This shows a kind of a reducibility result among reversible elements with memory, and gives an insight for the universality problem on them.

2   Preliminaries

A *sequential machine* (SM) M is defined by $M = (Q, \Sigma, \Gamma, \delta)$, where Q is a finite non-empty set of states, Σ and Γ are finite non-empty sets of input and output symbols, respectively. $\delta : Q \times \Sigma \rightarrow Q \times \Gamma$ is a mapping called a *move function*. M is called a *reversible sequential machine* (RSM) if δ is one-to-one (hence $|\Sigma| \leq |\Gamma|$). In an RSM, the previous state and the input are determined uniquely from the present state and the output of M. A *reversible logic elements with memory* (RLEM) is an RSM with small numbers of states and symbols.

Let $M = (\{q_0, q_1\}, \{a, b, c, d\}, \{w, x, y, z\}, \delta)$ be a 2-state 4-symbol RLEM. Since the move function $\delta : \{q_0, q_1\} \times \{a, b, c, d\} \rightarrow \{q_0, q_1\} \times \{w, x, y, z\}$ is one-to-one, it is specified by a permutation from the set $\{q_0, q_1\} \times \{w, x, y, z\}$. Hence, there are $8! = 40320$ RLEMs for $k = 4$. They are numbered by $0, \cdots, 40319$ in the lexicographic order of permutations. There are $6! = 720$ 2-state 3-symbol RLEMS and $4! = 24$ 2-state 2-symbol RLEMs, which are also numbered in this way [7]. To indicate k-symbol RLEM, the prefix "k-" is attached. In what follows, we sometimes call q_0 and q_1 by *state 0* and *state 1*, respectively. We also indicate $\delta(q_i, u) = (q_j, v)$ by $(i, u) \mapsto (j, v)$, and call it a *rule* of M.

Let us consider the move function of a 2-state 4-symbol RLEM given by Table 1. It defines the RLEM No. 4-289. We also use a pictorial representation for a 2-state RLEM as shown in Fig. 1. Note that in Fig. 1, an input signal (or a particle-like object) should be given at most one input line, because each input/output line corresponds to an input/output symbol of an RSM. Therefore, we should not confuse RLEMs with conventional logic gates.

Table 1. The move function of the 2-state 4-symbol RLEM 4-289

Present state	Input			
	a	b	c	d
q_0	$q_0\ w$	$q_0\ x$	$q_1\ w$	$q_1\ x$
q_1	$q_0\ y$	$q_0\ z$	$q_1\ z$	$q_1\ y$

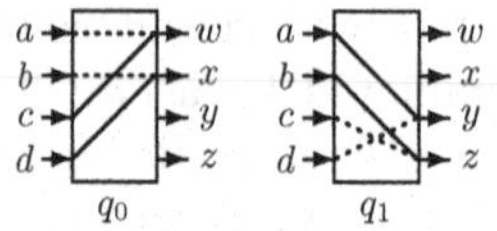

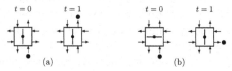

Fig. 1. Pictorial representation of the RLEM 4-289. Solid and dotted lines in a box describe the input-output relation for each state. A solid line shows the state changes to another, and a dotted line shows the state remains unchanged.

Fig. 2. Operations of an RE: (a) the parallel case, and (b) the orthogonal case

A *rotary element* (RE) [6] is a 2-state 4-symbol RLEM equivalent to the RLEM 4-289. It has two states called H-state (⊟) and V-state (⊞), four input lines $\{n, e, s, w\}$, and four output lines $\{n', e', s', w'\}$ corresponding to the input and output alphabets. We can interpret that an RE has a "rotating bar" to control the moving direction of an input signal (or a particle). When no particle exists, nothing happens on the RE. If a particle comes from a direction parallel to the rotating bar, then it goes out from the output line of the opposite side without affecting the direction of the bar (Fig. 2 (a)). If a particle comes from a direction orthogonal to the bar, then it makes a right turn, and rotates the bar by 90 degrees (Fig. 2 (b)). It is known that any reversible Turing machine can be simulated by a reversible logic circuit composed only of REs [6,8].

There are many 2-state RLEMs even if we limit $k = 2, 3, 4$, but we can regard two RLEMs are "equivalent" if one can be obtained by renaming the states and the input/output symbols of the other. The numbers of equivalence classes of 2-state 2-, 3- and 4-symbol RLEMs are 8, 24 and 82, respectively [7]. Fig. 3 shows representatives of the equivalence classes of 2- and 3-symbol RLEMs, where each representative is chosen as the one with the least index in its class.

Among them there are some "degenerate" elements that are either equivalent to connecting wires, or equivalent to a simpler element with fewer symbols. In Fig. 3, they are indicated at the upper-right corner of each box. Total numbers of equivalence classes of non-degenerate 2-state 2-, 3-, and 4-symbol RLEMs are 4, 14, and 55, respectively [7].

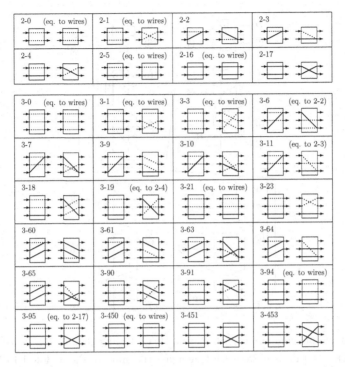

Fig. 3. Representatives of 8 equivalence classes of 2-state 2-symbol RLEMs, and those of 24 equivalence classes of 2-state 3-symbol ones [7]

3 Direct Simulation of an RE by 3-Symbol RLEMs

Here, we show that an RE is implemented by each of the 14 non-degenerate 2-state 3-symbol RLEMs in a systematic way. We first consider the construction method of an RE shown by Lee et al. [4]. They gave a circuit called a "C-D (coding-decoding) module", which is equivalent to the RLEM 3-10, using one RLEM 2-3 and one RLEM 2-4 as shown in Fig. 4. (In [4], the RLEMs 2-3 and 2-4 are called a "reading toggle" and an "inverse reading toggle", respectively.)

Lee et al. [4] then designed a circuit that simulates an RE using C-D modules. Fig. 5 gives the circuit, where C-D modules are replaced by RLEMs 3-10. Fig. 6 shows the operation of it for the parallel and the orthogonal cases.

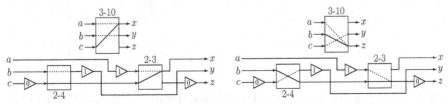

Fig. 4. Implementation of an RLEM 3-10 by RLEMs 2-4 and 2-3 [4]. A small triangle is a delay element, and the number in it shows the delay time.

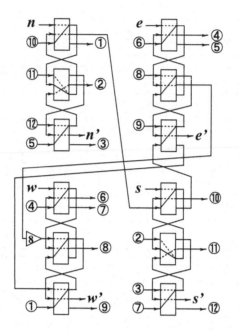

Fig. 5. A circuit composed of RLEMs 3-10 that simulates an RE [4]. It is in the H-state. The V-state is obtained by changing the state of the middle RLEM of each quarter. Each pair of lines with the same number should be connected. Additional delay elements of 8 units of time should be inserted at the positions of 2, 8 and 11. The total delay between the input and the output is 12.

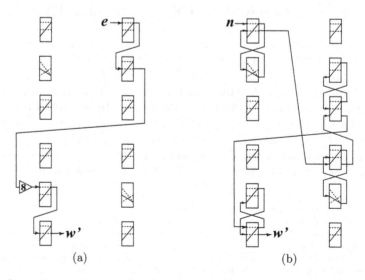

Fig. 6. Operations of the circuit of Fig. 5 in the H-state: (a) the parallel case where the input is from e, and (b) the orthogonal case where the input is from n

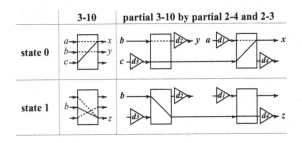

Fig. 7. A partial 3-10, and its implementation by partial 2-4 and 2-3

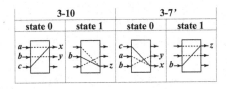

Fig. 8. Equivalence of a partial 3-10 and a partial 3-7. Note that the RLEM 3-7′ is obtained from 3-7 by exchanging the states 0 and 1

Table 2. Values of the delays $d_1, \cdots, d_4$ in Fig.7 for each of the 14 3-symbol RLEMs

	d_1 d_2 d_3 d_4		d_1 d_2 d_3 d_4		d_1 d_2 d_3 d_4		d_1 d_2 d_3 d_4
3-7	5 5 3 3	3-23	5 5 2 2	3-64	5 5 4 3	3-91	4 5 2 2
3-9	5 4 3 0	3-60	5 5 4 4	3-65	5 4 2 2	3-451	4 4 4 4
3-10	5 5 3 3	3-61	5 5 4 4	3-90	5 5 3 4	3-453	2 2 4 4
3-18	4 5 0 3	3-63	5 5 3 3				

From Fig. 6, we can see that not all the rules of the RLEM 3-10 are used, i.e., the set of the used rules is $R = \{(0, a) \mapsto (0, x),\ (0, b) \mapsto (0, y),\ (0, c) \mapsto (1, x),$ $(1, b) \mapsto (0, z)\}$. We call an RLEM with the rule set R a *partial 3-10*. We can further see that the partial 3-10 with the set R is implemented by a partial 2-4 and a partial 2-3, where only three rules are used in each of 2-3 and 2-4, as shown in Fig. 7. Note that the partial 3-10 with R is equivalent to a partial 3-7 under some permutation of input/output ports and states (see Fig. 8).

We now show that the partial 3-10 can be simulated by each of the 14 non-degenerate 3-symbol RLEMs by the method given in Fig. 7. It is shown in Fig. 9. By adding appropriate delays given in Table 2, these circuits acts as the partial 3-10 with the total delay of 6 units of time between the input and the output. Hence, replacing each occurrences of 3-10 in Fig. 5 by a circuit in Fig. 9, and changing the delay time of the delay element to 48 (= 6 × 8) in Fig. 5, we have a circuit made from each of the 14 3-symbol RLEMs that simulates an RE.

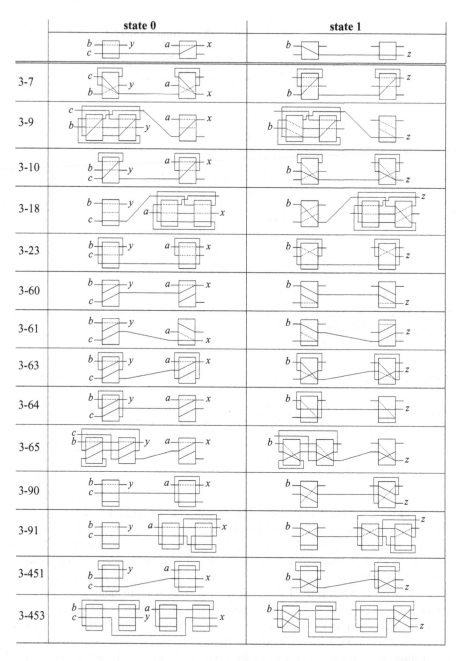

Fig. 9. Implementation of the partial 3-10 using the 3-symbol RLEMs by the method given in Fig. 7. Delay elements in Fig. 7 should be added according to the Table 2 so that the total delays between the input and the output are to be constant. Though RLEMs 3-7 and 3-10 apparently have the function of the partial 3-10, the circuits obtained by the method of Fig. 7 are shown above.

4   Concluding Remarks

Since reversibility is one of the important microscopic physical laws of Nature, future computing devices surely use this physical property directly. Hence, from a technical viewpoint, it is very important to know how simple a universal reversible logic element can be. In this paper, universality of all the 14 non-degenerate 2-state 3-symbol RLEM was shown by giving circuits that directly simulate an RE, in a systematic manner. Since it is known that any RSM can be implemented by REs [8], we can see all these 3-symbol RLEMs and an RE are mutually reducible. Here, all the constructed circuits work synchronously, i.e., the total delays between inputs and outputs are constant. On the other hand, we can also construct asynchronous circuits that may have smaller number of elements. This problem will be discussed in another paper. As for 2-state 2-symbol RLEMs, it is known that the set of RLEMs {2-3, 2-4} is universal [4]. But, it is an open problem whether there is a single universal 2-state 2-symbol RLEM.

Acknowledgement. This work was supported by JSPS Grant-in-Aid for Scientific Research (C) No. 21500015, and for JSPS Fellows No. 20·08364.

References

1. Bennett, C.H.: Logical reversibility of computation. IBM J. Res. Dev. 17, 525–532 (1973)
2. Bennett, C.H.: Notes on the history of reversible computation. IBM J. Res. Dev. 32, 16–23 (1988)
3. Fredkin, E., Toffoli, T.: Conservative logic. Int. J. Theoret. Phys. 21, 219–253 (1982)
4. Lee, J., Peper, F., Adachi, S., Morita, K.: An asynchronous cellular automaton implementing 2-state 2-input 2-output reversed-twin reversible elements. In: Umeo, H., Morishita, S., Nishinari, K., Komatsuzaki, T., Bandini, S. (eds.) ACRI 2008. LNCS, vol. 5191, pp. 67–76. Springer, Heidelberg (2008)
5. Morita, K.: A simple construction method of a reversible finite automaton out of Fredkin gates, and its related problem. Trans. IEICE Japan E-73, 978–984 (1990)
6. Morita, K.: A simple reversible logic element and cellular automata for reversible computing. In: Margenstern, M., Rogozhin, Y. (eds.) MCU 2001. LNCS, vol. 2055, pp. 102–113. Springer, Heidelberg (2001)
7. Morita, K., Ogiro, T., Tanaka, K., Kato, H.: Classification and universality of reversible logic elements with one-bit memory. In: Margenstern, M. (ed.) MCU 2004. LNCS, vol. 3354, pp. 245–256. Springer, Heidelberg (2005)
8. Morita, K.: Reversible computing and cellular automata — a survey. Theoretical Computer Science 395, 101–131 (2008)
9. Ogiro, T., Kanno, A., Tanaka, K., Kato, H., Morita, K.: Nondegenerate 2-state 3-symbol reversible logic elements are all universal. Int. Journ. of Unconventional Computing 1, 47–67 (2005)
10. Petri, C.A.: Grundsätzliches zur Beschreibung diskreter Prozesse. In: Proc. 3rd Colloquium über Automatentheorie, pp. 121–140. Birkhäuser Verlag, Basel (1967)
11. Toffoli, T.: Reversible computing. In: de Bakker, J.W., van Leeuwen, J. (eds.) ICALP 1980. LNCS, vol. 85, pp. 632–644. Springer, Heidelberg (1980)
12. Toffoli, T.: Bicontinuous extensions of invertible combinatorial functions. Mathematical Systems Theory 14, 12–23 (1981)

Pump Current as a Signal Transformation

Jun Ohkubo

Institute for Solid State Physics, University of Tokyo,
Kashiwanoha 5-1-5, Kashiwa-shi, Chiba 277-8581, Japan
ohkubo@issp.u-tokyo.ac.jp

Abstract. A signal transformation from cyclic inputs to particle current by a simple stochastic system is studied. The stochastic system describes a simple particle hopping, and it is known that it generates a unidirectional current of particles, so-called "pump current", under periodic perturbations. We give analytical results using a geometrical phase concept, and discuss the possibility of the pumping phenomenon as a natural computation; the stochastic system acts as a detector of the phase difference of two cyclic signals, or a converter of a cyclic signal into a unidirectional current of particles.

1   Introduction: Pump Current

'Fluctuations' is one of characteristics of living matter. Intrinsically, living matter has large fluctuations, and it has been revealed that such fluctuations play important roles in living matter [1]. Hence, it would be sometimes suitable to describe biological phenomena by stochastic interpretations. From the viewpoint of computation, how the system with large fluctuation performs signal processing or signal transformation is an important issue. The number of transmitter molecules in cells can be considerably small, and then there are large fluctuations. The cells would have mechanisms to control large fluctuations, or they may use such fluctuations in order to transmit information efficiently. Fluctuation in small systems is also one of important research topics in physics. For example, fluctuation plays an important role in molecular motors; there are many works related to Brownian ratchets [2,3]. In addition, many researchers try to find a general theoretical framework for fluctuating phenomena [4].

In the present paper, we focus on a signal transformation via a stochastic system; we show that a simple stochastic system with particle hopping can work as the signal transformation system. Figure 1 shows conceptual diagrams of the signal transformation. In Fig. 1(a), a stochastic system receives two cyclic signals as its inputs, and outputs a unidirectional current of particles. The direction of the current is determined by a phase difference of two input signals. Hence, the stochastic system acts as a detector of the phase difference of two input signals. In Fig. 1(b), a cyclic signal is converted into a unidirectional current. Note that a stochastic system under a cyclic perturbation evolves also cyclically, and hence the state of the stochastic system comes back to the initial one after one cycle period of the cyclic perturbation. Despite the cyclic behavior of the system, a

F. Peper et al. (Eds.): IWNC 2009, PICT 2, pp. 260–267, 2010.

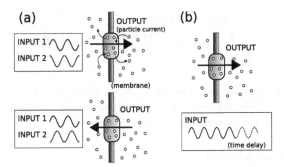

Fig. 1. Signal transformation from cyclic inputs to a unidirectional current. A membrane separates the whole system into two parts, and particles can move among the two parts with certain hopping rates. Inputs are cyclic perturbations affecting the hopping rates of the system, and the output is resultant particle current. (a) Detector of the phase difference of two cyclic inputs. According to the phase difference, the direction of the current changes. (b) Converter of a cyclic input into a unidirectional current.

unidirectional current is generated by a cyclic input. (We will explain Fig. 1 in more detail in Sec. 4.)

We will show a possibility for the signal transformations in Fig. 1 using a so-called "pump current". The pump current is a current induced by cyclic perturbations, and it has been already observed experimentally [5]. In the experiment, red blood cells are exposed to an oscillating electric field. The red blood cell maintains its intracellular sodium and potassium ion composition relatively constant because of the activity of its enzyme. It has been shown that the pumping mode of (Na,K)-ATPase is activated due to the oscillating electric field. The ion transfer would be effectively considered as a stochastic phenomenon, and the above experiment shows that cyclic perturbations can control particle currents using the stochasticity of the system. We will introduce a simple stochastic model which shows the pump current, and discuss how the pumping phenomenon can be used as a natural computation. Especially, we discuss a signal transformation from cyclic inputs to a unidirectional current of particles.

The present paper is organized as follows. In Sec. 2, we introduce a simple stochastic model for the pump current problem, and briefly explain the concept of counting statistics (for example, see [6]). The scheme of counting statistics is a powerful tool in order to study current and its fluctuation (including all statistics of the current). Section 3 gives the mathematical structure behind the pumping phenomena. It has been shown that a geometrical phase (Berry phase and Aharonov-Anandan phase) [7] is related to the pump current problem [8,9,10,11]. The interpretation of the pump current problem as the geometrical phase gives useful insights into the pump current. A new calculation result for a case with arbitrary phase differences of input signals is also given in Sec. 3. Finally, we discuss computational aspects of the pump current problem in Sec. 4; the pump current problem can be interpreted as a detector of phase differences of two oscillating inputs, or a converter of a cyclic input into a unidirectional current.

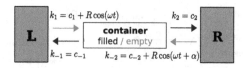

Fig. 2. A simple stochastic system which shows the pump current. [L] and [R] are particles baths, and the container [C] can contain at most one particle. k_1 and k_{-2} have time-dependent perturbations with different phases.

2   Stochastic Model

2.1   Simple Model for the Pumping Phenomenon

A simple stochastic system has been introduced in order to investigate the pump current problem [8]. Figure 2 is the stochastic model. The system consists of three parts. Two particle baths are denoted as [L] and [R]. The particle baths are assumed to be very large, and influx and efflux of particles do not affect them. These particle baths exchange particles via an intermediate container [C]. The container [C] can contain either zero or one particle in it. When the container is filled with one particle, the particle can escape from the container by jumping into one of the two particle baths [L] or [R]. On the contrary, when the container is empty, either of the particle baths can emit a new particle into the container. For simplicity, we here consider the kinetic rates in Fig. 2; each rate k_i has a constant contribution c_i, and only two kinetic rates (k_1 and k_{-2}) oscillate with time at frequency ω and amplitude R [12]. There is a phase difference α between k_1 and k_{-2}. So far, only a case with a phase difference $\alpha = -\pi/2$ has been investigated [8,10,11]. We here consider a more general case with arbitrary α.

A master equation for the system is as follows:

$$\frac{d}{dt}\boldsymbol{p}(t) = -\begin{bmatrix} k_1 + k_{-2} & -k_{-1} - k_2 \\ -k_1 - k_{-2} & k_{-1} + k_2 \end{bmatrix} \boldsymbol{p}(t), \tag{1}$$

where $\boldsymbol{p}$ is the state of the system defined as

$$\mathbf{p}(t) = \begin{bmatrix} P_e \\ P_f \end{bmatrix}. \tag{2}$$

Due to the normalization condition, we obtain $P_e + P_f = 1$.

We want to calculate the number of transitions from [C] into [R] during time T, where $T = 2\pi/\omega$ is the period of the rate oscillations. Neglecting discrete property of particles and assuming $\boldsymbol{p}(t)$ is always in equilibrium, it is easy to calculate the instantaneous [L] $\rightarrow$ [R] current from the master equation (1);

$$j(t) = \frac{\kappa_+(t) - \kappa_-(t)}{K(t)}, \quad K(t) \equiv \sum_{\{m\}} k_m(t), \quad \kappa_\pm(t) \equiv k_{\pm 1}(t)k_{\pm 2}(t). \tag{3}$$

One might think that the time average of the current over a cyclic perturbation is simply given by the time average of the current $j(t)$ in (3), but it is not true.

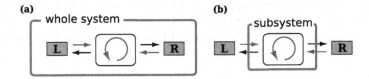

Fig. 3. (a) When we construct a master equation for the whole system, the state change is directly related to the dynamics of the particle hopping. (b) When a master equation describes only the subsystem, the state change does not give information about the particle hopping directly.

For example, when $c_1 = c_{-1} = c_2 = c_{-2}$, the time average of j is zero, but a net current is actually observed in numerical experiments. In addition, when one wants to calculate not only the average current, but also its fluctuation, a simple calculation based on the master equation (1) could not be enough. It has been shown that a concept of counting statistics is useful in order to calculate the pump current, and we explain it in the following subsection.

2.2 Counting Statistics

A master equation describes changes of states of a stochastic system. In some cases, the change of states directly gives information for dynamics. If one constructs a master equation for the whole system (Fig. 3(a)), a current [L] → [R] is obtained by observing the changes in the number of particles in [L] and [R]. That is, if the number of particles in [L] decreases and that in [R] increases, there is a net current [L] → [R]. In contrast, a current [R] → [L] is characterized by the decrease of particles in [R] (and then the increase of particles in [L]).

On the other hands, if a master equation describes only the subsystem (Fig. 3(b)), which is actually the case we want to treat, it would be difficult to extract infromation for dynamics from the change of states; after one cycle period, the subsystem comes back to the initial state of the cycle, and then the initial and final states are the same one.

In order to treat the case of Fig. 3(b), the concept of counting statistics is available. While it is possible to calculate an average current using other analytical methods, the counting statistics enables us to calculate higher cumulants (including fluctuation). Let P_n be the probability to have n net transitions from [C] into [R] during time T. A minus value of n means that there is a current from [R] to [L]. The characteristic function of P_n is given by [8,10,11]

$$Z(\chi) = e^{S(\chi)} = \sum_{s=-\infty}^{\infty} P_{n=s} e^{is\chi} = \mathbf{1}^{\dagger} \hat{T} \left(e^{-\int_0^T \hat{H}(\chi,t)dt} \right) \mathbf{p}(0), \qquad (4)$$

where $\mathbf{1}$ is the unit vector, $\hat{T}$ stands for the time-ordering operator, and

$$\hat{H}(\chi,t) = \begin{bmatrix} k_1 + k_{-2} & -k_{-1} - k_2 e^{i\chi} \\ -k_1 - k_{-2} e^{-i\chi} & k_{-1} + k_2 \end{bmatrix}. \qquad (5)$$

In (4), χ is called the counting field, and the derivatives of $S(\chi)$ give cumulants of P_n, e.g., $J_{\text{total}} = (-i/T)\partial S(\chi)/\partial \chi|_{\chi=0}$. Hence, the problem for calculating the current corresponds to the evaluation of the characteristic function $Z(\chi)$ (or $S(\chi)$).

3   Mathematical Structure Behind the Pumping Phenomenon

3.1   Interpretation as a Shrödinger-Like Equation

From (4), it is easy to see that the characteristic function $Z(\chi)$ is related to a solution of the following differential equation:

$$\frac{d}{dt}\mathbf{x}(t) = -\hat{H}(\chi, t)\mathbf{x}(t), \tag{6}$$

where $\mathbf{x}(t)$ is a state vector with two components. The final state $\mathbf{x}(T)$ at time T is formally calculated from

$$\mathbf{x}(T) = \hat{T}\left(e^{-\int_0^T \hat{H}(\chi,t)dt}\right)\mathbf{x}(0) \equiv \exp[i\mu(\chi)]\mathbf{x}(0). \tag{7}$$

Hence, we obtain the identity $S(\chi) = i\mu(\chi)$.

We here use an analogy between the classical stochastic system and a quantum mechanical formulation. Replacing the time evolution operator $\hat{H}(\chi, t)$ by $H \equiv -i\hat{H}(\chi, t)$, we obtain the following Shrödinger-like equation:

$$i\frac{d}{dt}|\phi(t)\rangle = H|\phi(t)\rangle. \tag{8}$$

When we take the initial state $|\phi(0)\rangle$ as a cyclic state, we have

$$|\phi(T)\rangle = U(T)|\phi(0)\rangle = e^{i\mu(\chi)}|\phi(0)\rangle, \tag{9}$$

where $\mu(\chi)$ is a phase factor.

Once one obtains the Shrödinger(-like) equation of (8), many analytical methods developed in quantum mechanics could be available in order to investigate the counting statistics in the *classical* stochastic system. Note that the 'Hamiltonian' H in (8) is a non-Hermitian operator, which is different from the usual quantum mechanics. As a consequence, the phase factor $\mu(\chi)$ can take a complex value. Although one may think that the complex phase is not realistic, there is no problem about the complex phase in our case; we here consider the characteristic function for a probability, and suitable derivatives of the phase give real values.

Due to the similarity to the quantum mechanics, we can expect that a system with periodic 'Hamiltonian' is related to geometrical phases, such as Berry phase and Aharonov-Anandan phase, and such relationships between the pump current problem and the geometrical phases have been actually investigated [8,9,10,11]. In the following subsection, we give a brief explanation for the scheme of the geometrical phase and a new analytical result for cases with arbitrary phase difference α.

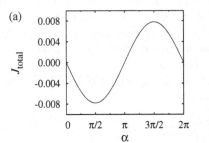
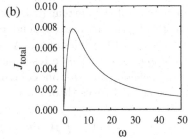

Fig. 4. Net current during one cycle period. We set $c_1 = c_2 = c_{-1} = c_{-2} = 1$ and $R = 0.5$. (a) α-dependence of the current. ω is set as 4.0. (b) ω-dependence of the current. α is set as $-\pi/2$.

3.2   Geometrical Phase Interpretation

According to discussions for the geometrical phase in quantum mechanics, it is possible to divide the (total) phase $\mu(\chi)$ in eq. (9) into two parts as follows [7]:

$$\mu(\chi) = \delta(\chi) + \gamma(\chi). \tag{10}$$

The phase $\delta(\chi)$ is called the dynamical phase, and $\gamma(\chi)$ is the Aharonov-Anandan phase. The dynamical phase is defined by [7]

$$\delta(\chi) = -\int_0^T \langle \widetilde{\phi}(t)|H|\phi(t)\rangle dt, \tag{11}$$

where $|\phi(t)\rangle$ is a state vector of a cyclic evolved state, and $\langle \widetilde{\phi}(t)|$ is related to a evolved state associated with the periodic adjoint Hamiltonian $H^\dagger$. The Aharonov-Anandan phase is calculated from [7,13]

$$\gamma(\chi) = \int_0^T \langle \widetilde{\phi}(t)|\frac{d}{dt}|\phi(t)\rangle dt. \tag{12}$$

It has been shown that the dynamical phase $\delta(\chi)$ gives the contribution from the simple discussion described in Sec 2.1 (equation (3)). The pump current, which is induced by the cyclic perturbation, is related to the phase $\gamma(\chi)$.

By using the Floquet theory, we can calculate the phases $\mu(\chi)$, $\delta(\chi)$, and $\gamma(\chi)$ perturbatively. While we consider the case with arbitrary phase difference α in the present paper, which is different from [10,11], the calculation scheme is similar to [10,11].

For simplicity, we here consider a case with $c_1 = c_2 = c_{-1} = c_{-2} = 1$. In this case, $\frac{\partial \delta(\chi)}{\partial \chi}|_{\chi=0} = 0$, and the net current stems only from the Aharonov-Anandan phase. In other words, we consider only the effect from the pump (cyclic perturbations). The net current in the case is as follows (up to second order in R):

$$J_{\text{total}} = -\frac{R^2}{4}\frac{\omega}{16+\omega^2}\sin(\alpha). \tag{13}$$

Figure 4 shows the analytical results from (13). We here set $R = 0.5$. Figure 4(a) is α-dependence of the current (ω is fixed at 4.0); the α dependence is given by the sin factor. Figure 4(b) is ω-dependence of the current. Here, we fix $\alpha = -\pi/2$, i.e., we set $k_{-2} = 1.0 + 0.5\sin(\omega t)$. As shown in Fig. 4(b), the ω-dependence of the current shows a stochastic resonance behavior.

Let us reconsider the Aharonov-Anandan phase of (12) from a different point of view. Using the Stokes theorem, we have

$$\gamma(\chi) = \oint_c A \cdot dk = \int\int_{S_c} dk_1 dk_{-2} F_{k_1, k_{-2}}, \qquad (14)$$

where $A_m = \langle \tilde{\phi}(k)|\partial_{k_m}|\phi(k)\rangle$, c the contour in the 2-dimensional parameter space (k_1-k_{-2} parameter space) drawn during one cycle, S_c the surface enclosed by the contour c, and

$$F_{k_1, k_{-2}} = \frac{\partial A_{-2}}{\partial k_1} - \frac{\partial A_1}{\partial k_{-2}}. \qquad (15)$$

Note that $\gamma(\chi)$ is given by the surface integral of $F_{k_1, k_{-2}}$. The above calculation shows that at least two cyclic perturbation are needed in order to make the pump current. If there is only one parameter change, a line is depicted in the parameter space and the area of the line is zero. In addition, if $\alpha = 0$ or π, a line is drawn on the parameter space, and hence there is no net current; this is consistent with (13). Actually, the sin factor is related to the area of an ellipse characterized by $x = \cos(\theta), y = \cos(\theta + \alpha)$ in polar coordinates.

4 Discussion: Pump Current as a Signal Transformation

In the previous section, it was revealed that the phase difference α is important to generate the pump current. Using understandings for the pump current, we can show that this mechanism for the pump current is considered as a "signal transformation".

We here explain Fig. 1 again. The whole system is separated by a membrane, and there is a container which transfers particles; one can see that Fig. 1 directly corresponds to Fig. 2. Figure 1(a) shows the detector of the phase difference of two inputs. According to the phase difference, the system outputs a unidirectional current. Although the correspondence between the phase difference and the current is not one-to-one (e.g., $\pi/8$ and $3\pi/8$ gives the same pump current), the stochastic system can roughly detect the phase difference. Hence, one could consider that this stochastic system performs a signal transformation from two cyclic inputs to one unidirectional current.

While two cyclic 'perturbations' are needed to make a pump current, it is possible to make a unidirectional current using only one cyclic 'input'. If a system has a compartment to move slowly, the system is considered to have two cyclic inputs effectively. For example, a protein folding or movement could be relatively slow compared with one cyclic input. Hence, the cyclic input may effect the system with time delay. If one of transition rates immediately responses the input, and the other transition rate responses it with a time delay, we have effectively two cyclic inputs. It enables the system to create the pump current, as explained above. The system in Fig. 1(b) is, therefore, considered as a converter of a cyclic input into a unidirectional current.

5   Concluding Remarks

We discussed how cyclic inputs are converted into a unidirectional current by using a simple stochastic system with particle hopping. Intuitively, the pumping phenomenon is understood as follows: because there is a restriction of the number of particles contained in the container, effects of the perturbations interfere each other and it causes the unidirectional current. It was shown that the geometrical phase interpretation is available to investigate the phenomenon, and it gives us an intuitive understanding for the condition to generate the pump current. Based on the understanding of the pumping phenomena, it is possible to construct signal transformation systems using such simple stochastic systems.

An information processing based on stochastic systems would exist widely in nature. In addition, such new mechanisms for information processing will be important in nanotechnology or biocomputing. It will be valuable to discuss fluctuating phenomena and topics in nonequilibrium physics from the viewpoint of natural computation.

References

1. Rao, C.V., Wolf, D.M., Arkin, A.P.: Nature (London) 420, 231 (2002)
2. Jülicher, F., Ajdari, A., Prost, J.: Rev. Mod. Phys. 69, 1269 (1997)
3. Reimann, P.: Phys. Rep. 361, 57 (2002)
4. Ritort, F.: Advances in Chemical Physics 137, 31 (2008)
5. Liu, D.S., Astumian, R.D., Tsong, T.Y.: J. Biol. Chem. 265, 7260 (1990)
6. Gopich, I., Szabo, A.: J. Chem. Phys. 122, 014707 (2005)
7. Borm, A., Mostafazadeh, A., Koizumi, H., Niu, Q., Zwanziger, J.: The Geometric Phase in Quantum Systems. Springer, Berlin (2003)
8. Sinitsyn, N.A., Nemenman, I.: Europhys. Lett. 77, 58001 (2007)
9. Sinitsyn, N.A., Nemenman, I.: Phys. Rev. Lett. 99, 220408 (2007)
10. Ohkubo, J.: J. Stat. Mech. P02011 (2008)
11. Ohkubo, J.: J. Chem. Phys. 129, 205102 (2008)
12. When there is a parameter mismatch in ω and/or R, we can also observe the pumping phenomenon. In: order to treat the model analytically, we here use the simple choice for the perturbations
13. Choutri, H., Maamache, M., Menouar, S.: J. Korean Phys. Soc. 40, 358 (2002)

Evaluation of Generation Alternation Models in Evolutionary Robotics

Masashi Oiso[1], Yoshiyuki Matsumura[2], Toshiyuki Yasuda[1], and Kazuhiro Ohkura[1]

[1] Graduate School of Engineering, Hiroshima University
1-4-1 Kagamiyama, Higashi-Hiroshima, Hiroshima, 739-8527, Japan
[2] Faculty of Textile Science, Shinshu University (Tokida Campus)
3-15-1, Tokida, Ueda City, Nagano, 386-8567, Japan
{oiso,yasuda,ohkura}@ohk.hiroshima-u.ac.jp,
matsumu@shinshu-u.ac.jp
http://www.ohk.hiroshima-u.ac.jp/en/index.html,
http://wimsl.shinshu-u.ac.jp/index.html

Abstract. For efficient implementation of Evolutionary Algorithms (EA) to a desktop grid computing environment, we propose a new generation alternation model called Grid-Oriented-Deletion (GOD) based on comparison with the conventional techniques. In previous research, generation alternation models are generally evaluated by using test functions. However, their exploration performance on the real problems such as Evolutionary Robotics (ER) has not been made very clear yet. Therefore we investigate the relationship between the exploration performance of EA on an ER problem and its generation alternation model. We applied four generation alternation models to the Evolutionary Multi-Robotics (EMR), which is the package-pushing problem to investigate their exploration performance. The results show that GOD is more effective than the other conventional models.

Keywords: Evolutionary Algorithms, Evolutionary Robotics, Generation Alternation Models, Neutral Networks.

1 Introduction

Evolutionary Robotics (ER)[1] is an effective method to design controllers of autonomous robots by Evolutionary Algorithms (EA). However, it takes a considerable amount of computational time because EA has to evaluate every candidate solution. To improve this situation, we focus on the parallel computation using many computers in a desktop grid computing environment[4]. A desktop grid environment potentially has a huge amount of computational power, but some problems may occur when applying traditional EA to a grid environment because of its heterogeneity, high parallelism, etc. For these reasons, we propose a new generation alternation model called Grid-Oriented-Deletion (GOD) for efficient implementation of EA to a desktop grid computing environment. GOD sends some individuals to each computational node at once so as to execute the sequential process composed of reproduction, evaluation, and selection

F. Peper et al. (Eds.): IWNC 2009, PICT 2, pp. 268–275, 2010.

independently in a generation on each node. This can suppress the overwork on the management node and avoid the problems caused by heterogeneity of the environment such as the synchronous waiting time.

In the previous research, generation alternation models are generally evaluated by using test functions. However, their exploration performance on the problem on real problems have been not made very clear yet. Moreover, in terms of generation alternation models, only few researches compare the performances of different models. Thus, we applied generation alternation models to an ER problem to investigate their exploration performance for ER.

In Section 2, four generation alternation models which are adopted in this paper are explained. In Section 3, experimental settings and results are shown. Finally, the conclusions are described in Section 4.

2   Generation Alternation Models

There are two important manipulation to design in EA, i.e., reproduction and selection. Designing appropriate reproduction processes can increase the exploration performance at neighboring points of each individual. However, reproduction processes normally do not affect for diversity of the population. By contrast, selection processes can adjust the diversity by adopting appropriate methods and parameters. Thus, we focus on generation alternation models which define methods of selection processes. In this paper, we aim to improve the exploration performance for ER problems applying a proper generation alternation model.

For designing generation alternation models, there are two kinds of selection method to define. One is the selection for reproduction; this selection is to choose parents to reproduce offspring. Another is the selection for survival; this selection is to choose individuals which survive to the next generation from the population. These two selection can improve the efficiency of global exploration.

In the beginning of EA, Simple GA, (μ, λ)-ES, etc. were proposed as generation alternation models. On these models, there are some problems such as premature convergence or stagnation in the genotype space. Today, many generation alternation models are proposed to overcome those problems. In the following, we introduce four generation alternation models adopted in this paper.

(μ, λ)[5]. This model is a classic discrete generation model of Evolution Strategies (ES).

- Selection for reproduction
  Use all μ individuals as parents.
- Selection for survival
  Select μ individuals from λ offspring.

Steady-State (SS)[6]. This model is a classic continuous generation model. Since this model has high selective pressure[7], premature convergence is sometimes occurred especially in a multimodal landscape.

- Selection for reproduction
  Select two individuals as parents from the population by Ranking Selection.

- Selection for survival
Remove two worst individuals from the population.
Then add two offspring to the population.

Minimal Generation Gap (MGG)[7]. This model is designed to minimize the distribution of the population between each generation. This model is compared its exploration performance with five traditional models by using some test functions and traveling salesman problems in [7]. The result shows that MGG can keep large diversity of the population to show better performance for those problems especially with a small population size.

- Selection for reproduction
Remove two individuals as parents from the population at random.

- Selection for survival
Select the best individual from the family (both parents and offspring).
Select an individual by roulette selection. Return them to the population.

Grid-Oriented-Deletion (GOD). This model is designed to apply EA to computational grid in order to minimize the execution time. In the following, the procedure of a generation of GOD is explained.

GOD has a parameter dr $(0 \leq dr \leq 1)$ which can adjust the selective pressure. GOD first decides one or two parents probabilistically according to this dr. Then GOD removes the individuals of the number as parents from the population, which are send to a computational node. In the computational node, the sequential process of reproduction, evaluation, and selection in a generation is executed. In the reproduction, each parent yields an offspring. And in the selection, best individuals of the same number of parents are selected from the family (both parents and offspring). Finally, the individuals of the same number of parents are returned to the population.

When the dr is set to a very small value, one parent is frequently selected (when dr is equal to zero, one parent is always selected). And dr is getting larger, the probability that two parents are selected is also getting higher (when dr is equal to one, two parents are always selected).

In the case that an individual is removed, the selection process is executed between a parent and its offspring. In this case, the characteristics of the parent is always kept in the population whether the parent or the offspring survives. By contrast, in the case that two individuals are removed, the selection is executed among two parents and two offspring. If the parent and its offspring have higher fitness than another parent and its offspring, the characteristics of the parent of lower fitness are lost from the population. Therefore the selective pressure in GOD is lower with a smaller dr, and higher with a larger dr.

As for preliminary experiments, we conducted exploratory experiments to investigate the effect of GOD on the execution time in the desktop grid environment. The results showed that GOD is effective in the grid environment compared with the other conventional models.

- Selection for reproduction
  Decide the number (one or two) of parents probabilistically according to the parameter dr.
  Remove one or two individuals as parents from the population at random.

- Selection for survival
  Select best individuals of the same number of individuals removed from the population from the family(both parents and offspring).
  Return them to the population.

3   Evaluation of Generation Alternation Models by Experiment

3.1   Experimental Settings

In this paper, we adopt the Evolutionary Multi-Robotics (EMR) package-pushing problem (Fig.1). There are nine autonomous robots and three kinds of packages on a square field. Each package should be pushed by robots of the number equal to the number written on the package or more. The task is to transport all packages to the goal line. A trial of the problem is finished when either the task is accomplished or 3,000 time steps are passed.

All robots have the same three-layered neural network as a controller. The neural network has sixteen input nodes, four hidden nodes, and two output nodes. All nodes are connected to all hidden nodes and output nodes (including themselves). Thus, there are 132 connections, and each weight is evolved by EA.

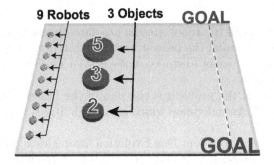

Fig. 1. EMR package-pushing problem environment (9 robots)

Table 1. Fitness table of EMR problem

State	Fitness
(a) A package reaches to the goal	+1000
(b) A package moves to right	+(X coordinate value)
(c) A robot touches a package	+0.05 per each step
(d) The number of time steps when a trial is finished	- time steps

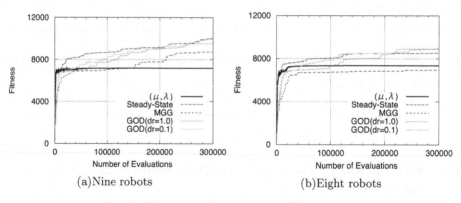

Fig. 2. Fitness transitions of the EMR problem

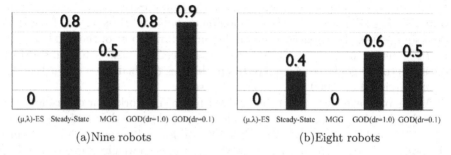

Fig. 3. Success rates of the task of each model

The input signals of the robot are the absolute direction($\sin\theta_1$, $\cos\theta_1$), eight IR sensors, the position of the nearest package (distance, $\sin\theta_2$, $\cos\theta_2$), and the position of the nearest robot (distance, $\sin\theta_3$, $\cos\theta_3$). The outputs are the rotation angles of two wheels.

A fitness value of the problem is the sum of the fitness values obtained according to Table.1. An individual which has a high fitness value means that the individual is good.

In this experiment, we adopt Fast Evolution Strategies [8] method for all generation alternation models to exclude the result from the influence on difference of reproduction process. The number of evaluation in a trial is 300,000.

In addition to this settings, we conducted another experiment using eight robots to make a difficult task to investigate the robustness of generation alternation models. These experiments were conducted ten times for each generation alternation models.

3.2 Experimental Results

Fig.2 shows the fitness transition of each model. Finally, GOD obtained higher fitness value than the other models. Fig.3 shows the success rates at the final

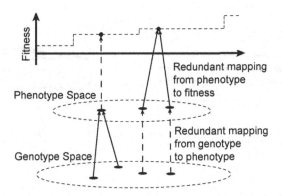

Fig. 4. Redundancy in genotype to phenotype or phenotype to fitness

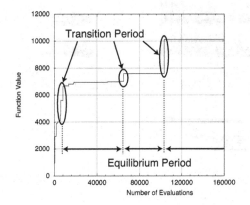

Fig. 5. Typical evolutionary dynamics on a problem with neutrality

generation. These results indicate that GOD also has higher success rate than the other models. GOD showed the good performance not depending on the parameter value of dr in this problem.

3.3   Discussion

Generally, in the design problem of a neural controller of ER, the mapping from genotype to phenotype and the mapping from phenotype to fitness are redundant (Fig. 4). This implies that the fitness landscape contains neutrality which is considered to be represented by neutral networks[2]. Evolutionary process exploring the problem is generally categorized into two period: transition period and equilibrium period (Fig.5). In the equilibrium period, the best fitness value is not improved although the population is moving around in the genotype space. In the transition period, the best fitness value improves rapidly. During the equilibrium period, the population is considered to form clusters in the genotype space and explore the neutral layer until it finds out the portal which guides the

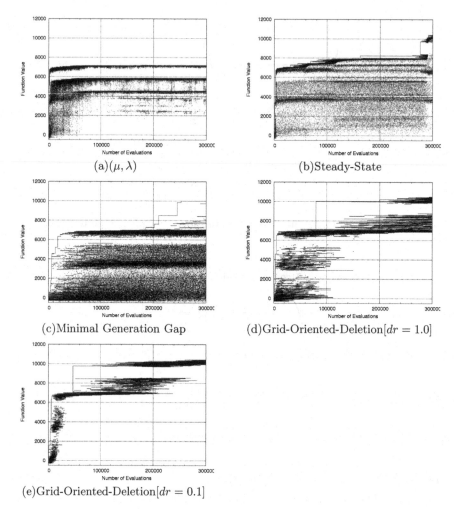

Fig. 6. Distribution of the population of a trial (nine robots)

population to an upper neutral layer. Then, the population transits to an upper layer in the transition period. Our research group has been analyzing evolutionary dynamics of problems with neutrality[3]. The results indicated that the more the population converges to a neutral layer, the more they can explore the layer efficiently. Fig.6 shows the distribution of the population in a trial with nine robots. Each point indicates an individual and its fitness value.

The individuals of GOD, which have the highest success rates of the task, converge on the best fitness value of the generation. This means that most of them are exploring the neutral layer where the best individual is on. As a result, GOD could explore most efficiently on the neutral layer while the equilibrium period compared with other three models. For this reason, GOD has the slightly higher success rate than other three models. We conjectured that this is because GOD adopts the elitist strategy and elitist preservation.

On the contrary, (μ, λ) do not adopt the elitist preservation. Therefore, the population did not converge to the highest neutral layer and the exploration of the neutral layer was insufficient.

SS and MGG also adopt the elitist preservation, however, they allow worse individuals to survive. In the case of problems with neutrality, the exploration around worse individuals can be wasteful in most cases. We consider that is the reason why the exploration performance are worse than GOD. In particular, it is clearly shown in MGG which aims to keep the high divergence in the population. The distribution of the individuals of MGG is hardly changed despite the best fitness value increases. Consequently, MGG has the lower success rate than the others.

4 Conclusions

In this paper, we proposed a new generation alternation model called GOD for parallel computation in the desktop grid environment in order to minimize the computational time for large scale problems. To investigate the exploration performance of GOD for ER problems, we adopted four generation alternation models to the experiment and the EMR package pushing problem. The results show that GOD outperforms the other three models not only on the fitness value but also the success rates. We will investigate to improve GOD to be more effective for other problems. Moreover, We found the relationship between the exploring performance of the EMR problem and the diversity of the population in the phenotype space. However, the relationship in the genotype space is not clear yet. In the future, we will investigate the distribution of the individuals in the genotype space.

References

1. Harvey, I., Husbands, P., Clif, D., Thompson, A., Jakobi, N.: Evolutionary Robotics: the Sussex Approach, Robotics and Autonomous Systems (1997)
2. Harvey, I., Thompson, A.: Through the Labyrinth Evolution Finds a Way: A Silicon Ridge. In: Proceedings of the First International Conference on Evolvable Systems: From Biology to Hardware, pp. 406–422. Springer, Heidelberg (1996)
3. Katada, Y., Ohkura, K., Ueda, K.: Measuring Neutrality of Fitness Landscapes Based on the Nei's Standard Genetic Distance. In: Proceedings of 2003 Asia Pacific Symposium on Intelligent and Evolutionary Systems: Technology and Applications, pp. 107–114 (2003)
4. Foster, I., Kesselman, C.: The grid: blueprint for a new computing infrastructure. Morgan Kaufmann Publishers Inc., San Francisco (1998)
5. Schwefel, H.-P.: Evolution and Optimum Seeking. John Wiley & Sons, Chichester (1995)
6. Syswerda, G.: Uniform Crossover in Genetic Algorithms. In: Proceedings of the third international conference on Genetic algorithms, pp. 2–9 (1989)
7. Sato, H., Yamamura, M., Kobayashi, S.: Minimal generation gap model for GAs considering both exploration and exploitation. In: Proceedings of 4th International Conference on Soft Computing, pp. 494–497 (1996)
8. Yao, X., Liu, Y.: Fast Evolution Strategies. Control and Cybernetics 26(3), 467–496 (1997)

Photonic Switching of DNA's Position
That Represents the Internal State
in Photonic DNA Automaton

Hiroto Sakai, Yusuke Ogura, and Jun Tanida

Department of Information and Physical Sciences,
Graduate School of Information Science and Technology,
Osaka University
1-5 Yamadaoka, Suita, Osaka 565-0871, Japan
{h-sakai,ogura,tanida}@ist.osaka-u.ac.jp

Abstract. Optical computing promises spatially parallel information processing, but the diffraction phenomenon prevents us from performing computations at a nanometer scale. To overcome the problem, the idea of DNA computing is useful because it offers methods for manipulating information at that scale. We have been studying on photonic DNA automaton, which is an implementation of automaton using light and DNA as information carriers. The internal state of the automaton is represented by the position of a specific DNA in a DNA nano-structure, and the state transition is executed by changing the position in the structure in accordance with photonic signals. In this paper, we describe a photonic switch of the position of a DNA strand in a DNA nano-structure by photonic control of other DNAs. The position is controlled through conformation-change of a pair of hairpin-DNAs that are responsive to visible and UV light irradiation. Experimental results show that the position of the DNA switches to the two position in response to visible and UV light. The method is applicable to state-transition of the photonic DNA automaton.

1   Introduction

DNA computing is a computational paradigm which offers autonomous information processing at a nanometer scale based on complementarity of DNA[1]. DNA automaton is an implementation example of computation using DNA. Since automaton is a fundamental model of computation, it is useful in various systems for information processing. A complicated processing is possible as a sequence of state-transitions depending on input and the current internal-state. Y. Benenson *et al.* demonstrated programmed control of release of a single-stranded DNA that works as a drug using DNA automaton[2].

In DNA computing, information is encoded into the base sequence, length, conformation, or other attributes of DNA. The longer the sequences used are, the more information can be encoded. However, use of long sequences force us a hard work for proper design and precise control of DNA reactions. Using another

F. Peper et al. (Eds.): IWNC 2009, PICT 2, pp. 276–289, 2010.

attribute of DNA is a practical strategy to deal with much information, and positional information such as the inter- and intra-molecular distance is promising in order to increase dimensional space into which information is encoded.

Light is a powerful information carrier because information that is encoded into optical signals is carried in parallel, locally and remotely. Diverse distributions of light can be generated easily using a variety of direct or external modulating methods. From the viewpoint of parallel computing, optical computing and DNA computing have several common features. For example, T. Head proposed photo-computing for solving 3-SAT using transparency of light. The problem is solved using an algorithm that is similar to the aqueous algorithm developed for DNA computing[3]. Optical control of DNA reactions offers usage of positional information of DNA at a micrometer scale. However, it is difficult to control the individual DNAs at a nanometer scale using propagating light because the spot size of light is limited to a micrometer scale owing to the diffraction limit of light.

At a nanometer scale, the position control of molecules is applicable to, for example, regulation of inter- or intramolecular interactions. DNA nano-structures are useful in controlling the position of DNA with nanometer order of precision. Various kinds of structures made of DNA have been demonstrated[4]. For example, S. Rinker *et al.* demonstrated the control of protein activities depending on position of two kinds of DNA aptamers on a DNA nano-structure[5].

We are investigating photonic DNA automaton to achieve information processing at a nanometer scale by using light and DNA[6,7]. We have proposed positional state representation and its transition for photonic DNA automaton[8]. In this scheme, the internal state is encoded into the position of a specific DNA, which can move in response to photonic signal, in a DNA structure as a working-basis. The state transition is implemented by moving the position through change of the conformation of a part of the working-basis. The next state, which is obtained as a result of conformation change, is determined according to optical signals as input and the positional information of DNA as the current state. Because the position is changed through control of annealing and denaturing of DNA strands in the order of nucleotides, nanoscale positioning of the DNA can be achieved. The method will be usable to control interactions of molecules depending on a nanoscale order of position.

We have designed four reaction schemes where conformation of a hairpin-DNA changes in response to visible light irradiation and ultraviolet (UV) light irradiation[9]. We also demonstrated controlling the position of a DNA strand in a DNA nano-structure using conformation-change of a pair of hairpin-DNAs which is induced by only DNAs[8]. These methods are important for realizing the photonic DNA automaton.

In this paper, we propose photonic switching of the position of a DNA strand using two hairpin-DNAs that are responsive to light irradiation as combination of the above mentioned methods. The method is applicable to implementation of photonic DNA automaton. The conformations of two kinds of hairpin-DNAs are changed inversely in response to light irradiation to achieve position control of the DNA. DNA strands used were tethered with azobenzene because the binding strength of an azobenzene-tethered strand and its complementary strand are controlled by light irradiation. Some experiments were performed to confirm the positional change based on the method. Section 2 explains the concept of photonic DNA automaton. In Section 3, we describe a design of reaction schemes of hairpin-DNAs and a method for photonic switching of position using the hairpin-DNAs. In Section 4, we show experimental results. In Section 5, we give conclusions.

2   Photonic DNA Automaton

Figure 1 shows the concept of photonic DNA automaton, which is a physical realization of automaton using photonics and DNA[9]. The internal state is encoded into DNA, and its state transition is performed by sending optical signals to the DNA as input. Light is a superior information carrier because the spatial and temporal distributions of light can be generated easily and they enable us to deal with much information in parallel. The features of light, including high-speed propagation, remote accessibility and flexible controllability, are also possible to be utilized. There are many automata that implement various state transition diagrams within a single light spot. The individual automata receiving the optical signal work simultaneously. The rules of state-transition are related to DNA reactions associated with the individual photonic DNA automata.

Although the feature size of the distribution of the optical signal is limited to a micrometer scale owing to diffraction limit of light, automata work at a nanoscale size by using DNA. As well as traditional DNA computing, the computational capability of DNA based on Watson-Crick's complementarity is used to implement the automaton in a nanometer scale. After broadcasting signal to automata through light, the automata work in parallel and independently according to a state-transition diagram associated to the individuals even if same input signals are transmitted to the automata in a light spot. This configuration provides nanoscale computing by use of propagating light.

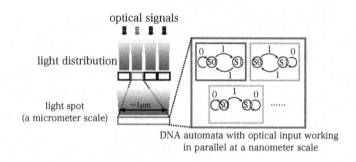

Fig. 1. The concept of photonic DNA automaton

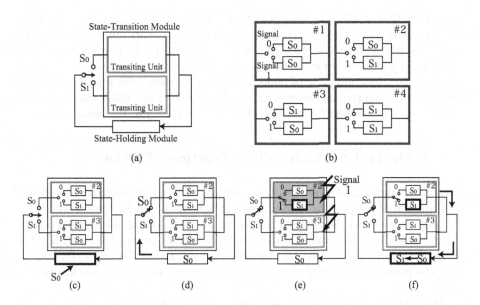

Fig. 2. The implementing method for photonic DNA automaton. (a) The overview of the system, (b) The four schemes for representing transiting rules of an automaton, (c) Setting of the initial state to the SHM, (d) Selecting of transiting unit depending on the initial state, (e) Transiting of the current state to next state in response to a signal, (f) Renewing of the current state to next state.

Figure 2 (a) shows a schematic illustration of the composition of a single photonic DNA automaton. The system consists of two modules: the state-transition module (STM) and the state-holding module (SHM). The STM receives optical signals and executes state-transitions in response to the signals. The SHM temporarily stores a current state, which is renewed after receiving information on the next state from the STM. These modules are actuated alternately. Let us consider an automaton with two internal-states, S_0 and S_1, and two input symbols, 0 and 1. The STM is constructed with two transiting units that correspond to each of the internal states, S_0 and S_1. The transiting units are used to determine the next state by receiving a photonic signal. One of the four schemes, which are used for four possible state-transiting patterns, is assigned to each of the transiting units. The four transition rules considered to implement two-state and two-symbol automaton are shown in Fig. 2 (b). The next state determined using a transiting unit depends on only an input symbol.

Figures 2 (c) through (f) show a procedure of operating an automaton. At first, an initial-state is loaded to the SHM (Fig. 2 (c)). The information on the current state is transferred from the SHM to the STM by selecting one of the transiting units depending on the state (Fig. 2 (d)). Only the selected transiting unit works in the following process. The state is transited to the next one by following the input signal (Fig. 2 (e)). The information on the next state is transferred from the STM to the SHM. In the SHM, the state is renewed to the next state (Fig. 2 (f)). The behavior of the photonic DNA automaton is regulated by repeating these operations. The system is possible to implement arbitrary state-transition diagrams with two states and two kind of input symbols by selecting the scheme used for each of the transiting units among four schemes. Among one transition cycle of photonic DNA automaton, the position of a DNA strand, which represents the internal state, is changed in a DNA nano-structure. The detail of a method for position control is described in the next section.

3   A Method for Controlling Position of DNA

The internal state of photonic DNA automaton is represented by positional information of DNA. This encoding method should be useful to deal with inter-actions of molecules depending on their positions or spatial information relating to molecules in a nanometer scale through computation using the automaton. State transition is executed by changing the position of the DNA. By control-ling the position using light, the state transition responding to optical signals is achieved.

We propose a method for controlling position of a DNA strand using the conformation of hairpin-DNAs that are responsive to light irradiation. Hairpin-DNAs take two different stable conformations: an open-state and a closed-state. Figure 3 shows switching between two states by using two kinds of DNA strands. Let us assume that the closed-state hairpin-DNA consists of a sticky-end (a), a stem (b and b'), and a loop (c). Here characters indicate a domain on a DNA sequence and the prime symbol indicates "complementary." In a solution of hairpin-DNA, H, and a linear strand L_1 consisting of b' , a' and d, the domain a of the hairpin-DNA binds with the domain a' of L_1, then the hairpin-DNA becomes the open-state as the result of branch-migration. If strand L_1 includes the sequence which is complement to another strand L_2, these strands can anneal together to form a duplex. As a result, L_1 is removed from the open-state hairpin-DNA, and it returns to the closed-state. In this paper, strands L_1 and L_2 are called opener DNA and closer DNA, respectively.

To control conformation of hairpin-DNAs using light, we use opener DNA ($O^{(azo)}$) and closer DNA ($C^{(azo)}$) tethered with azobenzene. Azobenzene is a

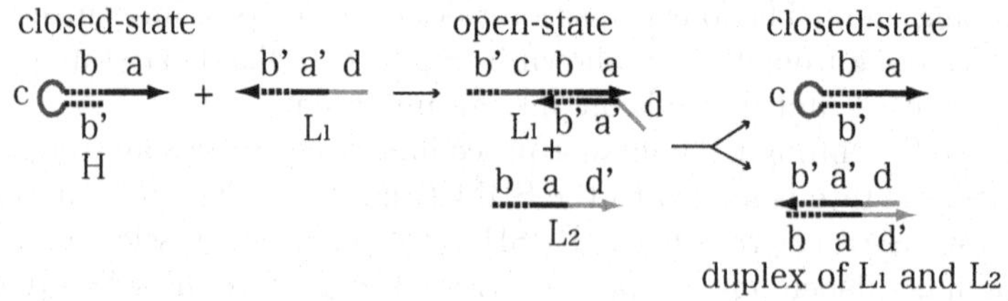

Fig. 3. Switching of the conformation of a hairpin-DNA using two kinds of DNA strands

photoisomerization molecule that switches to the trans-form by visible light irradiation and to the cis-form by UV light irradiation. A duplex consisting of DNA tethered with the cis-form azobenzene is less stable than that with the trans-form azobenzene, so that annealing between azobenzene-tethered DNA and its complementary DNA can be controlled using light[10]. We designed two reaction schemes ♯ 1 (Fig. 4(a)) and ♯ 2 (Fig. 4(b)) where conformation of a hairpin-DNA changes in response to visible light irradiation and UV light irradiation. The operation can be repeated; however, the operation efficiency decreases as repeating because the efficiency of photoisomerization of azobenzene decreases. The maximum number of operations is also limited owing to the decrease in photoisomerization efficiency. In the previous paper, 60% of a hairpin-DNA tethered with azobenzene switched to the open-state by visible light irradiation, and 70% of that switched to the closed state by UV light irradiation[9]. Introducing multiple azobenzene molecules into individual strands is expected to be effective to maintain high operation efficiency. The absorption peak wavelengths of azobenzene are around 350 nm and 450 nm, which DNA absorbs little, and therefore the light for switching the azobenzene makes no influence to maintain and functionality of the DNA structure.

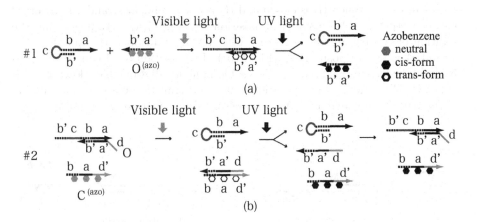

Fig. 4. Switching of the conformation of a hairpin-DNA using light. (a) Reaction scheme ♯ 1. The hairpin-DNA changes to open state using visible light and to closed state using UV light, (b) Reaction scheme ♯ 2. The hairpin-DNA changes to open state using UV light and to closed state using visible light.

In reaction scheme ♯ 1, $O^{(azo)}$ is utilized. The closed-state hairpin-DNA and $O^{(azo)}$ are mixed and the conformation of the hairpin-DNA is controlled by light irradiation. The azobenzene of $O^{(azo)}$ is switched to the trans-form by visible light irradiation, then the hairpin-DNA changes to the open state because $O^{(azo)}$ can hybridize to the hairpin-DNA. When the open-state hairpin-DNA is irradiated with UV light, $O^{(azo)}$ changes to the cis-form and moves away from the open-state hairpin-DNA; then the hairpin-DNA returns to the closed-state.

In reaction scheme ♯ 2, $C^{(azo)}$ and opener DNA (O) without azobenzene are utilized. Let us consider that $C^{(azo)}$ is added to a solution containing open-state hairpin-DNA annealing with O. When the solution is irradiated with visible light to switch the azobenzen of the $C^{(azo)}$ to the trans-form, $C^{(azo)}$ tethered with the trans-form azobenzen and O can anneal each other and become a duplex; then the open-state hairpin-DNA changes to the closed-state because O is removed from it. The duplex consisting of O and $C^{(azo)}$ is separated by UV light irradiation, then O rebinds to the closed-state hairpin-DNA, and it changes to the open-state. These reaction schemes enable us to switch the conformation of the hairpin-DNA using light. The conformation of the hairpin-DNA generated in reaction scheme ♯ 1 and that in ♯ 2 responsive to visible and UV light is inverse. The inverse relationship is useful to control position of DNA.

Figure 5 shows a method for controlling the position of a DNA strand in a DNA nano-structure as a working-basis using light. A solid line indicates a single-stranded DNA. The working-basis is a rectangular structure made of a number of single-stranded DNAs. A pair of hairpin-DNAs, one of which follows reaction scheme ♯ 1 and the other follows ♯ 2, are incorporated into the rectangular structure. Hairpin-DNAs H_1 and H_2 are used for reaction schemes ♯ 1 and ♯ 2, respectively. Single-stranded DNAs, O_1, O_2 and C_1 behave as O, $O^{(azo)}$ and $C^{(azo)}$ in Fig. 4. When H_1 is closed and H_2 is open, a linear strand, M, arranged between H_1 and H_2 is located at position P_1 (Fig. 5(a)). By irradiating the solution of the rectangular structure with UV light to open H_1 and close H_2, strand M moves to position P_0 (Fig. 5(b)). The position of M can be returned back to the initial position, P_1, by visible light irradiation that closes H_1 and opens H_2 (Fig. 5(a)). This operation is used for transforming the conformation of the pair of the hairpin-DNAs into positional information of strand M.

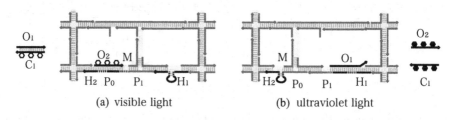

O_1
C_1
O_2 M
H_2 P_0 P_1 H_1
(a) visible light

M O_1
H_2 P_0 P_1 H_1
O_2
C_1
(b) ultraviolet light

Fig. 5. The scheme of converting conformation of a pair of hairpin-DNAs to the position of the DNA. (a) Under visible light, H_1 is closed and H_2 is open; then M is located at P_1, (b) Under UV light, H_1 is open and H_2 is closed; then M is located at P_0.

We can realize an automaton which records the last input signal by using the method with representing the internal state of the automaton as the position. Figure 6 shows the state-transition diagram of the automaton and the configuration used in this study. This automaton can receive input symbols, 0 and 1, which correspond to visible light and UV light (Fig 6(a)) and switches between two internal state, S_0 and S_1, which are represented by the position of M in P_0 or P_1. The automaton emulates the state-transition diagram shown in Fig 6(b).

Optical signal	Visible	Ultraviolet(UV)
Input symbol	0	1
Hairpin-DNA	Conformation of a hairpin-DNA	
H_1	closed	open
H_2	open	closed
Position	P_1	P_0
Internal state	S_1	S_0

(a)

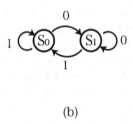

(b)

Fig. 6. (a) The relationship between input symbol/internal state and light/DNA. (b) The state-transition diagram implemented by the method shown in Fig. 4.

The number of internal states in the automaton can be increased by using more hairpin-DNAs on the working basis.

4   Experiments

The first experiment was aimed at confirming photonic switching of the position of a DNA strand through conformation-change of hairpin-DNAs. We used two kinds of hairpin-DNAs, which were incorporated into a DNA nano-structure with a rectangular shape, for reaction schemes ♯ 1 and ♯ 2. 4-arm DNA junctions[11] and 3-arm DNA junctions[12] were used to make the rectangular structure as shown in Fig. 7. The rectangular structure consists of 4-arm junctions, X1, X2, X3 and X4, a 3-arm junction Y, and a duplex T with two sticky-ends. In the structures of X1, X2, X3 and X4, the duplex part of the junction shape has the same sequence. A list of DNA sequences used in the experiment is shown in Appendix. The sequences of the DNA strands were designed using a software called NUPACK[13].

At first, the rectangular structure was made according to the following procedure. X1, X2, X3, X4, and T consist of four linear DNA strands, and Y consists of four linear strands and two hairpin-DNAs. Solutions of {X1, T}, {X2}, {X3, Y}, and {X4} were prepared by mixing necessary strands into individual tubes and they were incubated at 95 degrees C for 5 minutes, at 65 degrees C for 20 minutes, at 50 degrees C for 15 minutes, at 37 degrees C for 20 minutes, at 25 degrees C for 20 minutes, then at 4 degrees C for 20 minutes in a TAE/Mg buffer. Brackets { } indicate the composition of a solution. The concentration of each component was 10 μM. Solutions {X1, T} and {X2} were mixed in a tube and {X3, Y} and {X4} were mixed in another tube. These solutions were incubated at 4 degrees C for an hour, then these were mixed into a single tube and the tube was incubated at 4 degrees C for an hour. We confirmed the fabrication of the structure by electrophoresis[8].

The lower edge of the structure changes the position of strand M (indicated by "m" in Fig. 7), which is a part of Y, through conformations on the edge.

The upper edge was utilized to detect the position of strand M by fluorescence detection. Strand u2 on the upper edge was labeled with a fluorophore FAM and strand M was labeled with a quencher BHQ-1. FAM and BHQ-1 are a pair of fluorescence resonance energy transfer (FRET), which is energy transfer between two fluorophores, a donor molecule and an acceptor molecule. When the distance between two fluorophores is less than 10 nm, the donor in an excited state transfers energy to the acceptor without radiation, and the fluorescence from the donor decreases. The distance of the movement of strand M is estimated as 21 nm, which is the distance between strands u2 and u3. When strand M is at the position of u3, FRET between FAM and BHQ-1 doesn't occur and fluorescence of FAM is detectable. By measuring the fluorescence intensity, the position of strand M can be detected.

For controlling two hairpin-DNAs H_1 and H_2 using light, we used two kinds of strands tethered with three azobenzene molecules (C_1 and O_2) and an azobenzene-free strand (O_1). The individual sequences of three strands are 21-mer, and these duplexes are stable even if azobenzene molecules are in the cis-form. To make it possible to control the binding of these duplexes owing to the isomerization, these strands include a few base mismatches. Figures 8(a) and (b) show the assumed structures under visible light irradiation and UV light irradiation. When M is at P_1, high fluorescence intensity is obtained because FAM is away from BHQ-1 (Fig. 8(a)). In contrast, when the position of M is P_0, fluorescence of FAM is low because it is quenched by BHQ-1 (Fig. 8(b)).

To fabricate DNA structures whose initial position of strand M are P_0 and P_1, we used not only Y but also Y', which includes O_1 instead of O_2. Solutions of {X1, T, X2, X4, Y and X3} and {X1, T, X2, X4, Y' and X3} were incubated individually at 4 degrees C for an hour. The solution {X1, T, X2, X4, Y and X3} was used as the reference for P_1 before light irradiation (solution (i-1)), and the solution {X1, T, X2, X4, Y' and X3} was used as that for P_0 (solution (ii-1)). C_1 irradiated with visible light (450 nm) and O_1 were added to solution (i-1), then

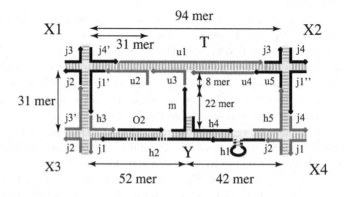

Fig. 7. The design of a DNA nano-structure using 4-arm DNA junctions and a 3-arm DNA junction

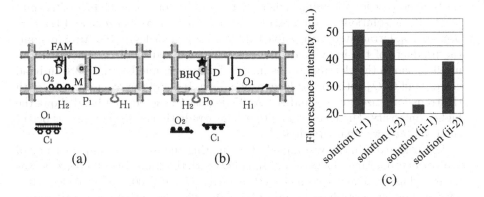

Fig. 8. Overview and result of the experiment on control of the position of M using light. The detection of the position using FRET when M is at (a) P_0 and (b) P_1. (c) The fluorescence intensity measured after adding strand D.

the solution was irradiated with UV light for 5 minutes (solution (i-2)). C_1 and O_2 that were irradiated with UV light (340 nm) were added to solution (ii-1), then the solution was irradiated with visible light for 5 minutes (solution (ii-2)). For detecting a FRET signal, strand D, which was complement to the sequence of the sticky-end of u2 and u3, was added to the individual solutions. This strand was used to fix the position of strand M in the upper edge. The strand binds with one of u2 and u3 depending on the position of strand M. At an hour after adding D, FRET signals of the solutions (i-1) to (ii-2) were measured.

Figure 8(c) shows the measured fluorescent intensity obtained from an examination for photonic switching of the position of strand M from the left to the right, and the reverse. The difference of the fluorescent intensity reflects the efficiency of FRET depending on the distance between FAM and BHQ-1. The intensity of solution (i-1) is a little higher than that of solution (i-2) and the intensity of solution (ii-1) is lower than that of solution (ii-2). This result suggests that the position of M moves to the intended direction.

The difference of intensities between solutions (i-1) and (i-2) is small in particular. The possible reasons include low efficiency of conformation-change of hairpin-DNAs owing to low operating temperature and low fabricating efficiency of rectangular structure. During fabrication of the rectangular structure and photonic switching of the conformation of hairpin-DNAs, the temperature was kept at 4 degrees C because it is easy to maintain the structure of DNA at low temperature. The main reason for low changing efficiency is that the operation was performed at low temperature. At higher temperature, the efficiency of photonic conformation-change can be improved. On the other hand, it is difficult to maintain the structure at high temperature, and redesign of the structure, for example lengthening the sequence of the duplex involved in the structure, is required to improve.

To confirm the effectiveness of the operation at higher temperature, we performed experiments on the reaction between O_2 and H_2. At first, we investigated the temperature for obtaining higher efficiency of conformation-change of hairpin-DNA. Three solutions containing O_2 with the cis-form azobenzene and H_2 were prepared and irradiated with visible light at 25, 30, or 35 degrees C. After an hour, the products of the individual solutions were analyzed by electrophoresis. Figure 9(a) shows the electrophoresis image. The upper, middle, and bottom band correspond to O_2 with the cis-form azobenzene, H_2, and the reactant of O_2 with the trans-form azobenzene and H_2. At 35 degrees C, the bottom band shows the highest intensity. The result shows that light irradiation at 35 degrees C is better than that at lower temperatures.

The next experiment was aimed at estimating the isomerization efficiency of azobenzene. The absorbance at 320 nm of O_2 with the trans-form azobenzene was measured using an absorptiometer (HitachiHigh-Tech; U-0080D) at 35 degrees C with changing the irradiation time of UV light. The absorption peak of the trans-form azobenzene is around 320 nm, and the absorbance of the trans-form at the wavelength is higher than that of the cis-form. Figure 9(b) shows the relationship between the measured absorbance of O_2 at 320 nm and the irradiation time of UV light. The relationship shows that 20 minutes irradiation is better than 5 minutes irradiation.

Considering on the above results, we confirmed changing H_2 by using O_2 irradiated with visible or UV light for 20 minutes at 35 degrees C. A solution of O_2 and H_2 were divided into four tubes, and they were irradiated with (i) visible light, (ii) visible light then UV, (iii) UV, and (iv) UV then visible. After the individual tubes were incubated at 35 degrees C for an hour, they were electrophoresed. Figure 9(c) shows the electrophoresis image. We confirmed the conformation of H_2 was changed depending on light irradiation. Using the improved condition of light irradiation, conformation-change of hairpin-DNAs will be improved and we will be able to obtain higher efficiency of position-change of strand M.

The necessary time for every transition of an automaton depends on the reaction time of a hairpin-DNA, which is about an hour at the present. The optimization of conditions for the reaction will enable us to shorten the transition time.

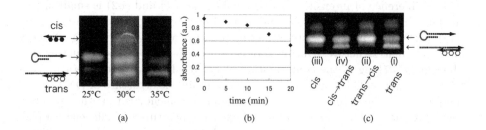

Fig. 9. The experimental results on the reaction of O_2 and H_2. (a) The electrophoresis image of the product in changing the temperature. (b) The absorbance of O_2 at 320 nm with changing the irradiation time of UV light. (c) The electrophoresis image of the product obtained with the improved condition.

Parallel driving of the automata is effective to increase the processing power, and it expected to be applied to control a molecular system.

5 Conclusions

Photonic DNA automaton is a new computing scheme for information processing at nanoscale. We studied a method for photonic switching of the position of DNA that represents the internal state of automaton. In this method, the position of a DNA strand is reversibly controlled by light. The method is realized by combining two of our previous achievements: conformation-change of hairpin-DNAs that are responsive to visible and UV light irradiation and position change of a DNA strand in a DNA structure induced by conformation-change of a pair of hairpin-DNAs. Experimental results showed that the position was able to be changed in response to the light. In the demonstrated automaton, the final state was determined by the optical signal input at the last, and the scheme itself emulates an automaton which memorizes the last input for no less than an hour. The memorized information can be renewed. The efficiency of position-change was not enough, but additional experiments showed that the efficiency was expected to increase by using improved reaction condition. Future issues include application of the method to the photonic DNA automaton capable of implementing arbitrary transition diagrams.

Acknowledgments

This work was supported by the Ministry of Education, Culture, Sports, Science and Technology, Japan, a Grant-in-Aid for Scientific Research (A), 18200022, 2006-2008 and for Young Scientist (B) 20700276, 2008-2009.

References

1. Benenson, Y., Paz-Elizur, T., Adar, R., Keinan, E., Livneh, Z., Shapiro, E.: Programmable and autonomous computing machine made of biomolecules. Nature 414, 430–434 (2001)
2. Benenson, Y., Gil, B., Ben-Dor, U., Adar, R., Shapiro, E.: An autonomous molecular computer for logical control of gene expression. Nature 429, 1–6 (2004)
3. Head, T.: Photocomputing: explorations with transparency and opacity. Parallel Processing Letters 17(4), 339–347 (2004)
4. Aldaye, F.A., Palmer, A.L., Sleiman, H.F.: Assembling Materials with DNA as the Guide. Science 321, 1794–1799 (2008)
5. Rinker, S., Ke, Y., Liu, Y., Chhabra, R., Yan, H.: Self-assembled DNA nanostructures for distance dependent multivalent ligand-protein binding. Nature nanotechnology 3, 418–422 (2008)
6. Ogura, Y., Horiguchi, Y., Tanida, J.: Stepwise State-Transition of DNA Structures Controlled Through Sequential Photonic Signals. In: Preliminary proceedings of the 14th international meeting on DNA computing, p. 197 (2008)

7. Sakai, H., Ogura, Y., Tanida, J.: An Implementation of a Nanoscale Automaton Using DNA Conformation Controlled by Optics Signals. In: International Topical Meeting on Information Photonics 2008, pp. 172–173 (2008)
8. Sakai, H., Ogura, Y., Tanida, J.: Positional state representation and its transition control for photonic DNA automaton. In: 15th International Meeting on DNA Computing and Molecular Programming, pp. 60–71 (2009)
9. Sakai, H., Ogura, Y., Tanida, J.: An Implementation of a Nanoscale Automaton Using DNA Conformation Controlled by Optical Signals. Japanese Journal of Applied Physics 48, 09LA01 (2009)
10. Asanuma, H.: Synthesis of azobenzene-tethered DNA for reversible photo-regulation of DNA functions: hybridization and transcription. Nature Protocols 2, 203–213 (2007)
11. Kallenbach, N.R., Ma, R., Seeman, N.C.: An immobile nucleic acid junction constructed from oligonucleotides. Nature 305, 829–831 (1983)
12. Ma, R., Kallenbach, N.R., Sheardy, R.D., Petrillo, M.L., Seeman, N.C.: Three-arm acid junctions are flexible. Nucleic Acids Research 14, 9745–9753 (1986)
13. http://www.nupack.org/

Appendix

The sequences of DNA used are shown below.
The left edge of the sequences is 5'-end, and the right edge is 3'-end.
The letter X is azobenzene.

j1 : CGCAATCCTGAGCACG

D :

j2 : CGTGCTCACCGAATGC

O1 : AGTGCATAGATAGTCATAGCCT

j3 : GCATTCGGACTATGGC

O2 : GAGTATGTCTCGCACGTGATTCACTTC

j4 : GCCATAGTGGATTGCG

C1 : GTGATXAGTCACTCGAXCCATAATGXCG
GAXAGATGXAATTCXACAGTXGCAGAXGAACAXTAACTC

h1 :
TCACACCAAATCACGTGCGAGACATACTCGCACGGCTATCACCTTCC

h2 :
AGTGCGGTGACGATGCCGTCCTCGCATATGGCGAGGACTTCACAAATCCACC

h3 : GCCATAGTGGATTGCGGGTGATTT

h4 : CGTGGTAGACCGCACTGGAAGGTGATAGC

h5 : TGGTGTGACGCAATCCTGAGCACGGTTAAAGGCGGGAA

m : GGCATCGTCCTACCACGAGGCTATGACTATC/BHQ-1/

u1 :
CGTAAGTCGCAGAGTATGCCATTGCCTCATCA
GCGTAGCATCGAGATCTAAGTTAGTAACTCTG

u2 :
/6-FAM/TATGCACTCGACTTACGCTTCCCCTAT
ATCC

u3 :
TATGCACTATGAGGCAATGGCATACTCTG

u4 :
TAGATCTCGATGCTACGCTG

u5 :
CGTGCTCACCGAATGCCAGAGTTACTAACT

j1" :
CGCAATCCTGAGCACGGTTAAAGGCGGGAA

j1' :
CGCAATCCTGAGCACGGTGAGGAGAAGAAG

j3' :
GCATTCGGACTATGGCCTTCTTCTCCTCAC

j3" :
GCATTCGGACTATGGCTTCCCGCCTTTAAC

j4' :
GCCATAGTGGATTGCGGGATATAGGGGAAG

Fluctuation Induced Structure in Chemical Reaction with Small Number of Molecules

Yasuhiro Suzuki

Nagoya University, Furocho Chikusaku Nagoya city Aichi 464 8603 Japan
ysuzuki@is.nagoya-u.ac.jp
http://sci.cs.is.nagoya-u.ac.jp

Abstract. We investigate the behaviors of chemical reactions of the Lotka-Volterra model with small number of molecules; hence the occurrence of random fluctuations modifies the deterministic behavior and the law of mass action is replaced by a stochastic model. We model it by using Abstract Rewriting System on Multisets, ARMS; ARMS is a stochastic method of simulating chemical reactions and it is based on the reaction rate equation. We confirmed that the magnitude of fluctuations on periodicity of oscillations becomes large, as the number of involved molecules is getting smaller; and these fluctuations induce another structure, which have not observed in the reactions with large number of molecules. We show that the underling mechanism through investigating the coarse grained phase space of ARMS.

Keywords: Artificial Chemistries, Lotka-Volterra Model, Small Number Effects, Chemical reactions with a small number of molecules.

Introduction

Biochemical reactions involve a small number of molecules; for example, the Transcription from DNA to RNA is performed by messenger RNA (mRNA) molecules and two copies of each gene. Hence in such chemical reactions occurrence of random fluctuations modifies the deterministic behavior and the law of mass action is replaced by a stochastic model (for example [7]). For considering such fluctuations in chemical reactions, Gillespie[2] proposed a stochastic method of simulating chemical kinetics, which has firmer physical basis than the deterministic formulation. In this method, stochastic chemical kinetics is defined by the chemical master equation. Since it is mathematically intractable, Monte Carlo procedure is used to simulate. It requires a great amount of computer time, several approximate procedures for this method have been proposed [3], [12].

Abstract Rewriting System on Multisets, ARMS

Abstract Rewriting system on MultiSets, ARMS was proposed[8] as an *Artificial Chemistry*[1], [5]. ARMS have been used in various subjects: the Systems biology (modeling the P53 signaling network [10], inflammatory response [11]),

F. Peper et al. (Eds.): IWNC 2009, PICT 2, pp. 290–297, 2010.

the Physical chemistry (modeling the *Belouzov-Zhabotinskii* reaction [9], [12]), Chemical Ecology [9] and so on. ARMS is a construct $\Gamma = (A, w, R)$, where A is an alphabet, w is the initial state and R is the set of reaction rules.

Let A be an *alphabet* (a finite set of abstract symbols). A *multiset* over A is a mapping $M : A \mapsto \mathbf{N}$, where $\mathbf{N}$ is the set of natural numbers; 0, 1, 2,.... For each $a_i \in A$, $M(a_i)$ is the *multiplicity* of a_i in M, we also denote $M(a_i)$ as $[a_i]$. We denote by $A^\#$ the set of all multisets over A, with the empty multiset, $\emptyset$, defined by $\emptyset(a) = 0$ for all $a \in A$. A multiset $M : A \mapsto \mathbf{N}$, for $A = \{a_1, \ldots, a_n\}$ is represented by the state vector $w = (M(a_1), M(a_2), \ldots, M(a_n))$, w. The union of two multisets $M_1, M_2 : A \mapsto \mathbf{N}$ is the addition of vectors w_1 and w_2 that represent the multisets M_1, M_2, respectively. If $M_1(a) \leq M_2(a)$ for all $a \in A$, then we say that multiset M_1 is included in multiset M_2 and we write $M_1 \subseteq M_2$.

A *reaction rule* r over A is defined as a couple of multisets, (s, u), with $s, u \in A^\#$. A set of reaction rules is expressed as R. A rule $r = (s, u)$ is also represented as $r = s \to u$. Given a multiset $w \subseteq s$, the application of a rule $r = s \to u$ to the multiset w produces a multiset w' such that $w' = w - s + u$. Note that s and u can also be zero vector (empty).

The *reaction vector*, ν_j denotes the change of the number of molecules produced by the rule r_j. For example ν for the reaction $a, b \to b, c$ is $(-1, 0, 1) \equiv (a, b, c)$. We employ multisets; such a multiset $X : A \mapsto \mathbf{R}$ for $A = \{a_1, \ldots, a_n\}$ is represented by the state vector $\mathbf{x} = (X(a_1), X(a_2), \ldots, X(a_n))$. $X(a_i)$ denotes the number of specie a_i. Let us assume that there are $N \geq 1$ molecular species $\{a_1, ..., a_n\}, a_i \in A$ that interact through reaction rules $R = \{r_1, ..., r_m\}$. As the time evolution of $\mathbf{x}$ unfolds from a certain initial state, let us suppose the state transition of the system to be recorded by marking on a time axis the successive instants $t_1, t_2, ...$ as $X(t_j)$ $(j = 1, 2, ...)$, where $j = 0$ denotes the initial state. We specify the dynamical state of $\mathbf{x}(t) \equiv (X(a_1(t), X(a_2(t)), ..., X(a_N(t)))$, where $X(a_i(t))$ is the number of a_i specie at time t. The time evolution of ARMS is given as $\mathbf{x}(t) = \mathbf{x}(t-1) + \nu_j$, when there are several rules can be applied for $\mathbf{x}(t)$, only one rule is selected for applying. We can define various ways of applying rules and in this study rules are applied sequentially according to the mass action law of chemical reactions. We will define it in the next section.

ARMS with Chemical Kinetics

We modify the ARMS for modeling chemical kinetics and assume that all chemical reactions take place in a well-stirred reactor; this assumption is required due to the strong dependence of the reaction rate on the concentration of the reagent species. For the dynamical state $\mathbf{x}(t)$, we define the probability of selecting $\nu_j \equiv P_{\nu_j}(\mathbf{x}(t))$ as

$$P_{\nu_j}(\mathbf{x}(t)) = \frac{c_j(\mathbf{x}(t)) \times k_j}{\sum_{j=1}^m c_j(\mathbf{x}(t))}, \tag{1}$$

where $c_j(\mathbf{x}(t))$ denotes the number of possible combination collisions of r_j reactant molecules on $\mathbf{x}(t)$, k_j is the reaction constant of r_j. The time evolution of

$\mathbf{x}(t)$ is a jump Markov process on the N-dimensional non-negative lattice. We define the function $f(\mathbf{x}(t))$, called the *propensity function* for $r_j \in R$ on $\mathbf{x}(t)$ by $f(\mathbf{x}(t)) = (P_{\nu_1}(\mathbf{x}(t)), P_{\nu_2}(\mathbf{x}(t)), ..., P_{\nu_m}(\mathbf{x}(t)))$.

Lotka-Volterra model. The Lotka-Volterra model, LV describes interactions between two species in an ecosystem, a predator and a prey. Reaction rules of the LV are given as;

$$X \xrightarrow{a} X, X : (r_1),$$
$$X, Y \xrightarrow{b} Y, Y : (r_2),$$
$$Y \xrightarrow{c} \bot : (r_3),$$

where a b and c are reaction constants and $\bot$ denotes an empty symbol, which represents decay of Y.

The reaction rate equation, RRE of the LV is

$$\dot{X} = aX - bXY = X(a - bY) \tag{2}$$
$$\dot{Y} = bXY - cY = Y(bX - c). \tag{3}$$

Since population equilibrium occurs in the model when neither of the population levels is changing, from $X(a - bY) = 0$ and $Y(bX - c) = 0$, equilibria of the LV are $X = Y = 0$ and $Y = a/b$, $X = c/b$. The first solution represents the extinction of both species and the second solution represents a equilibrium point at which both populations sustain their current, non-zero numbers. The LV shows oscillations around this equilibrium point.

We analyze the behavior of the LV around the equilibrium point. The RRE of the LV can be rewritten into

$$\left(\frac{c}{x} - d\right)\dot{x} + \left(\frac{a}{y} - b\right)\dot{y} = 0,$$

and we obtain

$$\frac{d}{dx}[c \log x - dx + a \log y - by] = 0.$$

We define

$$H(x) = \bar{x} - x, \quad G(y) = \bar{y} \log y - y,$$

where $\bar{x} = c/d$ and $\bar{y} = a/b$. Then we obtain the Lyapnov function,

$$V(x, y) = dH(x) + bG(y).$$

Since

$$\frac{d}{dt}V(x(t), y(t)) = 0, \quad V(x(t), y(t)) = constant,$$

the time evolution of LV is preodic. Hence if fluctuations displace the time evolutions from the equilibrium point, the system should show periodic time evolutions and these fluctuations never lead the time evolutions to the equilibrium point [4]. Hence, we will observe a variety of periodic oscillations, when the magnitude of fluctuations are large.

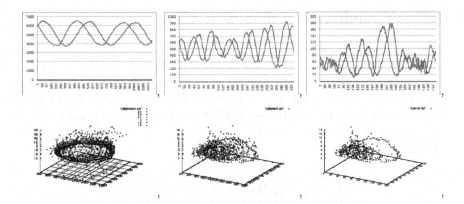

Fig. 1. The time evolution and its phase space and the distribution of existence probability are arranged one above the other; from left to right, the equilibrium points and initial states are (5000, 5000) and (6000,6000), (500,500) and (600, 600) and (50,50) and (60, 60); the Z-axis denotes the value of the existence probability. respectively

Comparison of Schematic Views

We compared the periodic oscillations by changing equilibria; we examined the case when equilibria $(X, Y) = (5000, 5000)$, $(500, 500)$ and $(50, 50)$, for each equilibrium point we set the initial state as $(X, Y) = (6000, 6000)$, $(600, 600)$ and $(60, 60)$, respectively (top row in the fig.1).

We confirmed that the magnitude of fluctuations on periodicity of oscillations became large, as the equilibrium point was getting smaller. When the equilibrium point $(X, Y) = (5000, 5000)$, the magnitude of fluctuations was small and when $(X, Y) = (500, 500)$, fluctuations became large, while keeping periodic oscillations. When $(X, Y) = (50, 50)$, fluctuations became more larger and a variety of oscillations were observed; especially small period oscillations near the equilibrium point was observed.

Phase Space

We compared phase spaces near to the each of equilibrium point (bottom raw in the fig.1) and existence probability; the existence probability of $\mathbf{x}_i \in \mathbf{x}(t), t = 1, 2, \dots$ was obtained by dividing the number of times of visiting $\mathbf{x}_i$ by the total number of visiting states in the time evolution. When the equilibrium point $(X, Y) = (5000, 5000)$, cyclic phase structure was clearly observed and the distribution of existence probability was mostly homogenous around the cyclic phase structure. When $(X, Y) = (500, 500)$, a variety of cyclic phase structures were observed; the distribution of existence probability was mostly homogeneous and these structures generate a variety of periodic oscillations. When $(X, Y) = (50, 50)$, the phase structure was skewed; the distribution of existence probability was inhomogeneous and most of them were attracted near to the equilibrium point.

Coarse Graining of Probabilistic Field

From comparison of schematic views, it was shown that the number of molecules in reactions effect behaviors of the time evolutions. Next, we will focus on effect of the number of molecules in reactions to the probability of selecting reaction rules. Since the probability of selecting a reaction rule was given by the propensity function on $\mathbf{x}$, we examine the probabilistic field of the propensity function. The propensity function of the LV was $f(\mathbf{x}) = (\frac{aX}{M}, \frac{bXY}{M}, \frac{cY}{M})$, where $M = aX + bXY + cY$.

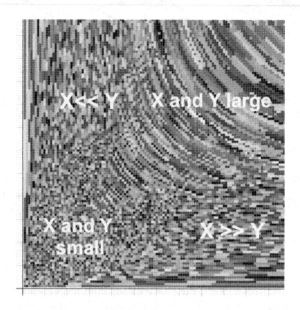

Fig. 2. The coarse grained probabilistic field of the LV; the equilibrium point was $(X, Y) = (10, 10)$ and from $(1, 1) to (200, 200)$ were observed, where the same color illustrates the same closure. The vertical axis illustrates Y and the horizontal axis, X, respectively.

Closure. A closure is defined as a sub space in a state space, where every value of a propensity function is approximately the same. The closure of $\mathbf{x}_i$ is defined as the set of vectors $\mathbf{x}_j, (j = 1, 2, 3, ...)$, where the maximal absolute component values in the differences of vector $\mathbf{x}_i$ and $\mathbf{x}_j$ are less than ϵ, ϵ is a very small number and given in advance; $\{ \mathbf{x_j} \mid MAX \mid f(\mathbf{x_j}) - f(\mathbf{x_i}) \mid < \epsilon \}$, where the function $MAX(\mathbf{x})$ returns the maximal value in $\mathbf{x}$. We investigated the coarse grained phase space of equilibrium point $(X, Y) = (10, 10)$, while $(X, Y) = (1, 1)$ to $(200, 200)$. When the value of X and Y were small, the size of closures was 1 and the size of a closure was proportion to the value of X and Y, where the size of closure i is given by the cardinality of $\mathbf{x}_i$. The shape of closures were different according to the value of X and Y; when X and Y were large, the shape of closures were hyperbolic; when X was considerably larger than Y, the shape of closures were horizontally, on the other hand, when Y was considerably larger than X, the shape of closures were vertically (fig 2). Because, when X or/and Y is/are so large that $\frac{bXY}{M}$ is considerably larger than $\frac{aX}{M}$ and $\frac{cY}{M}$, we can ignore $\frac{aX}{M}$ and $\frac{cY}{M}$ in a propensity function, so the propensity function can be regarded as $f(\mathbf{x}) \simeq (0, \frac{bXY}{M}, 0)$, approximately. Hence, if X and Y are large the shape of closure becomes hyperbolic. In case X is large and Y is considerably smaller than X, even if X is changed largely, XY would not change so much. So the shape of closures would along the X-axis and horizontally long. On the other hand, in case Y is large and X is considerably smaller than Y, even if Y is changed largely, XY would not change so much. So the shape of closures would along the Y-axis and vertically long.

Size of closure and time evolutions. If the probability of selecting a rule is homogeneous, the directions of time evolutions from the closure are also same. So when the size of closures are large, even if a time evolution fluctuates and bifurcates into several states, those of bifurcated states tend to be covered by a closure and the directions of time evolutions from these bifurcated states would be kept the same. On the other hand, when the size of closure are small, if a time evolution fluctuates and bifurcates into several states, those of bifurcated states not tend to be covered by a closure and the directions of time evolutions from the states would be different. Therefore, the time evolution of the reactions with small molecules should be suffered large fluctuations. In this study, because of a reaction rule is applied sequentially, a bifurcation may occur, after the time evolution returns to the visited state.

Small Numbers Effects, SNE. We showed that in the LV, fluctuations never leads the time evolution to the equilibrium point; however in reactions with small numbers, we observed the fluctuations lead the system to the equilibrium point (fig. 3). When equilibrium point was $(50, 50)$ and initial state, $(60, 60)$, its time evolution was trapped near to the equilibrium point and oscillated in small periods (in the right of the fig.3). The coarse grained phase space shows that the reactions with $X, Y \simeq 50$, closures are small. So its time evolution would be suffered large fluctuations and trapped. We call the characteristic changes

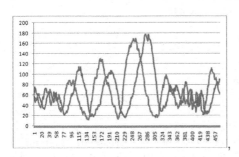 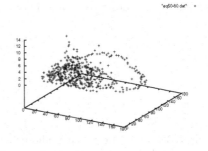

Fig. 3. Small Number Effects: the time evolution was trapped near to the equilibrium point, $(50, 50)$, where the initial state was $(60, 60)$ (during $t = 1 - 77$ and $324 - 438$)

in reactions caused by the small number molecules, the Small Number Effects (SNE). We also observed the SNE in the Brusselator model[6]. The Brusselator is a model of the Belousov Zhabotinskii (BZ) reaction;

$$A \xrightarrow{k_1} X : r_1,$$

$$B + X \xrightarrow{k_2} Y + D : r_2,$$

$$2X + Y \xrightarrow{k_3} 3X : r_3,$$

$$X \xrightarrow{k_4} E : r_4.$$

The Brusselator shows a limit-cycle oscillation; when a time evolution in the limit-cycle oscillation, even if a time evolution fluctuates, it returns to the limit-cycle oscillation and does not show a variety of periodic oscillations, which as we have seen in the LV.

We investigated the behaviors of limit-cycle oscillations in the Brusselator with changing the number of molecules [12]. We defined the size of system s by $[x] + [y] \leq s$, where $[x]$ and $[y]$ denote the total number of each molecular of x and y. When the size reaches s, r_1 cannot be applied because the other rules do not change the total number of molecules. This s was only one parameter and all other parameters were fixed; $k_1 = 100, k_2 = 3, k_3 = 10^{-3}$, $k_4 = 1$ and the initial state was $(X, Y) = (100, 100)$. We changed $s = 1000, 530$ and 500.

When s was 1000, the time evolution showed the limit-cycle (left in the fig.4), which is in good agreement with the kinetics of the differential equation model despite appreciable fluctuations [12]. As the number of involved molecules decreased to 530, the fluctuations in the amplitude of oscillation increase and the kinetics began to differ from those of the differential equation; however the periodicity was maintained and the time evolution remains quasi-periodic (Fig.4 in the middle).

When $s = 500$, the magnitude of fluctuations grew large and the system was trapped in an unreactive state where $X = 0$ (Fig.4 in the right); however the time evolution remains quasi-periodic oscillation and a variety of periods in oscillations could not be observed. These results illustrates that there would be various types of the SNE existing and their behaviors would be related to its dynamical characteristics. The coarse grained phase spaces would show underlying mechanisms among them and it is our future work.

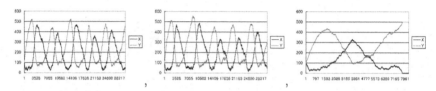

Fig. 4. Disappearance of periodic behavior of ARMS for the Brusselator system. In every simulation, parameters are $k_1 = 100$, $k_2 = 3$, $k_3 = 10^{-3}$, $k_4 = 1$ and the initial state is $(X, Y) = (100, 100)$. From left to right, the size of system is 1000, 530 and 500, respectively.

Acknowledgment. This research was partly supported by Information Initiative Center, Hokkaido University JAPAN. Visualization of the probabilistic field were designed by Kenji Tsunoda.

References

1. Dittrich, P., Ziegler, J., Banzhaf, W.: Artificial chemistries, a review. Artif. Life 7(3), 225–275 (2001)
2. Gillespie, D.T.: Exact Stochastic Simulation of Coupled Chemical Reactions. J. Phys. Chem. 81(25), 2340–2361 (1977)
3. Gillespie, D.T.: Approximate accelerated stochastic simulation of chemically reacting systems. J. Chem. Phys. 115(4), 1716–1733 (2001)
4. Hofbauer, J., Sigmund, K.: The Theory of Evolution and Dynamical Systems. Cambridge Univ. Press, Cambridge (1988)
5. Manca, V.: String rewriting and metabolism: a logical perspective, in Computing with Bio-Molecules. In: Theory and Experiments, pp. 36–60. Springer, Singapore (1998)
6. Nicolis, G., Prigogine, I.: Exploring Complexity, An Introduction. Freeman and Company, San Francisco (1989)
7. Ohkubo, J., Shnerb, N., Kesselar, D.A.: Transition Phenomena Induced by Internal Noise and Quasi-Absorbing State. J. Phisical Society Japan 77(4), 044002-1 – 042005 (2008)
8. Suzuki, Y., Tsumoto, S., Tanaka, H.: Analysis of Cycles in Symbolic Chemical System based on Abstract Rewriting System on Multisets. In: Artificial Life V, pp. 482–489. MIT Press, Cambridge (1996)
9. Suzuki, Y., Fujiwara, Y., Takabayashi, J., Tanaka, H.: Artificial Life Applications of a Class of P Systems: Abstract Rewriting System on Multisets. In: Calude, C.S., Pun, G., Rozenberg, G., Salomaa, A. (eds.) Multiset Processing. LNCS, vol. 2235, p. 299. Springer, Heidelberg (2001)
10. Suzuki, Y., Tanaka, H.: Modeling P53 signaling network by using multiset processing. In: Applications of Membrane Computing, pp. 203–215. Springer, Tokyo (2006)
11. Suzuki, Y.: An investigation of the Brusselator on the mesoscopic scale. Inter. J. of Parallel, Emergent and Distributed Sys. 22, 91–102 (2007)
12. Umeki, M., Suzuki, Y.: A Simple Membrane Computing Method For Simulating Bio-chemical Reactions 27(3), 529–550 (2008)

Parallel Retrieval of Nanometer-Scale Light-Matter Interactions for Nanophotonic Systems

Naoya Tate[1,2], Wataru Nomura[1,2], Takashi Yatsui[1,2], Tadashi Kawazoe[1,2], Makoto Naruse[1,2,3], and Motoichi Ohtsu[1,2]

[1] Department of Electrical Engineering and Information Systems, School of Engineering, The University of Tokyo, 2-11-16 Yayoi, Bunkyo-ku, Tokyo, 113-8656, Japan
[2] Nanophotonics Research Center, School of Engineering, The University of Tokyo, 2-11-16 Yayoi, Bunkyo-ku, Tokyo, 113-8656, Japan
[3] National Institute of Information and Communications Technology, 4-2-1 Nukuikita, Koganei, Tokyo, 184-8795, Japan

Abstract. Exploiting the unique attributes of nanometer-scale optical near-field interactions in a completely parallel manner is important for innovative nanometric optical processing systems. In this paper, we propose the basic concepts necessary for parallel retrieval of light–matter interactions on the nanometer-scale instead of the conventional one-dimensional scanning method. One is the *macro-scale observation* of optical near-fields, and the other is the *transcription* of optical near-fields. The former converts effects occurring locally on the nanometer scale involving optical near-field interactions to propagating light radiation, and the latter magnifies the distributions of optical near-fields from the nanometer scale to the sub-micrometer one. Those techniques allow us to observe optical far-field signals that originate from the effects occurring at the nanometer scale. We numerically verified the concepts and principles using electromagnetic simulations.

1 Introduction

Nanophotonics is a novel technology that utilizes the optical near-field, which is the electromagnetic field that mediates the interactions between closely spaced nanometric matter [1,2]. By exploiting optical near-field interactions, nanophotonics has broken the integration density restrictions imposed on conventional optical devices by the diffraction limit of light. This higher integration density has enabled realization of *quantitative* innovations in photonic devices and optical fabrication technologies [3,4]. Moreover, *qualitative* innovations have been accomplished by utilizing novel functions and phenomena made possible by optical near-field interactions, which are otherwise unachievable with conventional propagating light [5,6].

One of the most important technological vehicles that has contributed to the study of nanophotonics so far is high-quality optical near-field probing tips, such

F. Peper et al. (Eds.): IWNC 2009, PICT 2, pp. 298–307, 2010.

as those based on optical fiber probes [7]. They have achieved high spatial resolution and high energy efficiency, up to 10% optical near-field generation efficiency in some cases. For instance, near-field optical microscopes (NOMs) have been widely applied for obtaining ultrahigh-resolution images [8]. However, methods of characterizing optical near-fields using probing tips require one-dimensional (1D) scanning processes, which severely limit the throughput in obtaining two-dimensional (2D) information on the nanometer scale. Additionally, precision technologies are indispensable in fabricating probe tips and also in controlling their position during measurement; such technologies are large obstacles to developing more practical and easy characterization or utilization of optical near-fields. Therefore, eliminating one-dimensional scanning processes, or in other words, *probe-free nanophotonics*, is an important step toward further exploiting the possibilities of light–matter interactions on the nanometer scale.

In fact, we have successfully utilized optical near-field interactions without any scanning processes in nano-optical fabrication: For instance, optical near-field lithography, utilizing the near-field interactions between photomasks and photoresists, has already been developed [9]. Furthermore, non-adiabatic processes, meaning that optical near-field interactions activate conventionally light-insensitive materials, are additional novelties that are available in fabricating nano-structures by nanophotonics [10]. Optical near-field etching is another example of probe-free nanophotonics, where photochemical reactions are selectively excited in regions where optical near-fields are generated; this approach has been successfully demonstrated in flattening the rough surfaces of optical elements [11].

We consider that eliminating optical near-field probe tips is also important for characterization of optical near-fields and information processing applications. In this paper, we discuss the concept of parallel processing in nanophotonics, and we propose two techniques that allow parallel retrieval of optical near-fields: *macro-scale observation* of optical near-field interactions, which involves quadrupole–dipole transformation of charge distributions in engineered nanostructures, and *magnified transcription* of optical near-fields based on the photoinduced structural changes in metal complexes that exhibit photoinduced phase transitions.

In related applications, we have proposed shape-engineering of metal nanostructures so that optical near-field interactions between two nanostructures control the resultant far-field radiation [12]. Here we can also find probe-free utilization of optical near-field interactions. Although they are well matched with some applications, such as optical security [12], they are not applicable to nanostructures in general. They still require well-controlled shape-engineering processes, as well as precise alignment between planar nanostructures. In fact, another motivation of this study is to overcome such stringent alignment issues that are unavoidable in typical situations involving nanostructures. The concept of transcription of optical near-fields proposed in this paper would solve such problems by inducing unique processes enabled by the materials themselves—metal complexes in the particular demonstrations described below.

2   Nanometric Optical Processing Based on Nanophotonics

Nanometric optical processing systems, whose features include high integration density, low-energy operation, and innovative functions, are practical embodiments of such qualitative innovations offered by nanophotonics. Several fundamental proposals have been developed for their implementation, such as nanophotonic devices [13,14,15] and interconnect technologies [16]. However, regarding the interfaces between those nanometric systems and their associated surrounding systems, existing concepts that apply to conventional optical systems are not applicable since the physical basis is completely different between conventional propagating light and optical near-fields. Appropriate interfacing concepts and techniques that are well matched with the features of nanophotonics are strongly demanded. Eliminating the conventional scanning processes is also important to significantly improve the throughput in obtaining information on the nanometer scale.

Figures 1(a) and (b) show schematic diagrams that conceptually illustrate two kinds of representative nanometric processing systems. The system shown in Fig. 1(a) is operated by the host processing system located in the center. All of the processing nodes around the center are precisely controlled by the host system. In the case of nanometric processing, however, such a centralized architecture is difficult to apply since it is technically difficult to precisely align the nanometric components and to precisely and flexibly interconnect each component. On the other hand, we propose the system schematically shown in Fig. 1(b) for the implementation of nanometric optical processing systems. The processing components are connected in an autonomous manner, and the global control signals do not specify the detailed operations of each component. The local interactions between components in the system, involving optical near-field interactions, bring about certain behavior in the system as a whole.

In order to realize the parallel processing system shown in Fig. 1(b), it is necessary to retrieve the information located at the nanometer scale and expose it at the sub-micrometer scale so that the outputs of the system are obtainable in the optical far-field. For this purpose, here we propose two fundamental techniques: one is the *macro-scale observation* of optical near-field interactions, and the other is the *transcription* of optical near-field interactions, as schematically shown in Fig. 1(c). These techniques eliminate one-dimensional scanning processes; in other words, *probe-free nanophotonics* is accomplished.

3   Nanophotonic Matching Utilizing Macro-scale Observation of Optical Near-Field Interactions

For the macro-scale observation of optical near-fields, here we demonstrate that two nanostructures can be designed to exhibit far-field radiation only under the condition that the shapes of the two structures are appropriately combined and

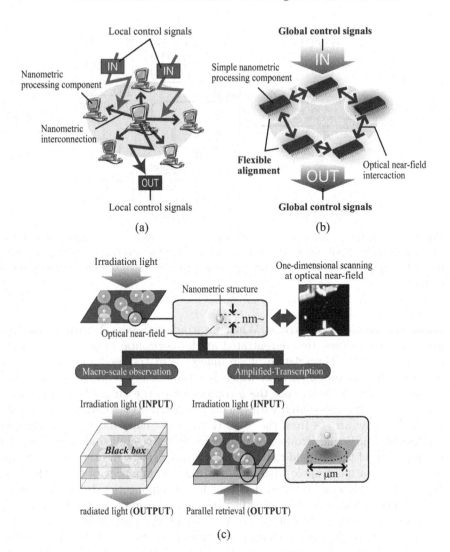

Fig. 1. Basic concept of nanometric processing systems (a) based on existing processing protocol and (b) utilizing nanophotonics. (c) Proposed fundamental processes of spatially parallel data retrieval for the nanometric optical processing system.

closely stacked. This function of the two nanostructures can be regarded effectively as a lock and a key, because only an appropriate combination of a lock and a key yields an output signal, namely, far-field radiation [12]. Figure 2 shows a schematic diagram of the nanophotonic matching system based on this function.

We design two nanostructural patterns, called *Shape A* and *Shape B* hereafter, to effectively induce a quadrupole–dipole transform via optical near-field interactions. Shape A and Shape B are designed as rectangular units on the xy-plane with constant intervals horizontally (along the x-axis) and vertically

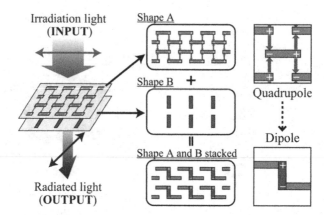

Fig. 2. Schematic diagram of nanophotonic matching system. The function is based on a quadrupole–dipole transform via optical near-field interactions, and it is achieved through shape-engineered nanostructures and their associated optical near-field interactions.

(along the y-axis), respectively. When we irradiate Shape A with x-polarized light, surface charges are concentrated at the horizontal edges of each of the rectangular units. The relative phase difference of the oscillating charges between the horizontal edges is π, which is schematically represented by $+$ and $-$ marks in Fig. 2. Now, note the y-component of the far-field radiation from Shape A, which is associated with the charge distributions induced in the rectangle. We draw arrows from the $+$ mark to the $-$ mark along the y-axis. We can find that adjacent arrows are always directed oppositely, indicating that the y-component of the far-field radiation is externally small. In other words, Shape A behaves as a quadrupole with regard to the y-component of the far-field radiation. It should also be noted that near-field components exist in the vicinity of the units in Shape A. With this fact in mind, we put the other metal nanostructure, Shape B, on top of Shape A. Through the optical near-fields in the vicinity of Shape A, surface charges are induced on Shape B. What should be noted here is that the arrows connecting the $+$ and $-$ marks along the y-axis are now aligned in the same direction, and so the y-component of the far-field radiation appears; that is, the stacked structure of Shape A and Shape B behaves as a dipole. Also, Shape A and Shape B need to be closely located to invoke this effect since the optical near-field interactions between Shape A and Shape B are critical. In other words, a quadrupole–dipole transform is achieved through shape-engineered nanostructures and their associated optical near-field interactions.

In order to verify this mechanism of the quadrupole–dipole transform by shape-engineered nanostructures, we numerically calculated the surface charge distributions induced in the nanostructures and their associated far-field radiation based on a finite-difference time-domain (FDTD) electromagnetic simulator (*Poynting for Optics*, a product of Fujitsu, Japan). Figure 3(a) schematically represents the design of (i) Shape A only, (ii) Shape B only, and (iii) a stacked structure of Shape A and Shape B, which consist of arrays of gold rectangular units. The length of each of the rectangular units is 500 nm, and the width and height are 100 nm. As the material, we assumed a Drude model of gold with a refractive index of 0.16 and an extinction ratio of 3.8 at a wavelength of 688 nm [17].

When irradiating these three structures with continuous-wave x-polarized input light at a wavelength of 688 nm, Figs. 3(b), (c) and (d) respectively show the induced surface charge density distributions (simply called surface charge hereafter) by calculating the divergence of the electric fields. For the Shape A only structure (Fig. 3(b)), we can find a local maximum and local minimum of the surface charges, denoted by + and − marks. When we draw arrows from the + marks to the − marks between adjacent rectangular units, as shown in

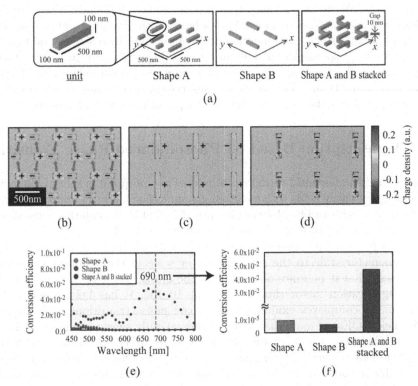

Fig. 3. (a) Specifications of the three types of nanostructures used in numerical evaluation of the conversion efficiency based on the FDTD method: Shape A only, Shape B only, and a stacked structure of Shapes A and B. (b,c) Surface charge density distributions induced in (b) Shape A only and (c) Shape B only. (d) Surface charge density distribution induced in Shape B only when it is stacked on top of Shape A. (e) Calculated performance figures of the quadrupole–dipole transform, namely, polarization conversion efficiency, with the three types of nanostructures. (f) Selective comparison at a wavelength of 690 nm.

Fig. 2, we can see that the arrows are always directed oppositely between the adjacent ones, meaning that the Shape A only structure behaves as a quadrupole for the y-components of the far-field radiation. For the Shape B only structure (Fig. 3(c)), the charges are concentrated at the horizontal edges of each of the rectangular units, and there are no y-components that could contribute to the far-field radiation. Figure 3(d) shows the surface charge distributions induced in Shape B when it is stacked on top of Shape A. We can clearly see that the charges are induced at the vertical edges of each of the rectangular units, and they are aligned in the same direction. In other words, a dipole arrangement is accomplished with respect to the y-component, leading to a drastic increase in the far-field radiation.

Now, one of the performance figures of the quadrupole-dipole transform is $I_{\mathrm{conv}} = I_{y-\mathrm{OUT}}/I_{x-\mathrm{IN}}$, where $I_{x-\mathrm{IN}}$ and $I_{y-\mathrm{OUT}}$ represent the intensities of the x-component of the incident light and the y-component of the radiated light, respectively. We radiate a short optical pulse with a differential Gaussian form whose width is 0.9 fs, corresponding to a bandwidth of around 200 to 1300 THz. Figure 3(e) shows I_{conv} as a function of the input light wavelength, and Fig. 3(f) compares I_{conv} specifically at 690 nm. I_{conv} appears strongly with the stacked structure of Shapes A and B, whereas it exhibits a small value with Shape A only and Shape B only. We can clearly observe the quadrupole–dipole transform in the optical near-fields as the change of I_{conv} in the optical far-fields.

4   Transcription Based on Photoinduced Phase Transition

The second fundamental technique to realize our proposal is spatial magnification of optical near-fields with a certain magnification factor so that the effects of optical near-fields can be observed in optical far-fields. We call this transcription of optical near-fields. It is schematically shown in Fig. 4.

In our proposal, it is necessary to spatially magnify the optical near-field from the nanometer scale to the sub-micrometer scale in the resultant transcribed pattern so that it becomes observable in the optical far-field. It turns out that the magnification factor should be about 10–100. It has been reported that some metal complexes exhibit photoinduced phase transitions with quantum efficiencies (QEs) of more than ten, as schematically shown in the upper half of Fig. 4. This suggests the possibility of transcription of optical near-fields with a certain magnification factor using such materials. Once the spatial pattern is detectable in the far-field, various concepts and technologies common in parallel optical processing will be applicable.

A photoinduced phase transition has been observed in several cyano-bridged metal complexes [18]. They exhibit bistable electronic states at room temperature. The energy barrier between these bistable states maintains a photoinduced state even after irradiation is terminated. Also, the state can easily be reset via either optical irradiation or temperature control. Moreover, typical phase transitions are excited in a cascaded manner, meaning that they exhibit high quantum efficiencies. This transition between the high-temperature (HT) phase and the

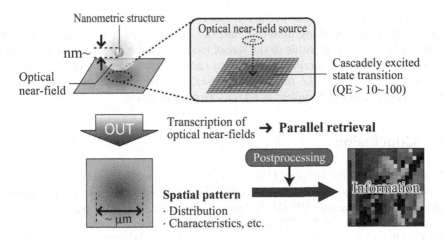

Fig. 4. Fundamental process of optical parallel data retrieval based on the transcription of optical near-fields. In our proposal, it is necessary to spatially magnify the optical near-field from the nanometer scale to the sub-micrometer scale in the resultant transcribed pattern so that it is observable in the optical far-field.

low-temperature (LT) phase is accompanied by a structural change from a cubic to a tetragonal structure due to Jahn–Teller distortion, and their physical properties are strongly changed by the transition. Rubidium manganese hexacyanoferrate [19] is a suitable material for the transcription medium.

To validate the transcription effect numerically, we simulated the difference of the optical response between the LT- and HT-phase materials using FDTD simulation. A nanometric fiber probe with a radius of 20 nm and an illumination light source with an operating wavelength of 635 nm were included in the calculation model, as shown in Fig. 5(a). Figure 5(b) compares the electric field intensities at the surfaces of the LT- and HT-phase materials.

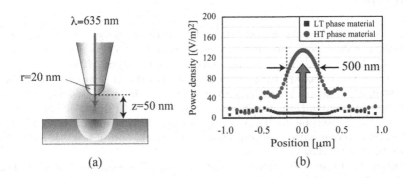

Fig. 5. (a) Schematic diagrams of calculation model for numerical validation of the transcription. (b) Numerical results. The signal intensity in the near-field is obviously increased by the irradiation.

The power density at the surface of the HT-phase material was much stronger than that of the LT-phase material, and the scale of the distribution was magnified from 20 nm, comparable to the size of the fiber probe, to 500 nm, comparable to the size of the transcribed area. From these results, we can confirm the fundamental principle of the magnified transcription of optical near-fields using a cyano-bridged metal complex as a transcription medium.

5 Conclusions

In summary, we have proposed the concepts of macro-scale observation and transcription of optical near-fields as fundamental techniques for nanometric optical processing. We numerically demonstrated the principles with practical calculation models based on FDTD methods. These are key, fundamental techniques for parallel retrieval in nanometric optical processing. We will further investigate the principles and show experimental demonstrations for exploiting the unique physical processes originating at the nanometer scale and making them available as valuable system-level functions for implementation of nanometric processing systems.

The authors thank Prof. Ohkoshi, Dr. Tokoro, and Dr. Takeda of the Department of Chemistry, School of Science, the University of Tokyo, for their valuable contributions to our collaborative research.

This work was supported in part by a comprehensive program for personnel training and industry–academia collaboration based on NEDO projects funded by the New Energy and Industrial Technology Organization (NEDO), Japan, as well as the Global Center of Excellence (G-COE) "Secure-Life Electronics" and Special Coordination Funds for Promoting Science and Technology sponsored by the Ministry of Education, Culture, Sports, Science and Technology (MEXT), Japan.

References

1. Ohtsu, M., Kobayashi, K., Kawazoe, T., Sangu, S., Yatsui, T.: Nanophotonics: Design, Fabrication, and Operation of Nanometric Devices Using Optical Near Fields. IEEE J. Sel. Top. Quantum Electron. 8(4), 839–862 (2002)
2. Naruse, M., Miyazaki, T., Kawazoe, T., Kobayashi, K., Sangu, S., Kubota, F., Ohtsu, M.: Nanophotonic Computing Based on Optical Near-Field Interactions between Quantum Dots. IEICE Trans. Electron E88-C(9), 1817–1823 (2005)
3. Nishida, T., Matsumoto, T., Akagi, F., Hieda, H., Kikitsu, A., Naito, K.: Hybrid recording on bit-patterned media using a near-field optical head. J. Nanophotonics B, 011597 (2007)
4. Ozbay, E.: Plasmonics: Merging Photonics and Electronics at Nanoscale Dimensions. Science 311, 189–193 (2006)
5. Ohtsu, M.: Nanophotonics in Japan. J. Nanophotonics 1, 0115901–7 (2007)
6. Naruse, M., Kawazoe, T., Yatsui, T., Sangu, S., Kobayashi, K., Ohtsu, M.: Progress in Nano-Electro-Optics V. Ohtsu, M. (ed.). Springer, Berlin (2006)

7. Mononobe, S.: Near-Field Nano/Atom Optics and Technology, pp. 31–69. Springer, Berlin (1998)
8. Maheswari Rajagopalan, U., Mononobe, S., Yoshida, K., Yoshimoto, M., Ohtsu, M.: Nanometer level resolving near field optical microscope under optical feedback in the observation of a single-string deoxyribo nucleic acid. Jpn. J. Appl. Phys. 38(12A), 6713–6720 (1999)
9. Inao, Y., Nakasato, S., Kuroda, R., Ohtsu, M.: Near-field lithography as prototype nano-fabrication tool. Science Direct Microelectronic Engineering 84, 705–710 (2007)
10. Kawazoe, T., Ohtsu, M., Inao, Y., Kuroda, R.: Exposure dependence of the developed depth in nonadiabatic photolithography using visible optical near fields. J. Nanophotonics 1, 0115951–9 (2007)
11. Yatsui, T., Hirata, K., Nomura, W., Tabata, Y., Ohtsu, M.: Realization of an ultra-flat silica surface with angstrom-scale. Appl. Phys. B 93, 55–57 (2008)
12. Naruse, M., Yatsui, T., Kawazoe, T., Tate, N., Sugiyama, H., Ohtsu, M.: Nanophotonic Matching by Optical Near-Fields between Shape-Engineered Nanostructures. Appl. Phys. Exp. 1, 112101 1-3 (2008)
13. Kobayashi, K., Sangu, S., Kawazoe, T., Shojiguchi, A., Kitahara, K., Ohtsu, M.: Excitation dynamics in a three-quantum dot system driven by optical near field interaction: towards a nanometric photonic device. J. Microscopy 210, 247–251 (2003)
14. Kawazoe, T., Kobayashi, K., Akahane, K., Naruse, M., Yamamoto, N., Ohtsu, M.: Demonstration of a nanophotonic NOT gate using near-field optically coupled quantum dots. Appl. Phys. B 84, 243–246 (2006)
15. Yatsui, T., Sangu, S., Kawazoe, T., Ohtsu, M., An, S.-J., Yoo, J.: Nanophotonic switch using ZnO nanorod double-quantum-well structures. Appl. Phys. Lett. 90(22), 223110-1-3 (2007)
16. Nomura, W., Yatsui, T., Kawazoe, T., Ohtsu, M.: Observation of dissipated optical energy transfer between CdSe quantum dots. J. Nanophotonics 1, 011591-1-7 (2007)
17. Lynch, D.W., Hunter, W.R.: Comments on the Optical Constants of Metals and an Introduction to the Data for Several Metals. In: Palik, E.D. (ed.) Handbook of Optical Constants of Solids, pp. 275–367. Academic Press, Orlando (1985)
18. Sato, O., Hayami, S., Einaga, Y., Gu, Z.Z.: Control of the Magnetic and Optical Properties in Molecular Compounds by Electrochemical, Photochemical and Chemical Methods. Bull. Chem. Soc. Jpn. 76(3), 443–470 (2003)
19. Tokoro, H., Matsuda, T., Hashimoto, K., Ohkoshi, S.: Optical switching between bistable phases in rubidium manganese hexacyanoferrate at room temperature. J. Appl. Phys. 97, 10M508 (2005)

A Compressible Fluid Model for Traffic Flow and Nonlinear Saturation of Perturbation Growth

Akiyasu Tomoeda[1,2], Daisuke Shamoto[2], Ryosuke Nishi[2], Kazumichi Ohtsuka[2], and Katsuhiro Nishinari[2,3]

[1] Meiji Institute for Advanced Study of Mathematical Sciences,
Meiji Univeristy, Japan
[2] Research Center for Advanced Science and Technology,
The University of Tokyo, Japan
[3] School of Engineering, University of Tokyo, Japan
[4] PRESTO, Japan Science and Technology Corporation
atom@isc.meiji.ac.jp

Abstract. In this paper, we have proposed a new compressible fluid model for the one-dimensional traffic flow taking into account the reaction time of drivers, which is based on the actual measurements. This model is a generalization of Payne model by introducing a density-dependent function of reaction time. The linear stability analysis of this new model shows the instability of homogeneous flow around a critical density of vehicles. Moreover, the condition of the nonlinear saturation of density against small perturbation is derived from the analysis by using reduction perturbation method.

1 Introduction

Among various kinds of jamming phenomena, a traffic jam of vehicles is a very familiar phenomenon and causes several losses in our daily life such as decreasing efficiency of transportation, waste of energy, serious environmental degradation, etc. In particular, highway traffic dynamics has attracted many researchers and has been investigated as a nonequilibrium system of interacting particles for the last few decades [1,2]. A lot of mathematical models for one-dimensional traffic flow have ever been proposed [3,4,5,6,7,8,9,10,11,12] and these models are classified into *microscopic* and *macroscopic* model in terms of the treatments of particles. In *microscopic* model, e.g. car-following model [3,4] and cellular automaton model [5,6,7,8], the dynamics of traffic flow is described by the movement of individual vehicles . Whereas in *macroscopic* model, the dynamics is treated as an effectively one-dimensional compressible fluid by focused on the collective behavior of vehicles [9,10,11,12]. Moreover, it is widely known that some of these mathematical models are related with each other, which is shown by using mathematical method such as ultra-discretization method [13] or Euler-Lagrange transformation [14]. That is, rule-184 elementary cellular automaton model [5] is derived from Burgers equation [10] by ultra-discretization method . The specific case of optimal velocity (OV) model [4] is formally derived from the

F. Peper et al. (Eds.): IWNC 2009, PICT 2, pp. 308–315, 2010.

rule-184 cellular automaton model via the Euler-Lagrange transformation. Quite recently, more noteworthy are that the ultra-discrete versions of OV model are shown by Takahashi *et al.* [15] and Kanai *et al.* [16].

In contrast to practically reasonable microscopic models, macroscopic models with reasonable expression have almost never been proposed, even in various traffic models based on the hydrodynamic theory of fluids [9,10,11,12]. In previous fluid models, one does not have any choice to introduce the diffusion term into the models such as *Kerner* and *Konhäuser* model [12], in order to represent stabilized density wave, which indicates the formation of traffic jam. However, it is emerged as a serious problem that some vehicles move backward even under heavy traffic, since the diffusion term has a spatial isotropy. As mentioned by *Daganzo* in [17], the most essential difference between traffic and fluids is as follows:

"*A fluid particle responds to the stimulus from the front and from behind, however a vehicle is an anisotropic particle that mostly responds to frontal stimulus.*"

That is, traffic vehicles exhibits an *anisotropic* behavior, although the behavior of fluid particles with simple diffusion is *isotropic*. Therefore, unfortunately we would have to conclude that traffic models which include the diffusion term are not reasonable for the realistic expression of traffic flow. Given these factors, we suppose that traffic jam forms as a result of the plateaued growth of small perturbation by the nonlinear saturation effect.

Now let us return to Payne model [11], which is one of the most fundamental and significant fluid models of traffic flow without diffusion term. Payne model is given by

$$\frac{\partial \rho}{\partial t} + \frac{\partial}{\partial x}(\rho v) = 0, \tag{1}$$

$$\frac{\partial v}{\partial t} + v\frac{\partial v}{\partial x} = \frac{1}{\tau}\left(V_{\text{opt}}(\rho) - v\right) + \frac{1}{2\tau\rho}\frac{dV_{\text{opt}}(\rho)}{d\rho}\frac{\partial \rho}{\partial x}, \tag{2}$$

where $\rho(x,t)$ and $v(x,t)$ corresponds to the spatial vehicle density and the average velocity at position x and time t, respectively. τ is the reaction time of drivers, which is a positive *constant* value, and $V_{\text{opt}}(\rho)$ is the optimal velocity function, which represent the desired velocity of drivers under the density ρ.

Payne employs the results of observations by *Greenshield* [18] as the optimal velocity function:

$$V_{\text{opt}}(\rho) = V_0\left(1 - \frac{\rho}{\rho_{\text{max}}}\right), \tag{3}$$

where V_0 is the maximum velocity in free flow phase and ρ_{max} corresponds to the maximum density when all cars become completely still.

The linear stability analysis for Payne model gives us the dispersion relation as follows [11]:

$$\omega = -kV_{opt}(\rho_0) + \frac{i}{2\tau}\left\{1 \pm \sqrt{1 + 4a_0^2\tau^2(2k\rho_0 i - k^2)}\right\}, \tag{4}$$

$$\text{where} \quad a_0^2 = -\frac{1}{2\tau}\frac{dV_{opt}(\rho)}{d\rho}\bigg|_{\rho=\rho_0} > 0. \tag{5}$$

Furthermore, the linear stability condition is calculated from dispersion relation (4),

$$\frac{1}{2\tau} > -\rho_0^2\frac{dV_{opt}(\rho)}{d\rho}\bigg|_{\rho=\rho_0}. \tag{6}$$

If one applies velocity-density relation of Greenshield (3) into this stability condition, the following linear stability condition is obtained

$$\frac{\rho_{max}}{2\tau V_0} > \rho_0^2. \tag{7}$$

Here, let us define the stability function $(S(\rho_0))$ as

$$S(\rho_0) = \frac{\rho_{max}}{2\tau V_0} - \rho_0^2. \tag{8}$$

In this function, the condition $S(\rho_0) > 0$ $(S(\rho_0) < 0)$ corresponds to the stable (unstable) state.

Fig. 2 shows the stability plots for several constant value of τ. From this figure, we see that the instability region of homogeneous flow occurs beyond a critical density of vehicles.

However, Payne model shows the condensation of vehicles due to the momentum equation (2). That is, as the density increases, the value of optimal velocity in the first term of right-hand-side becomes zero and the value of second term also becomes zero. Thus, the vehicles gather in one place due to the nonlinear effect vv_x and the small perturbation blows up without stabilization. Therefore, Payne model is also incomplete to describe the realistic dynamics of traffic flow.

Thus, in this paper we propose a new compressible fluid model, by extending the Payne model in terms of the reaction time of drivers based on the following actual measurements.

2   New Compressible Fluid Model Based on Experimental Data

We have performed car-following experiment on a highway. The leading vehicle cruises with legal velocity and following vehicle pursues the front one. The time-series data of the velocity and position (latitude and longitude) of each vehicle are recorded every 0.2 seconds (5Hz) by a global positioning system (GPS) receiver on-board with high-precision (< 50 centimeters precision).

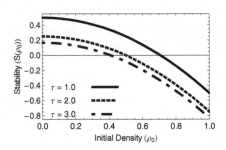

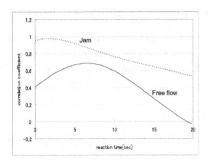

Fig. 1. The plots of stability in the case of the Greenshield relation (3) with $\rho_{\max} = 1.0$ and $v_0 = 1.0$

Fig. 2. The plots of correlation coefficient based on (9) for given each reaction time (τ)

By dividing the time-series data into two phases, i.e. *free-flow phase* and *jam phase*, based on the velocity, and assuming that the reaction time of drivers is considered as a slight delay of behavior, we calculate the correlation coefficient which is denoted by

$$r_{i+1}(\tau) = \langle v_i(t)v_{i+1}(t+\tau)\rangle_t \tag{9}$$

where $v_i(t)$ shows the velocity of i-th car at time t. Note that, i-th car drives in front of $(i+1)$-th car. The symbol $\langle * \rangle$ indicates an ensemble average . Finally we have obtained the correlation coefficient for each given τ which is shown in Fig. 2. From this figure, we have found that the peak of correlation coefficient shifts according to the situation of the road. Here, since the reaction time of a driver is considered as τ, the reaction time of a driver is not constant, but obviously changes according to the situation of the road. That is, if the traffic state is free (jam), the reaction time of a driver is longer (shorter). Note that, this experimental results (Fig. 2) is quite important to verify the changing of reaction time of drivers, which is considered as a constant value in previous models.

As a reasonable assumption based on this result, the reaction time of drivers would depend on the density on the road. Under this assumption, we have extended Payne model and proposed a new compressible fluid model as follows,

$$\frac{\partial \rho}{\partial t} + \frac{\partial}{\partial x}(\rho v) = 0, \tag{10}$$

$$\frac{\partial v}{\partial t} + v\frac{\partial v}{\partial x} = \frac{1}{\tau(\rho)}\left(V_{\text{opt}}(\rho) - v\right) + \frac{1}{2\rho\tau(\rho)}\frac{dV_{\text{opt}}(\rho)}{d\rho}\frac{\partial \rho}{\partial x}. \tag{11}$$

The difference between this model and Payne model is only that the reaction time of drivers is changed from constant value to density-dependent function $\tau(\rho)$.

2.1 Linear Stability Analysis

Now let us perform the linear stability analysis for our new dynamical model to investigate the instability of homogeneous flow. The homogeneous flow and small perturbation are given by

$$\rho = \rho_0 + \varepsilon\rho_1, \tag{12}$$
$$v = V_{\text{opt}}(\rho_0) + \varepsilon v_1. \tag{13}$$

One obtains the form of dispersion relation as

$$\omega = -kV_{\text{opt}} + \frac{i}{2\tau(\rho_0)}\left\{1 \pm \sqrt{1 + 4a_0^2\tau(\rho_0)^2(2k\rho_0 i - k^2)}\right\}, \tag{14}$$

$$\text{where} \quad a_0^2 = -\frac{1}{2\tau(\rho_0)}\frac{dV_{\text{opt}}(\rho)}{d\rho}\bigg|_{\rho=\rho_0} > 0. \tag{15}$$

Hence, the stability conditions

$$\frac{1}{2\tau(\rho_0)} > -\rho_0^2\frac{dV_{\text{opt}}(\rho)}{d\rho}\bigg|_{\rho=\rho_0}, \tag{16}$$

are obtained. The difference between (6) and (16) is only the reaction time, which changes from τ to $\tau(\rho_0)$. Although $\tau(\rho)$ is variable value, $\tau(\rho_0)$ is the constant value which is decided by only the initial density of homogeneous flow. Therefore, this stability condition (16) and the stability condition of Payne model (6) are essentially equivalent, that is, our new model also shows the instability of homogeneous flow. Substituting the relation of Greenshield (3) into (16), the stability condition leads to

$$\frac{\rho_{\text{max}}}{2\tau(\rho_0)V_0} > \rho_0^2. \tag{17}$$

The most important point of our new model is that it is possible to stabilize the perturbation by the nonlinear effect which is created by the function $\tau(\rho)$, which was failed in the Payne model. In order to show this nonlinear effect, the evolution equation of small perturbation is derived in the next subsection.

2.2 Reductive Perturbation Analysis

Let us define the new coordinates X and T by *Galilei transformation* as

$$X = \varepsilon(x - c_g t) \quad (X \sim \varepsilon), \tag{18}$$
$$T = \varepsilon^2 t \qquad\quad (T \sim \varepsilon^2), \tag{19}$$

where $c_g = d\omega/dk$ is the group velocity. Then, we assume that $\rho(x,t), v(x,t)$ can be given in terms of the power series in ε, i.e.,

$$\rho \sim \rho_0 + \varepsilon\rho_1 + \varepsilon^2\rho_2 + \varepsilon^3\rho_3 + \cdots, \tag{20}$$
$$v \sim v_0 + \varepsilon v_1 + \varepsilon^2 v_2 + \varepsilon^3 v_3 + \cdots. \tag{21}$$

Substituting (20) and (21) into (10) and (11), for each order term in ε, we have, respectively,

$$\varepsilon^2: \quad \frac{\partial}{\partial X}\Big(\rho_1(v_0 - c_g) + \rho_0 v_1\Big) = 0, \tag{22}$$

$$\varepsilon^3: \quad \frac{\partial \rho_1}{\partial T} + \frac{\partial}{\partial X}\Big(\rho_2(v_0 - c_g) + \rho_1 v_1 + \rho_0 v_2\Big) = 0, \tag{23}$$

$$\varepsilon^4: \quad \frac{\partial \rho_2}{\partial T} + \frac{\partial}{\partial X}\Big(\rho_3(v_0 - c_g) + \rho_2 v_1 + \rho_1 v_2 + \rho_0 v_3\Big) = 0, \tag{24}$$

and

$$\varepsilon^0: \quad V_{\text{opt}} = v_0, \tag{25}$$

$$\varepsilon^1: \quad V'_{\text{opt}} = \frac{v_1}{\rho_1}, \tag{26}$$

$$\varepsilon^2: \quad (v_0 - c_g)\frac{\partial v_1}{\partial X} = \frac{1}{2\tau(\rho_0)}\Big(V''_{\text{opt}}\rho_1^2 + 2V'_{\text{opt}}\rho_2 - 2v_2\Big) + \frac{1}{2\tau(\rho_0)\rho_0}V'_{\text{opt}}\frac{\partial \rho_1}{\partial X}, \tag{27}$$

$$\varepsilon^3: \quad \frac{\partial v_1}{\partial T} + (v_0 - c_g)\frac{\partial v_2}{\partial X} + v_1\frac{\partial v_1}{\partial X}$$
$$= \frac{1}{\tau(\rho_0)}\Big(\frac{1}{6}V'''_{\text{opt}}\rho_1^3 + V''_{\text{opt}}\rho_1\rho_2 + V'_{\text{opt}}\rho_3 - v_3\Big) - \frac{\rho_1\tau'(\rho_0)}{\tau(\rho_0)^2}\Big(\rho_2 V'_{\text{opt}} + \frac{1}{2}\rho_1^2 V''_{\text{opt}} - v_2\Big)$$
$$+ \frac{1}{2\tau(\rho_0)}\Big(V'_{\text{opt}}\frac{1}{\rho_0}\frac{\partial \rho_2}{\partial X} + V''_{\text{opt}}\frac{\rho_1}{\rho_0}\frac{\partial \rho_1}{\partial X} - V'_{\text{opt}}\frac{\rho_1}{\rho_0^2}\frac{\partial \rho_1}{\partial X} - \frac{\rho_1\tau'(\rho_0)V'_{\text{opt}}}{\rho_0\tau(\rho_0)}\frac{\partial \rho_1}{\partial X}\Big). \tag{28}$$

Note that the prime means the abbreviation of each derivation.

Putting $\phi_1 = \rho_1$ as a first-order perturbation quantity and eliminating the second-order quantities (ρ_2, v_2) in (23) and (27), we have obtained the Burgers equation with constant coefficients

$$\frac{\partial \phi_1}{\partial T} = \Big(\frac{2(v_0 - c_g)}{\rho_0} - V''_{\text{opt}}\rho_0\Big)\phi_1\frac{\partial \phi_1}{\partial X} + \Big(\frac{v_0 - c_g}{2\rho_0} - \tau(\rho_0)(v_0 - c_g)^2\Big)\frac{\partial^2 \phi_1}{\partial X^2}, \tag{29}$$

as a evolution equation of first-order quantity. Moreover, eliminating the third-order quantities (ρ_3, v_3) in (24) and (28) and defining the perturbation Φ included the first- and second-order perturbation as $\Phi = \phi_1 + \varepsilon\phi_2$, the higher-order Burgers equation with constant coefficients

$$\frac{\partial \Phi}{\partial T} = \frac{2(v_0 - c_g)}{\rho_0}\Phi\frac{\partial \Phi}{\partial X} + \Big(\frac{v_0 - c_g}{2\rho_0} - \tau(\rho_0)(v_0 - c_g)^2\Big)\frac{\partial^2 \Phi}{\partial X^2}$$
$$+ \varepsilon\Big\{-\Big(\frac{2(v_0 - c_g)^2\tau(\rho_0)}{\rho_0} + (v_0 - c_g)^2\tau'(\rho_0)\Big)\frac{\partial}{\partial X}\Big(\Phi\frac{\partial \Phi}{\partial X}\Big)$$
$$- \Big(\frac{\tau(\rho_0)(v_0 - c_g)^2}{\rho_0} - 2(v_0 - c_g)^3\tau(\rho_0)^2\Big)\frac{\partial^3}{\partial X^3}\Phi\Big\}, \tag{30}$$

is obtained. Note that, in this derivation, we put $V''_{\text{opt}} = V'''_{\text{opt}} = 0$ due to the relation (3).

Although the first-order equation (29) of our model is essentially equivalent to the one of Payne model, the second-order equation (30) is differ from that of Payne model in terms of the coefficient of the third term of right-hand side.

In order to analyze the nonlinear effect of our model, let us consider the coefficient of the diffusion term of second-order equation. Let us put the coefficient of the second term of right-hand side in (30) as

$$P = \frac{v_0 - c_g}{2\rho_0} - \tau(\rho_0)(v_0 - c_g)^2, \tag{31}$$

and also put the coefficient of the third term as

$$Q = \frac{2(v_0 - c_g)^2 \tau(\rho_0)}{\rho_0} + (v_0 - c_g)^2 \tau'(\rho_0). \tag{32}$$

Thus, diffusion term of (30) is described as

$$\left(P - \varepsilon Q\Phi\right)\frac{\partial^2 \Phi}{\partial X^2}. \tag{33}$$

Since $P = 0$ corresponds to the neutrally stable condition, we assume the value P is negative, which corresponds to the linear unstable case. In the case of Payne model, Q is always positive because τ is constant, i.e. τ' is always zero. Therefore, the diffusion coefficient is always negative under the linear unstable condition of Payne model, which makes the model hard to treat numerically. However, in the case of our model, $\tau'(\rho)$ is always negative since $\tau(\rho)$ is considered as monotonically decreasing function. If Q is negative, the diffusion coefficient becomes positive value as Φ increases even under the linear unstable condition. In this situation, the small perturbation will be saturated by nonlinear effect created by the density-dependent function of reaction time of drivers. The conditions for nonlinear saturation corresponds to $P < 0$ and $Q < 0$, which are transformed into the following expressions,

$$\tau(\rho_0) > \frac{1}{2\rho_0(v_0 - c_g)}, \tag{34}$$

$$\tau'(\rho_0) < -\frac{2\tau(\rho_0)}{\rho_0}. \tag{35}$$

All conditions which include the other cases are summarized as following Tab. 1.

Table 1. Classification table based on the coefficient of diffusion term

P	Q	$P - \varepsilon Q\Phi$	Time evolution
(Linear unstable) $P<0$	$Q<0$	$P - \varepsilon Q\Phi > 0$	Saturation
		$P - \varepsilon Q\Phi < 0$	Amplification
	$Q>0$	$P - \varepsilon Q\Phi < 0$	Amplification
(Linear stable) $P>0$	$Q<0$	$P - \varepsilon Q\Phi > 0$	Damping
	$Q>0$	$P - \varepsilon Q\Phi > 0$	Damping
		$P - \varepsilon Q\Phi < 0$	Amplification

3   Conclusion

In this paper, by introducing the density-dependent function $\tau(\rho)$ about reaction time of drivers based on the actual measurements, we have proposed a new compressible fluid model for one-dimensional traffic flow. Our new model does not include the diffusion term which exhibits the unrealistic isotropic behavior of vehicles, since vehicles mostly responds to the stimulus from the front one. The linear stability analysis for our new model gives us the existence of instability of homogeneous flow and we have found that the stability condition is essentially equivalent to Payne model.

Moreover, the behavior of small perturbation is classified according to the diffusion coefficient of the higher-order Burgers equation which is derived from our new model by using reductive perturbation method taking into account both first- and second-order perturbation quantities. From this classification, we have obtained the special condition where the small perturbation is saturated by nonlinear effect.

References

1. Chowdhury, D., Santen, L., Schadschneider, A.: Phys. Rep. 329, 199 (2000)
2. Helbing, D.: Rev. Mod. Phys. 73, 1067 (2001)
3. Newell, G.F.: Oper. Res. 9, 209 (1961)
4. Bando, M., Hasebe, K., Shibata, A., Sugiyama, Y.: Phys. Rev. E 51, 1035 (1995a)
5. Nishinari, K., Takahashi, D.: J. Phys. A: Math. Gen. 31, 5439 (1998)
6. Matsukidaira, J., Nishinari, K.: Int. Mod. Phys. C 15, 507 (2004)
7. Kanai, M., Nishinari, K., Tokihiro, T.: Phys. Rev. E 72, 035102 (2005)
8. Sakai, S., Nishinari, K., Iida, S.: J. Phys. A: Math. Gen. 39, 15327 (2006)
9. Lighthill, M.J., Whitham, G.B.: Proc. R. Soc. A 299, 317 (1955)
10. Whitham, G.B.: Linear and Nonlinear Waves. John Wiley and Sons, New York (1974)
11. Payne, H.J.: Mathematical Models of Public Systems 1. In: Bekey, G.A. (ed.) Simulation Council, La Jolla, p. 51 (1971)
12. Kerner, B.S., Konhauser, P.: Phys. Rev. E 48, R2355 (1993)
13. Tokihiro, T., Takahashi, D., Matsukidaira, J., Satsuma, J.: Phys. Rev. Lett. 76, 3247 (1996)
14. Matsukidaira, J., Nishinari, K.: Phys. Rev. Lett. 90, 088701 (2003)
15. Takahashi, D., Matsukidaira, J.: J. SIAM Letters 1, 1 (2009)
16. Kanai, M., Isojima, S., Nishinari, K., Tokihiro, T.: arXiv:0902.2633 (2009)
17. Daganzo, C.F.: Trans. Res. B 29B, 277 (1995)
18. Greenshields, B.D.: Proceedings of the Highway Research Board, Washington, D.C., vol. 14, p. 448 (1935)

Functional Sized Population Magnetic Optimization Algorithm

Mehdi Torshizi[1] and M. Tayarani-N.[2]

[1] University of Payam-e-Noor, Iran
mtorshizi@yahoo.com
[2] Azad University of Mashhad, Iran
tayarani@ieee.org

Abstract. Magnetic Optimization Algorithm (MOA) is a recently novel optimization algorithm inspired by the principles of magnetic field theory whose possible solutions are magnetic particles scattered in the search space. In order improve the performance of MOA, a Functional Size population MOA (FSMOA) is proposed here. To find the best function for the size of the population, several functions for MOA are considered and investigated and the best parameters for the functions will be derived. In order to test the proposed algorithm and operators, the proposed algorithm will be compared with GA, PSO, QEA and sawtooth GA on 14 numerical benchmark functions. Experimental results show that the proposed algorithm consistently has a better performance than those of other algorithms in most benchmark function.

Keywords: Optimization Method, Magnetic Optimization algorithm, Particle Swarm Optimization, Genetic Algorithms, Numerical Function Optimization.

1 Introduction

Magnetic Optimization Algorithm (MOA) is a novel optimization algorithm [1] inspired by the magnetic field theory and the attraction between the magnetic particles. In order to improve the performance of optimization algorithms several operators are introduced one of which controls the size of the population. The size of the population is an effective parameter of the evolutionary algorithms which has a great role on the performance of EAs. Several researches investigate the effect of population size and try to improve the performance of EAs with controlling the size of the population. A functional sized population GA with a periodic function of saw-tooth function is proposed in [2]. In order to improve the performance of QEA, [3] proposes a sinusoid sized population for QEA. The best population size for genetic algorithms is found in [4]. Inspired by the natural features of the variable size of the population, an improved genetic algorithm with variable population-size is suggested in [5]. In [6], an adaptive population size for the population is proposed for a novel evolutionary algorithm. To adjust the population size, a scheme has been proposed which results in a balance between exploration and exploitation, see [7].

This paper proposes a functional size population for MOA to improve its exploration/exploitation tradeoff. Several functions for the population are proposed here to

F. Peper et al. (Eds.): IWNC 2009, PICT 2, pp. 316–324, 2010.

find the best function for the population. This paper is organized as follows: section 2 proposes the functional size for the population. Section 3 finds the best parameters for the proposed algorithm. Section 4 analyzes the proposed algorithm on 14 numerical functions and finally section 5 concludes the proposed operator.

2   Functional Sized Population MOA (FSMOA)

Another approach which maintains the diversity of the population and improves the performance of the evolutionary algorithms is the use of a variable size for the population. In [2] a variable size population is proposed for GA that improves the performance of GA. They use a saw-tooth function for the size of the population with partially re-initialization of the particles. Here to improve the performance of MOA, this paper uses a functional population size for this algorithm. In addition to the saw-tooth function, this paper uses some other functions for the proposed MOA; the functions are saw-tooth, inverse saw-tooth, triangular, sinusoid and square functions. Fig. 1 shows the functions which are examined in this paper. The pseudo code of the proposed Functional Size MOA (FSMOA) is described as below:

Procedure FSMOA
begin
      $t = 0$
1.   initialize X^0 with a structured population
2.   **while** not termination condition **do**
        begin
            $t = t + 1$
3. $n(t) = f(t)$
4.   if $n(t) < n(t - 1)$ delete particles with worst fitness
5.   and put them in the reserve set
6.   if $n(t) > n(t - 1)$ if reserve set is empty create random particles,
               else put the particles of reserve set in the population
7.   evaluate the particles in X^t and store their performance in magnetic
               fields B^t
8.   normalize B^t according to (1)
9.   evaluate the mass M^t for all particles according to (2)
10.   **for** all particles x_{ij}^t in X^t **do**
            begin
11. $F_{ij} = 0$
12.   find neighborhood set N_{ij}
13.   **for** all x_{uv}^t in N_{ij} **do**
14. $F_{ij,k} = F_{ij,k} + \frac{(x_{uv,k}^t - x_{ij,k}^t) \times B_{uv}^t}{D(x_{ij,k}^t, x_{uv,k}^t)}$
15.   **for** all particles x_{ij}^t in X^t **do**
               begin
16. $v_{ij,k}^{t+1} = \frac{F_{ij,k}}{M_{ij,k}} \times R(l_k, u_k)$
17. $x_{ij,k}^{t+1} = x_{ij,k}^t + v_{ij,k}^{t+1}$
            end

end

     end

end

The pseudo code of FSMOA is described as below:

1. In this step all of the particles in the population X^t (for $t = 0$) are initialized randomly.
2. The while loop is terminated when the termination condition is satisfied.
3. The size of the population is calculated as a function.
4. If the size of the population is decreased comparison to size of the population in the previous iteration, some particles should be deleted from the population. In the proposed algorithm, the worst particles are deleted until the size of the population is adjusted. The deleted particles are deleted and inserted in a reserve population for the future needs.
5. If the size of the population is increased, some new particles should be inserted in the population. In the proposed algorithm the particles that are inserted in the reserve set are used to inserts in the population. In this step, if the size of the population is increased, the particles of the reserve set are inserted in the population until the size of the population is adjusted. In inserting the particles in the population, if the reserve set is empty, the new random particles are created and inserted in the population.

The functions that are used in this paper are:

saw-tooth [2]:

$$n(t) = \text{Round}\left[\bar{n} - \frac{2A\bar{n}}{T-1}\left(t - T \times \text{Round}\frac{t-1}{T} - 1\right)\right] \tag{1}$$

inverse saw-tooth:

$$n(t) = \text{Round}\left[\bar{n} + \frac{2A\bar{n}}{T-1}\left(t - T \times \text{Round}\frac{t-1}{T} - 1\right)\right] \tag{2}$$

Triangular:

$$n(t) = \text{Round}\left[\bar{n} - A\bar{n} + 2A\bar{n} \max\left(\min\left(\frac{2\text{mod}(t,T)}{T}, 2 - \frac{2\text{mod}(t,T)}{T}\right), 0\right)\right] \tag{3}$$

sinusoid:

$$n(t) = \text{Round}\left[\bar{n} + A\bar{n} \sin\left(\frac{2\pi}{T}t\right)\right] \tag{4}$$

square:

$$n(t) = \text{Round}\left[\bar{n} + A\bar{n} - 2A\bar{n} \times \text{Round}\left(\frac{\text{mod}(t,T)}{T}\right)\right] \tag{5}$$

Where $n(t)$ is the size of the population in generation t, $\bar{n}$ is the average size of the population, A is the amplitude of the periodic function of population size, T is the period of the functional population, Round(.) is the round function (rounds its input to nearest integer), and mod(., .) is modulus after division function. Fig 1 shows the functions which are used in this paper. The best values for T and A are found in the following of this section.

Table 1. The best parameters for the proposed FSMOA

	saw-tooth				inverse-saw				sinusoid				square				triangular				MOA
	A	T	Best	Ttest	A	T	Best	Ttest	A	T	Best	Ttest	A	T	Best	Ttest	A	T	Best	Ttest	Best
f_1	0.9	250	**23000**	0.19	0.4	25	22528	0.55	0.9	500	22743	0.33	0.2	25	22921	0.26	0.9	250	23031	0.18	21881
f_2	0.6	25	-78.79	0.03	0.4	25	-80.51	0.05	0.9	250	**-70.91**	0.002	0.9	250	-78.74	0.03	0.9	250	-79.62	0.05	-105
f_3	0.9	25	**-0.80**	7e-5	0.6	25	-1.03	0.002	0.9	100	-0.84	1e-4	0.9	1000	-0.82	6e-4	0.9	25	-0.86	0.0002	-1.71
f_4	0.9	25	**0.60**	2e-6	0.9	25	0.52	7e-5	0.9	25	0.58	1e-5	0.9	25	0.49	6e-4	0.9	25	0.60	1.7e-6	0.29
f_5	0.1	25	**61.73**	0.22	0.4	25	60.36	0.55	0.2	25	59.37	0.86	0.2	250	61.47	0.30	0.1	250	60.48	0.55	51.26
f_6	0.9	25	-28.91	1e-4	0.9	25	-34.9	7e4	0.9	25	-32.15	3e-4	0.9	500	**-27.93**	1e-4	0.9	25	-36.96	0.002	-108
f_7	0.2	25	30.44	0.98	0.1	25	31.60	0.38	0.4	1000	31.56	0.28	0.2	100	31.86	0.28	0.2	25	**32.47**	0.07	31.28
f_8	0.4	25	93.87	0.29	0.9	25	**94.55**	0.04	0.4	25	94.01	0.27	0.6	1000	93.94	0.31	0.9	250	93.93	0.28	91.59
f_9	0.9	25	-79.2	3e-4	0.9	25	-91	7e4	0.4	25	-121	0.006	0.9	25	-175	0.21	0.9	100	-97.54	0.001	-362
f_{10}	0.4	25	**-0.05**	8e-6	0.9	25	-0.06	0.01	0.9	25	-0.059	0.002	0.9	25	-0.07	0.73	0.9	25	-0.05	8.5e-5	-0.07
f_{11}	0.9	25	-7.44	1e-5	0.9	25	-7.67	2e-5	0.9	25	**-7.30**	1e-5	0.9	25	-7.44	2e-5	0.9	100	-7.67	5e-5	-14.44
f_{12}	0.9	25	**-68**	0.004	0.9	25	-85	0.03	0.9	100	-82	0.01	0.9	25	-100	0.07	0.9	100	-83	0.02	-187
f_{13}	0.9	25	**-100**	6e-4	0.9	100	-102	0.002	0.9	25	-102	0.002	0.9	25	-106	0.01	0.9	25	-101	0.001	-117
f_{14}	0.1	1000	-0.033	0.10	0.1	250	-0.033	0.34	0.2	500	-0.033	0.24	0.6	250	**-0.032**	5e-10	0.2	1000	-0.033	0.01	-0.033

6. In this step the fitness of each particle x_{ij}^t in X^t is calculated and stored in the magnetic field, B_{ij}^t.

7. The normalization is performed on the B^t. The normalization is performed as:

$$B_{ij} = \frac{B_{ij} - Min}{Max - Min} \tag{6}$$

Where: $Min = \underset{i,j=1}{\overset{S}{\text{Minimum}}}(B_{ij}^t)$, $Max = \underset{i,j=1}{\overset{S}{\text{Maximum}}}(B_{ij}^t)$.

8. In this step the mass of all particles is calculated and stored in M^t:

$$M_{ij}^t = \alpha + \rho \times B_{ij}^t \tag{7}$$

Where α and ρ are constant values.

9. In this step in the "for" loop, the resultant force of all forces on each particle is calculated.

10. At first the resultant force which is applied to particle $x_{ij}^t (F_{ij})$ is set to zero.

11. In this step the neighbors of x_{ij}^t is found. The set of neighbors for particle x_{ij} is defined as: $N_{ij} = \left\{ x_{i^1 j}, x_{i j^1}, x_{i^2 j}, x_{i j^2} \right\}$
Where:

$$i^1 = \begin{cases} i-1 & i \neq 1 \\ S & i = 1 \end{cases}, j^1 = \begin{cases} j-1 & j \neq 1 \\ S & j = 1 \end{cases} i^2 = \begin{cases} i+1 & i \neq S \\ 1 & i = S \end{cases}, j^2 = \begin{cases} j+1 & j \neq S \\ 1 & j = S \end{cases}$$

Table 2. Median and standard deviation(Std) of the best parameters for the proposed FSMOA

	A		T	
	Median	Std	Median	Std
Saw-tooth	0.9	0.33	25	262
inverse-saw	0.9	0.33	25	81
sinusoid	0.9	0.31	25	289
square	0.9	0.25	175	340
triangular	0.9	0.35	100	270

Table 3. Experimental results on 14 numerical benchmark functions for $m = 100$ and $m = 500$. the results are averaged over 30 runs. Ttest shows Ttest between the results of each algorithm with FSMOA.

	FSMOA		MOA			GA			PSO			saw-tooth GA		
$m-100$														
	Mean	Std	Mean	Std	Ttest	Mean	Std	Ttest	Mean	Std	Ttest	Mean	Std	Ttest
f_1	23067	664	22803	2139	0.71	**39249**	350	0	9634	1568	2e-15	22413	855	0.59
f_2	-74	15.63	-114	19.75	9e-5	-134.28	7.62	2e-9	-842	51.71	0	-743	42.48	0
f_3	**-0.86**	0.38	-1.95	0.77	8e-4	-5.05	0.24	2e-16	-8.93	0.43	0	-14.92	0.32	0
f_4	**0.60**	0.04	0.16	0.17	3e-7	-0.32	0.04	0	-0.90	0.11	0	-5.72	0.78	7e-15
f_5	**58.25**	6.0	34.25	9.3	2e-6	51.46	7.30	0.03	-782	784	0.003	-29493	3105	0
f_6	**-26.06**	4.13	-141	38.55	2e-8	-162.33	28.18	1e-11	-600	169	3e-9	-5625	525.9	0
f_7	30.94	2.1	31.38	3.70	0.74	79.11	1.27	0	13.91	2.06	5e-13	35.22	2.08	0.01
f_8	**93.72**	1.26	87.71	1.75	6e-8	92.99	0.66	0.12	58.78	2.01	0	65.20	1.84	2e-16
f_9	**-66.96**	12.4	-517.61	396	0.002	-3824	368	0	-11624	1704	3e-14	-61051	5514	0
F_{10}	**-0.06**	0.01	-0.09	0.03	0.002	-0.32	0.02	0	-0.81	0.08	2e-16	-1.87	0.08	0
F_{11}	**-8.02**	1.28	-19.42	2.9	2e-9	-51.88	3.58	0	-27.62	3.07	3e-13	-75.13	3.08	0
F_{12}	**-66.66**	23	-176	140	0.02	-12387	3049	2e-10	-12696	2291	1e-12	-4.8e5	1.2e5	2e-10
F_{13}	**-100**	0.45	-125	15.51	7e-5	-657	74.54	5e-15	-542.03	66.04	4e-14	-4409	618	2e-14
F_{14}	**-0.0323**	7e-18	-2.29	0.48	2e-11	-0.13	0.01	5e-15	-10.92	3.15	2e-9	-3.73	0.67	3e-5
$m=500$														
	Mean	Std	Mean	Std	Ttest	Mean	Std	Ttest	Mean	Std	Ttest	Mean	Std	Ttest
f_1	7.2e4	2602	6.0e4	6615	5e-5	**1.17e5**	824	0	2.2e4	3183	0	5.4e4	3508	0.02
f_2	-2233	112	-2460.6	125	4e-4	-3551	83	0	-4710	124	0	-6583	139	0
f_3	**-5**	0.29	-7	0.46	5e-10	-14	0.17	0	-9	0.22	0	-17	0.05	0
f_4	**-1.15**	0.06	-3.03	0.18	0	-29.89	0.98	0	-5.70	0.25	0	-80.94	2.06	0
f_5	**-420**	325	-3735	1433	1e-6	-1.41e5	1e4	0	-8119	2497	1e-8	-4.5e5	1.8e4	0
f_6	**-983**	97	-2709	267	2e-13	-2.9e4	1526	0	-3588	268	2e-16	-8.9e4	2770	0
f_7	63.46	6.06	62.99	6.27	0.86	**197.18**	5.05	0	32.34	2.30	1e-11	70.75	2.88	0.002
f_8	**365**	7.86	318	5.17	6e-12	338	3.91	2e-8	259	4.76	0	223.92	4.64	0
f_9	**-1.1e4**	733	-3.6e4	5270	2e-11	-3.17e5	1.6e4	0	-6.5e4	3218	0	-9.0e5	2.8e4	0
F_{10}	**-0.39**	0.01	-0.54	0.02	2e-13	-1.77	0.04	0	-0.88	0.03	0	-3.45	0.06	0
F_{11}	**-20**	1.12	-31	2.29	1e-10	-87.38	1.31	0	-32.94	2.93	3e-10	-94.76	0.87	0
F_{12}	**-4.0e4**	1.0e4	-2.2e5	7.3e4	2e-7	-1.13e7	8e5	0	-5.9e5	7.2e4	4e-15	-9e+7	7.6e6	0
F_{13}	**-984**	36	-2012	117	6e-16	-1.9e4	1635	0	-3406	236	0	-8.8e4	4015	0
F_{14}	**-24**	1.31	-58	2.34	0	-32.5	1.95	1e-9	-132	2.64	0	-92.63	4.95	1e-13

12. In this step, in the "for" loop the force which is applied to particle x_{ij}^t from its neighbor's $x_{uv}^t (\forall x_{uv}^t \in N_{ij})$ is calculated.

13. The force which is applied from x_{uv}^t to x_{ij}^t is related to the distance between two particles and is calculated as:

$$F_{ij,k} = \frac{\left(x_{uv,k}^t - x_{ij,k}^t\right) \times B_{uv}^t}{D\left(x_{ij,k}^t, x_{uv,k}^t\right)} \tag{8}$$

Where $D(.,.)$ is the distance between each pair of neighboring particles and defined as:

$$D\left(x_{ij}^t, x_{uv}^t\right) = \sqrt{\frac{1}{m} \sum_{k=1}^{m} \left(\frac{x_{ij,k}^t - x_{uv,k}^t}{u_k - l_k}\right)^2} \tag{9}$$

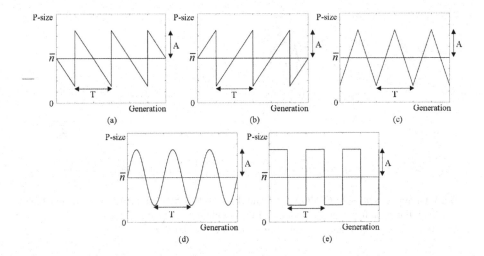

Fig. 1. a) The functions which are used for the population size. a) saw-tooth b) inverse saw-tooth c) triangular d) sinusoid e) square. T is the period of the functions, A is the amplitude and P-size is the size of the population in generation t.

14. In the "for" loop, the movement of all particles is calculated.

15,16 In these two steps the velocity and the location of the particles is updated:

$$v_{ij,k}^{t+1} = \frac{F_{ij,k}}{M_{ij,k}} \times R(l_k, u_k) \qquad x_{ij,k}^{t+1} = x_{ij,k}^t + v_{ij,k}^{t+1} \tag{10}$$

Indeed in the proposed algorithm, in the decreasing step of the algorithm, the particles are not deleted, they only leave the population and in the increasing step they return to the population. This policy is used because if the random particles be inserted in the population, the randomness of the algorithm is overflowed and the performance of the algorithm is decreased. With this policy, the knowledge that exists in the worst particles is not ignored and is used in the future iterations.

The proposed functional size population has two cycles. The first cycle is increasing the size of population. In the increasing cycle, the new particles are inserted in the population. Inserting the new particles in the population increases the diversity of the population and improves the exploration performance of the algorithm. The other cycle is the decreasing cycle. In this cycle, the worst particles of the population are eliminated. This treatment improves the exploitation of the algorithm by exploiting the best solutions and ignoring the inferior ones. This means that the proposed algorithm has two cycles: exploration cycle and exploitation cycle.

3   Finding the Best Parameters

As it can be seen in Fig. 1, the proposed functions have some parameters that are A, the amplitude and T the period of the functions. In order to find the best values for these

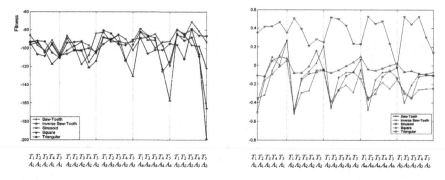

Fig. 2. Parameter setting of FSMOA for T and A for (a) Rastrigin's Function and (b) Generalized Griewank Function for several functions for the population. The parameters are set to $T_1 \cdots T_5 = (25, 100, 250, 500, 1000)$ and $A_1 \cdots A_5 = \bar{n} \times (0.1, 0.2, 0.4, 0.6, 0.9)$.

parameters some experiments are done. Fig. 2 shows the finding of the best parameters for the proposed FSMOA for Generalized Griewank Function. The best parameters for the other 14 numerical benchmark functions are found similar to the Generalized Griewank function. The best parameters and the best functions for the size of the population are summarized in Table 1. According to Table 1 the saw-tooth function has the best results for 8 numerical benchmark functions, the sinusoid and square for 2 benchmark functions, triangular function for 1 benchmark functions, and inverse saw-tooth for 1 benchmark function, so the best function for the size of the population is saw-tooth function. Table 1 shows that the functional size for the population improves the performance of MOA for all the benchmark functions. Table 2 shows the median and standard deviation of the best parameters for 5 proposed functions. According to this table the best amplitude for the proposed functions is 0.9 and best period T is 25. Table 3 shows the Ttest between the FSMOA and MOA. The Ttest for most of results is so small; it shows that the results of Table 1 are valid.

4   Experimental Results

The proposed FSMOA is compared with 4 algorithms, GA, PSO, MOA and saw-tooth GA. In order to compare the algorithms with their best parameters the best parameters for each algorithm is found independently. The experimental results are performed for several dimensions ($m = 5, 10, 25, \mathbf{100}, 250, 500, 1000, 1500, 2000$ and 3000) of 14 numerical benchmark functions. The average population size of all algorithms for all of the experiments is set to 25; termination condition is set for a maximum of 1000 generations and the structure of population is considered as cellular. Due to statistical nature of the optimization algorithms, all results are averaged over 30 runs and Ttest analysis is performed on results. The best parameters of MOA, GA, PSO and saw-tooth GA are found and the parameters of FSMOA are set to the best parameters which are found in previous section.

Table 4. The effect of dimension on the performance of FSMOA. By increasing the dimension of problems, FSMOA dominates the other algorithms.

Dimension	FSMOA	MOA	GA	PSO	saw-tooth
5	3	6	0	0	5
10	3	6	5	0	0
25	2	7	5	0	0
100	12	0	2	0	0
250	12	0	2	0	0
500	12	0	2	0	0
1000	12	0	2	0	0
1500	12	0	2	0	0
2000	12	0	2	0	0

Table 3 summarizes the experimental results on MOA, FSMOA, GA, PSO and saw-tooth GA for 14 benchmark functions (The results for some dimensions are not summarized in Table 3 because of small space of the paper). The last 3 algorithms are some well known and state of the art algorithms for the comparison aim. In Table 3 the best results are bolded. As it seen the Ttest value for most of experiments is so small and this means that each algorithm reaches a distinct performance. The Ttest is performed between the results of FSMOA and several algorithms. For the size of problem $m = 5$ the best algorithm is saw-tooth GA. This algorithm has the best results for 4 benchmark functions. After saw-tooth is FSMOA with 3 best results and MOA with 2 and QEA with 1 best result in 14 numerical benchmark functions. Standard GA and PSO could not yield best result for any functions. For $m = 5$, for 7 functions the best results is for the proposed algorithms in this paper and for 7 functions the best results are for other well known algorithms.

Increasing the dimension of the problem changes the performance of the algorithms. For $m = 10$ saw-tooth GA gives up its superiority and FSMOA and GA reach the best performance for 4 functions and after them are QEA and EMOA with 2 and finally SRMOA and saw-tooth GA with 1 best result among 14 numerical functions and 8 compared algorithms. By increasing the dimension of the problem FSMOA dominates the other algorithms. For $m = 25$ FSMOA has the best results for 8 functions, GA for 4 functions and QEA and MOA for 1 functions. For $m = 100$ FSMOA is the best algorithm. Because it has the best results for 10 functions, after FSMOA is GA with 3 and QEA with 1 best result. By increasing the dimension of algorithm to 500 the FSMOA retains its domination. According to Table 3 the best algorithm is FSMOA, but FSMOA can not work well on Schwefel 2.26, Goldberg, Michalewicz and Kennedy functions. The best algorithm for Schwefel 2.26 is QEA which has the best performance for this function in most dimensions. saw-tooth GA has a fine performance only for low dimensions and for problems with high dimensions saw-tooth GA can not work well.

5 Conclusion

One of the main parameters of evolutionary algorithms is the population of the algorithm. This paper proposes a novel functional sized population magnetic optimization algorithm called FSMOA. The functional sized population of FSMOA makes a find trade-off between exploration and exploitation in FSMOA and makes it possible to escape from local optima. This paper examines 5 functions for the population on 14 numerical objective functions and finds the best parameters for the 14 objective functions. In order to examine the proposed algorithm several experiments are performed on several dimensions of several objective functions. Experimental results show that the proposed functional sized population improves the performance of MOA.

In order to compare the performance of the algorithm through several dimensions the experiments are performed over several dimensions. Table 4 shows the domination of each algorithm for several functions and dimensions. We can claim that by increasing the dimension of the problem, the performance of FSMOA dominates other algorithms.

References

1. Tayarani-N, M.H., Akbarzadeh-T, M.R.: Magnetic Optimization Algorithm, A New Synthesis. In: IEEE World Conference on Computational Intelligence (2008)
2. Koumousis, V.K., Katsaras, C.P.: A saw-tooth Genetic Algorithm Combining the Effects of Variable Population Size and Reinitialization to Enhance Performance. IEEE Trans. Evol. Comput. 10, 19–28 (2006)
3. Tayarani-N, M.H., Akbarzadeh-T, M.R.: A sinusoid Size Ring Structure Quantum Evolutionary Algorithm. In: IEEE International Conference on Cybernetics and Intelligent Systems Robotics, Automation and Mechanics (2008)
4. Wang, D.L.: A Study on the Optimal Population Size of Genetic Algorithm. In: Proceedings of the 4th World Congress on Intelligent Control and Automation (2002)
5. Shi, X.H., Wan, L.M., Lee, H.P., Yang, X.W., Wang, L.M., Liang, Y.C.: An Improved Genetic Algorithm with Variable Population Size and a PSO-GA Based Hybrid Evolutionary Algorithm. In: International Conference on Machine Learning and Cybernetics (2003)
6. Jun, Q., Li-Shan, K.: A Novel Dynamic Population Based Evolutionary Algorithm for Revised Multimodal Function Optimization Problem. In: Fifth World Congress on Intelligent Control and Automation (2004)
7. Zhong, W., Liu, J., Xue, M., Jiao, L.: A Multi-agent Genetic Algorithm for Global Numerical Optimization. IEEE Trans. Sys., Man and Cyber. 34, 1128–1141 (2004)
8. Khorsand, A.R., Akbarzadeh-T, M.R.: Quantum Gate Optimization in a Meta-Level Genetic Quantum Algorithm. In: IEEE International Conference on Systems, Man and Cybernetics (2005)

Appendix

In order to compare the proposed algorithm with the well-known and state of the art algorithms, the proposed algorithm is compared with other algorithms on 14 numerical benchmark functions. The functions are f_1:Schwefel 2.26 [7], f_2:Rastrigin [7], f_3:Ackley [7], f_4:Griewank [7], f_5:Penalized 1 [7], f_6:Penalized 2 [7], f_7:Michalewicz [8], f_8:Goldberg [2], f_9:Sphere Model [7], f_{10}:Schwefel 2.22 [7], f_{11}:Schwefel 2.21 [7], f_{12}:Dejong [8], f_{13}:Rosenbrock [2], and f_{14}:Kennedy [2].

Emergence and Collapse of Order
in Ad Hoc Cellular Automata

Soichiro Tsuda

International Center of Unconventional Computing
University of the West of England
Soichiro.Tsuda@uwe.ac.uk

Abstract. A system that shows life-like properties requires moderate complexity, so-called the "edge of chaos" state. To investigate how such state is maintained by a system in which agents have limited access to the information about other agents, a simple cellular automata-based model, called ad hoc cellular automata (ACA) is designed and its behaviour is examined. This model shows complex Class 3-like pattern and follows a power law relation. It is also found that the classification of the model is totally different from elementary cellular automata in terms of entropy.

1 Introduction

Complexity is essential for life. However, complexity science revealed high complexity would result in utter chaos, and therefore modest complexity is required for a system to be complex and emergent (cf. [1]). Numerous complex system models are so far proposed to answer how a system can maintain the complexity of the system at an appropriate level, i.e., at the "edge of chaos" [2]. Some models adopt two coupled dynamics to achieve complex emergent systems. A classical example of such model is Turing's chemical morphogenesis [3]. More recently, self-organised criticality (SOC) models have been proposed by Bak and his co-workers [4] in the context of complex systems. In a SOC system, two, often opposing, dynamics maintain the system in the edge-of-chaos state.

The aim of this paper is to discuss how the edge-of-chaos state is achieved with a cellular automata (CA)-like multi-agent system employing two cross-scale dynamics. As agents in the real world (e.g., traders in the stock market) cannot access to all the information available at the present moment, agents in the model are assumed to have limited access to the information about other agents. Of particular interest here is the effect of the interaction length between agents in the system and how it contributes to the maintenance of complex and emergent structures in the system.

2 Ad Hoc Cellular Automata with Intra-agent Dynamics

The CA-like model, called ad hoc cellular automata (ACA), is based on one-dimensional two-state and three-neighbourhood cellular automata (elementary

F. Peper et al. (Eds.): IWNC 2009, PICT 2, pp. 325–332, 2010.

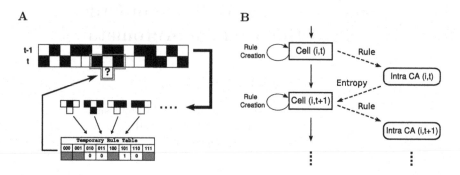

Fig. 1. The schematic diagrams of ACA. Each cell creates a temporary rule table from past transitions of other cells at time t-1 and applies it to compute its next state at time t+1 (A). Each cell in ACA interacts with its intra-agent automata by giving a temporarily created rule and taking a entropy value of the intraCA. The entropy value is used to create a rule table at time t+1 (B).

cellular automata, ECA). Different from the ECA, however, cells in ACA do not share a fixed transition rule with others. Instead, each cell in ACA mimics behaviour of other cells by copying their past transitions (including itself) at time t-1 and create a temporary rule from those collected information. The temporary rule is then applied to compute the state of the cell at time $t + 1$ (Fig. 1A). The number of transitions a cell can copy depends on a lookup range parameter E, which is determined by the intra-agent dynamics as described below. For example, if $E = 5$, transitions from the $(i$-2)-th to $(i$+2)-th cell are collected and used to make a table for the i-th cell. If two or more transitions conflict with each other, e.g. there exist 010→1 and 010→0 in a collected transition data set, a transition from the rightmost cell triplet is adopted. If the temporary rule table is incomplete, blank rules is filled in by random number generators to complete the table. The created temporary rule is then applied to the current cell at time t to compute the next state of the cell. Every time step each cell repeats this rule creation process (called inter-agent dynamics) creates a temporary rule table, which is effectively different from that of other cells.

The intra-agent dynamics of a cell in ACA is defined as one-dimensional two-state three-neighbourhood ECA, called intraCA. A temporary rule table created by the i-th ACA cell at time t is handed over to the intraCA and used to compute the time development of intraCA over T_{in} steps (Fig. 1B). During the computation of the time development, the metric entropy $H(t)$ of bit sequence of cells in intraCA is calculated for each time step by:

$$H(t) = -\frac{1}{X} \sum_{i=0}^{2^X - 1} \frac{Q_t^i}{N} log(\frac{Q_t^i}{N}) \tag{1}$$

where X is neighbourhood size, N is the number of cells. Q_t^i is the appearance frequency of bit pattern i and therefore Q_t^i/N represents the probability of

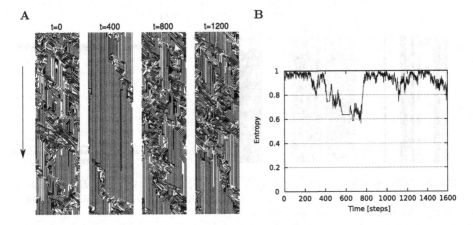

Fig. 2. **(A)** A spatio-temporal pattern of ACA. Black sites indicate the state is one. A globally stable state can be observed around t=600, followed by a collapse of the order as a bundle of chaotic region spreads over the entire system. **(B)** Time course of metric entropy for (A). Entropy increases or decreases as bundles of chaotic region appear or disappear, respectively.

pattern i^1. After T_{in} steps, mean entropy $\overline{H}$ is returned to the i-th ACA cell at time $t+1$ and used to determine the lookup range parameter E, as:

$$E = \alpha\overline{H} \quad (\text{mod } 10) \tag{2}$$

where α is a scaling parameter. As mentioned above, depending on the parameter E, the i-th ACA cell at time $t+1$ collects the past transitions at time t to create a new rule. Accordingly, two dynamics of an ACA cell, rule generating dynamics and intraCA dynamics, are coupled together to compute the time development of the ACA cell.

In this study, we fix the parameters as $X = 3$, $N = 100$, $T_{in} = 200$, and $\alpha = 10$. The number of ACA and intraCA cells are set to 100 and 50, respectively. The spatio-temporal pattern of ACA with periodic boundary condition is observed over 1600 steps.

3   Results

3.1   Emergence and Collapse of Complex/Periodic Patterns

We have carried out 20 runs with random initial condition, periodic boundary condition for ACA cells, and random initial rules for intraCA. It is particularly worth noting that, in all 20 runs, ACA showed similar pattern as shown in Fig. 2A. The cells in ACA are organised into chaotic region or periodic stripe region shortly after the start of computation. Those regions dynamically evolve

---

[1] i is decimal representation of bit pattern, e.g. pattern 5 represents a bit pattern 101.

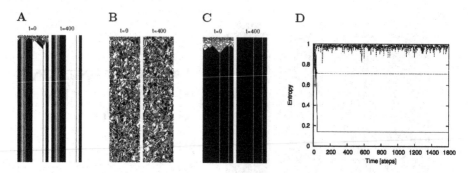

Fig. 3. Spatio-temporal patterns of ACA with fixed lookup parameter E (A), random E (B), and the case E is equivalent to the system size (C). Time courses of metric entropy for these three cases are plotted in (D).

in time and a bundle of chaotic region travels through the stripe background. In general, the bundle expands from left to right. This is possibly because the rightmost cell triplet is adopted when a temporary rule table is created. In this example case, bundles slowly diminish after $t = 200$ steps and completely disappear around at $t = 600$. However, it should be emphasised that a chaotic region re-appears from a point even after the whole system completely fell into a periodic phase, and it forms a bundle and spreads all over the cells ($t = 1000$). The chaotic region started to re-diminish at the end of the run (around $t = 1600$). All 20 runs showed such repeating pattern of the chaotic and periodic phase, although the timing of the first appearance of a periodic phase was different. It has been confirmed that ACA systems started with periodic initial conditions and single point initial condition (all zeros except single one for one ACA cell) showed similar time development patterns.

To quantify these pattern transitions, the metric entropy by Eq. (1) is calculated for a bit sequence of ACA at each time step[2]. Fig. 2B shows the time course of the entropy for the spatio-temporal pattern of Fig. 2A. Decrease in entropy is observed as the chaotic regions shrink, and finally the entropy becomes constant when the entire system is in a periodic phase around at $t = 600$, followed by a sharp increase in entropy along with the spread of the chaotic bundle.

For control experiments, several models based on the ACA are examined. As mentioned in the previous section, the effect of the interaction length between cells, i.e., the lookup range parameter E, is of interest here. Three cases are considered: (1) E is fixed, (2) E is determined randomly, and (3) E is equal to the ACA system size N. Note that the computation of intraCA is not required in these three cases.

For the first case that the lookup range parameter E was fixed to a constant value ranging from zero to ten throughout the computation, all the spatio-temporal pattern fallen into Class 1 (all black or all white state) or Class 2

---

[2] Note that this is calculated over ACA cells, not for intraCA cells.

pattern (periodic state) after several hundreds of time steps and the pattern never changed once it is fallen into a state (Fig. 3A). The second case, when a random number from one to ten is assigned for E of each ACA cell at every time step, there was no specific structure but fragile small periodic regions in the pattern (Fig. 3B). In the last case, an ACA cell can obtain all the past transitions made at a previous time step. A temporary rule for each agent is created in a 'democratic' way: The frequencies of transitions to zero and one from a triplet are compared. For example, the number of $110 \rightarrow 0$ transition is greater than that of $110 \rightarrow 1$ in the collection of all transitions made at a previous time step, $110 \rightarrow 0$ is adopted to compute the next state of an ACA cell (and vice versa). This is applied to all the triplets to fill a temporary rule table. If no transition to both zero and one from a triplet in the collected transition data set, a random value, either zero or one, is used to complete the rule table. As all ACA cells share the same transition data set (all the past transitions made at a previous time step), temporary rule tables for ACA cells created by this process are identical. 100 runs are carried out with this model. Most of observed spatio-temporal pattens are Class 2 periodic patterns as shown in Fig. 3C, although Class 1 and 3-like patterns are observed sporadically.

Regarding the metric entropy of these models, it settled down to a specific level after several hundreds of time steps in the first and last case (Fig. 3D middle and bottom lines, respectively), and remained high throughout the time development in the second case (Fig. 3D top curve).

Additionally, we investigated the effect of random bit filling for temporal rule generation. Specifically, empty rule cells are filled either with all zeros or all ones, instead of random bits to make a complete rule table. The observed patterns were mostly either Class 1 or Class 2 (no data shown). This suggests that the random bit filling secures a driving force for the whole ACA system to deviate itself from a globally-ordered mode, whereas the past transition mimicking for temporal rule generation organises neighbouring ACA cells to behave similarly.

3.2   Emergence and Collapse of Clusters in Rule-Entry Space

Figure 4A–C are plots of the temporary rule that passed from each ACA cell to its intraCA against the entropy returned from it using the same example run for Fig. 2A. Each point in the plots corresponds to an ACA cell at a time step. When the whole ACA system is in the chaotic phase, ACA cells in the rule-entropy space is uniformly scattered ($t = 100$, Fig. 4A). Cells slowly gather as the chaotic region in ACA disappears, and eventually form four clusters around at time $t = 600$ (Fig. 4B). However, the clusters start to dissipate within a hundred steps with the re-appearance of a chaotic region and finally come back to the scattered distribution again after 400 steps ($t = 1000$, Fig. 4C).

Fig. 4D shows the time development of entropy of a cell at site = 9. Even if the whole system is in a periodic phase (around $t = 600$), in most cases an ACA cell is required to fill empty rule cells with random bits to make a complete CA rule table. Thus the created CA rule slightly differs from that of the previous time step and the resulting entropy values are always fluctuating in time. More

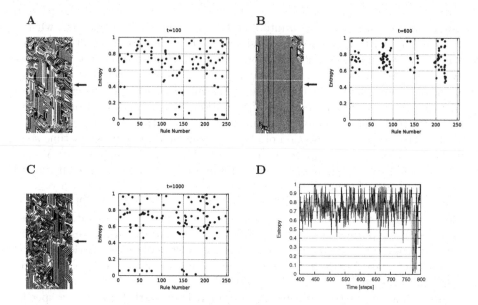

Fig. 4. Emergence and collapse of clusters in entropy-rule space. The entropy is plotted against rule numbers (A)-(C) based on the same data from Fig. 2A . Spatio-temporal patterns on the left of the plots are taken from Fig. 2A. Arrows by the patterns indicate the time in which entropy values and rule numbers are sampled for the plots. The time course of the entropy value returned from the intraCA of ACA cell at site = 9 (D).

importantly, this sporadically gives a low entropy value, which could cause a big change in the spatio-temporal pattern of the ACA system. In fact, a low entropy around at $t = 670$ triggered a collapse of the periodic phase and resulted in the widespread chaotic phase (see Fig. 2A). This is because the low entropy gives a small lookup range parameter E (see eq. (refeq:2)), which allows an ACA cell to collect only a few past transitions of other cells. As a result, a new temporary rule for the ACA cell is mostly filled with random bits and largely differs from those for the the other cells as well as its past rule at previous time steps. This worked as a "seed" for the entire collapse of the globally-ordered structure.

3.3   Classifying ACA

Wuensche proposed a method to classify complex Class 4 rules from the other Classes using the metric entropy [5] and found complex rules can be separated by plotting the standard deviation (SD) of the entropy time series against the mean. We here use this method to allocate a position for ACA in the mean-SD space. The SD and mean for entropy time series, such as Fig. 2B, are calculated for each 20 runs of normal ACA and a control experiment with random lookup parameter E (the second case, termed "random ACA" for short). Additionally, those for each ECA rule (256 rules) are calculated for comparison. Fig. 5A is the

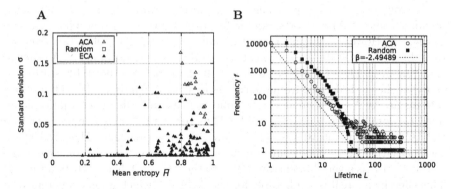

Fig. 5. Classification of ACA by plotting mean entropy against the standard deviation of the entropy (A) and lifetime of stationary states (B). Plotted data for each model is taken from one of the 20 runs. All the cases showed similar distribution for each model. A power law can be found in the distribution of normal ACA. A dashed regression line is the least square fitting of lifetime from $L = 2$ to 50 by $\ln f = \alpha + \beta \ln(L)$.

plot of the SD against the mean. Clearly, these three types of CA models are located in different areas. The random ACA concentrates in a high-mean low-SD spot, while normal ACA extends over a wider range from lower to higher SD values in a higher mean entropy region. More importantly, all runs of normal ACA took higher SD values than any of ECA. This is rather obvious because ACA has more complex configuration than ECA, but at least it indicates that the configuration of normal ACA allows the system to have several complexity levels, and at the same time it keeps its complexity high on average.

3.4   Power Law in Stationary State Distribution

Lifetime of an ACA cell in the stationary state is counted. This indicates how long the cell stay in the same state. We did not distinguish different states (zero or one) and simply focused on how many time steps zeros and ones are preserved throughout the time evolution of a cell. The distribution of the frequency of lifetime against the length of lifetime for normal ACA follows a power law with exponent -2.5, whereas that for ACA without the intraCA does not (Fig. 5B).

4   Discussion and Conclusion

We introduced a simple rule-changing cellular automaton, called ah hoc cellular automata (ACA), each cell of which continually creates temporary rules using the intra-agent dynamics for computation of the next state. The spatio-temporal pattern of the system showed a complex, Class 3-like behaviour that the chaotic regions travel through the periodic background. This complex pattern is due to the coupling of two dynamics in the ACA cell. As each ACA cell follows past transitions of other ACA cells, the inter-agent dynamics (temporary rule generation) orchestrate the whole system into a stable state. The intra-agent dynamics (intraCA) supports this orchestration into a periodic phase, as seen in Fig. 2A. However, the intra-agent dynamics can trigger a deviation from the periodic phase as well. As mentioned in Fig. 4D, the entropy of intraCA is always changing even when the whole ACA system is in the periodic stable state. This constant change is derived from the rule generating dynamics. When a temporary rule for an ACA cell is created at a time step, blank cells in the rule table are filled randomly to complete the table. Thus temporary rules handed over from the ACA cell to its intraCA is not the same at every time step. This gives rise to the fluctuation in the time course of the intraCA entropy, and in some cases, the intraCA returns a very low entropy value to the ACA cell, like $t = 670$ in Fig. 4D. This effectively creates a rule table for the ACA cell which is very different from other neighbouring ACA cells and also from the previous rule table for the cell because the ACA cell received the low entry value can obtain only one or two past transitions to create a rule for the next step. This eventually leads the whole system to deviate from a periodic phase. As illustrated by the entropy analysis in section 3.3, the interdependency of two dynamics maintains the whole ACA system in the complex state (high mean entropy) and allows it to have wider range of complexity levels (high SD of the entropy). This means each ACA cell is open to change due to the intrinsic randomness in the rule generating dynamics, while the same dynamics lead the cell to behave the same as neighbouring cells. In fact, other models proposed as control experiments showed either simple or chaotic behaviour and could not take the wide range of complex states as they rely only on the rule generating dynamics.

Acknowledgements

The author wishes to thank Prof Yukio-Pegio Gunji and Dr Stefan Artmann for their helpful comments.

References

1. Wolfram, S.: A New Kind of Science. Wolfram Media (May 2002)
2. Langton, C.G.: Computation at the edge of chaos: phase transitions and emergent computation. Phys. D 42(1-3), 12–37 (1990)
3. Turing, A.M.: The chemical basis of morphogenesis. Philosophical Transactions of the Royal Society (B) 237, 37–72 (1952)
4. Bak, P., Tang, C., Wiesenfeld, K.: Self-organized criticality: An explanation of the 1/f noise. Phys. Rev. Lett. 59(4), 381–384 (1987)
5. Wuensche, A.: Classifying cellular automata automatically: Finding gliders, filtering, and relating space-time patterns, attractor basins, and the z parameter. Complexity 4(3), 47–66 (1999)

A Transition Rule Set for the First 2-D Optimum-Time Synchronization Algorithm

Hiroshi Umeo, Kaori Ishida, Koutarou Tachibana, and Naoki Kamikawa

Univ. of Osaka Electro-Communication,
Neyagawa-shi, Hastu-cho, 18-8, Osaka, 572-8530, Japan
umeo@cyt.osakac.ac.jp

Abstract. The firing squad synchronization problem on cellular automata has been studied extensively for more than forty years, and a rich variety of synchronization algorithms have been proposed for two-dimensional cellular arrays. In the present paper, we reconstruct a *real-coded* transition rule set for an optimum-time synchronization algorithm proposed by Shinahr [11], known as the first optimum-time synchronization algorithm for two-dimensional rectangle arrays. Based on our computer simulation, it is shown that the proposed rule set consists of 28-state, 12849 transition rules and has a validity for the synchronization for any rectangle arrays of size $m \times n$ such that $2 \le m, n \le 500$.

1 Introduction

We study a synchronization problem that gives a finite-state protocol for synchronizing a large scale of cellular automata. The synchronization in cellular automata has been known as a firing squad synchronization problem (FSSP, for short) since its development, in which it was originally proposed by J. Myhill in Moore [6] to synchronize all parts of self-reproducing cellular automata. The problem has been studied extensively for more than 40 years and a rich variety of synchronization algorithms have been proposed for one- and two-dimensional arrays. In the present paper, we reconstruct a *real-coded* transition rule set for an optimum-time synchronization algorithm proposed by Shinahr [11], known as the first optimum-time synchronization algorithm for two-dimensional rectangle arrays. Not only internal states of each cell but also the number of transition rules are very important factors in the evaluation of synchronization algorithms. We attempt to answer the following questions:

- How can we get a *real-coded* transition rule set from a set of local rules in a wild-card format?
- How many state transition rules are needed for synchronizing rectangle arrays?
- Can we convert transition rules from wild-card type to *real-coded* one automatically?

No such examination on the complexity of rule numbers has been made in the study of 2-D synchronization algorithms. Based on our computer simulation, it

F. Peper et al. (Eds.): IWNC 2009, PICT 2, pp. 333–341, 2010.

is shown that the proposed rule set consists of 28-state, 12849 transition rules and has a validity for the synchronization for any rectangle arrays of size $m \times n$ such that $2 \leq m, n \leq 500$.

2   Firing Squad Synchronization Problem

2.1   FSSP on Two-Dimensional Cellular Arrays

A finite two-dimensional (2-D) cellular array consists of $m \times n$ cells. Each cell is an identical (except the border cells) finite-state automaton. The array operates in lock-step mode in such a way that the next state of each cell (except border cells) is determined by both its own present state and the present states of its north, south, east, and west neighbors. All cells (*soldiers*), except the north-west corner cell (*general*), are initially in the quiescent state at time $t = 0$ with the property that the next state of a quiescent cell with quiescent neighbors is the quiescent state again. At time $t = 0$, the north-west upper corner cell C_{11} is in the *fire-when-ready* state, which is the initiation signal for the array. The firing squad synchronization problem is to determine a description (state set and next-state function) for cells that ensures all cells enter the *fire* state at exactly the same time and for the first time. The tricky part of the problem is that the same kind of soldier having a fixed number of states must be synchronized, regardless of the size $m \times n$ of the array. The set of states and transition rules must be independent of m and n.

Several synchronization algorithms on 2-D arrays have been proposed by Beyer [2], Grasselli [4], Shinahr [11], Szwerinski [12], Umeo, Maeda, Hisaoka and Teraoka [17], and Umeo and Uchino [18]. As for the time optimality of the two-dimensional firing squad synchronization algorithms, it has been shown by Beyer [2] and Shinahr [11] independently that there exists no two-dimensional cellular automaton that can synchronize any 2-D array of size $m \times n$ in less than $m+n+\max(m,n)-3$ steps. In addition they first proposed an optimum-time synchronization algorithm that can synchronize any 2-D array of size $m \times n$ in optimum $m+n+\max(m,n)-3$ steps. Shinahr [11] gave a 28-state implementation.

2.2   Overview of Shinahr's Time-Optimum Algorithm

Beyer [2] and Shinahr [11] presented an optimum-time synchronization scheme for rectangle arrays of size $m \times n$ operating in $m + n + max(m, n) - 3$ optimum-steps. The scheme is based on the generalized synchronization algorithm for one-dimensional (1-D) arrays, which allows the general to be located at any position in the array. It has been shown, in Moore and Langdon [7], to be impossible to synchronize any array of length n less than $n+\max(k, n - k + 1) - 2$ steps, where the general is initially on C_k. Moore and Langdon [7] also developed firstly a generalized optimum-time synchronization algorithm with 17 internal states that can synchronize any array of length n at exactly $n+\max(k, n - k + 1) - 2$ steps.

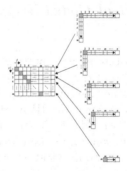

Fig. 1. Two-dimensional array synchronization scheme employed in Beyer [2] and Shinahr [11]

The first optimum-time synchronization algorithm $\mathcal{A}_{BS}$, developed independently by Beyer [2] and Shinahr [11], is based on the rotated L-shaped mapping which maps configurations of generalized FSSP solutions on 1-D L-shaped arrays onto 2-D arrays. A rectangular array of size $m \times n$ is regarded as $min(m,n)$ rotated L-shaped 1-D arrays, where each rotated L-shaped 1-D array is synchronized independently using the generalized firing squad synchronization algorithm. Configurations of the generalized synchronization processes on 1-D array can be mapped on the rotated L-shaped array. We refer it as L-array. See Fig. 1. We overview the algorithm operating on an array of size $m \times n$. At time $t = 0$, the north-west cell C_{11} is in general state and all other cells are in quiescent state. For any i such that $1 \leq i \leq min(m,n)$, the cell C_{ii} will be in the general state at time $t = 3i - 3$. A special signal which travels towards a diagonal direction is used to generate generals on the cells $\{C_{ii}|1 \leq i \leq min(m,n)\}$. For each i such that $1 \leq i \leq min(m,n)$, the cells $\{C_{ij}|i \leq j \leq n\}$ and $\{C_{ki}|i \leq k \leq m\}$ constitute the ith L-shaped array. Note that the ith general generated at time $t = 3i - 3$ is on the $(m - i + 1)$th cell from the left end of the ith L-array. The length of the ith L-array is $m + n - 2i + 1$. Thus, based on the time complexity on the generalized firing squad synchronization algorithm for one-dimensional array given above, the ith L-array can be synchronized at exactly $t_i = 3i - 3 + m + n - 2i + 1 - 2 + max(m - i + 1, n - i + 1) = m + n + max(m,n) - 3$, which is independent of i. In this way, all of the L-arrays can be synchronized simultaneously. Thus, an $m \times n$ array synchronization problem is reduced to independent $min(m,n)$ 1-D generalized synchronization problems such that:

$$\begin{cases} \mathcal{P}(m, m+n-1), \mathcal{P}(m-1, m+n-3), ..., \mathcal{P}(1, n-m+1), & m \leq n, \\ \mathcal{P}(m, m+n-1), \mathcal{P}(m-1, m+n-3), ..., \mathcal{P}(m-n+1, m-n+1), & m > n. \end{cases}$$

Here $\mathcal{P}(k, \ell)$ means the 1-D generalized synchronization problem for ℓ cells with a general on the kth cell from left end. Shinahr [11] has given a 28-state implementation of the algorithm $\mathcal{A}_{BS}$.

3   Construction of *Real-Coded* Transition Rule Set for $\mathcal{A}_{BS}$

Shinahr [11] presented a 28-state implementation for 2-D synchronization algorithm $\mathcal{A}_{BS}$, where most (97%) of the transition rules had *wild cards* which can match any state. We denote the rule set by $\mathcal{R}_{Shi}$. Note that 17 pairs of rules in $\mathcal{R}_{Shi}$ were doubly listed in the Appendix III in Shinahr [11]. See the details in Shinahr [11] and H. Umeo, K. Ishida, K. Tachibana, and N. Kamikawa [14]. It is easily seen that a simple expansion of those transition rules with wild cards into ones having no wild cards would yield a vast amount of redundant rules and some inconsistency in the definition of transition rule set. In this section we reconstruct a *real-coded* transition rule set for the Shinahr's optimum-time synchronization algorithm. Here the word "*real-coded*" means that the rule set reconstructed includes no wild cards.

The set of internal states $\mathcal{Q}$ in the Shinahr's 28-state implementation is as follows:

$$\mathcal{Q} = \{\text{Q0, P0, Q1, A1, S1, A2, S2, M1, Q2, A3, S3, A4, S4, M2, K0,}$$
$$\text{K1, Y1, Y2, Q6, A5, A6, S5, S6, K2, K3, P1, M0, F0}\},$$

where the symbols P0, Q0 and F0 denote the general, quiescent and firing states, respectively.

Each rule in the state transition table in $\mathcal{R}_{Shi}$ is expressed by 6 tuples: Y1 Y2 Y3 Y4 Y5 Y6, where Y1 denotes the present state of a cell, Y2 the west state, Y3 the east state, Y4 the north state, Y5 the south state, and Y6 the next state of the cell. The symbol "X0" denotes a boundary state for virtual cells acting as a neighbor of border cells. The symbol "- -" denotes a wild card which matches any state in $\mathcal{Q} \cup \{\text{X0}\}$.

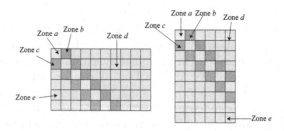

Fig. 2. Zones on the 2-D rectangle array, where some specific transition rules with wild cards are applied

Shinahr [11] defined three subsets of internal states: $\mathbb{U}_1$, $\mathbb{U}_2$, and $\mathbb{U}_3$, in order to make the construction of transition rule set $\mathcal{R}_{Shi}$ transparent. The three subsets are as follows:

$U_1 = \{A5,\ A6,\ F0,\ K2,\ K3,\ M0,\ P0,\ P1,\ Q0,\ Q6,\ S5,\ S6,\ Y1,\ Y2\},$
$U_2 = \{A1,\ A2,\ F0,\ K0,\ K1,\ M1,\ Q0,\ Q1,\ S1,\ S2\},$
$U_3 = \{A3,\ A4,\ F0,\ K0,\ K1,\ M2,\ Q0,\ Q2,\ S3,\ S4\}.$

Those states appear in specific zones shown in Fig. 2. Zone a consists of cells on the principal diagonal of the 2-D array. Cells in the upper half of the array are in Zone b and d. Cells in Zone c and e are in the lower half of the array. The cells included in Zone b and c are adjacent to the cells in Zone a. The states in U_1 are used for cells in Zone a, states in U_2 are for cells in Zone b and d, and states in U_3 are for cells in Zone c and e.

The transition rules applied to the cells in Zone a, b, c, d, and e have the following forms specified below:

- Zone a: $U_1\ U_3\ U_2\ U_2\ U_3 \rightarrow U_1$,
- Zone b: $U_2\ U_1\ U_2\ U_2\ U_1 \rightarrow U_2$,
- Zone c: $U_3\ U_3\ U_1\ U_1\ U_3 \rightarrow U_3$,
- Zone d: $U_2\ U_2\ U_2\ U_2\ U_2 \rightarrow U_2$,
- Zone e: $U_3\ U_3\ U_3\ U_3\ U_3 \rightarrow U_3$.

We expand the rules with two wild cards in $\mathcal{R}_{Shi}$ so that each rule has the form above depending on the first state in U_1, U_2, and U_3. We give an example. Consider a transition rule KOM1A1----A1 in $\mathcal{R}_{Shi}$. The first state K0 is in U_2 and U_3, and the west state M1 is in U_2. With these local constraints, the rule KOM1A1----A1 should be applied only to the cells in Zone d. We expand the rule KOM1A1----A1 by taking any state in U_2 as the forth and fifth states so that it has the form: $U_2\ U_2\ U_2\ U_2\ U_2 \rightarrow U_2$. This yields the following transition rules:

$$\text{KOM1A1A1A1} \rightarrow \text{A1},$$
$$\text{KOM1A1A1A2} \rightarrow \text{A1},$$
$$\text{KOM1A1A1F0} \rightarrow \text{A1},$$
$$\vdots$$
$$\text{KOM1A1S2S2} \rightarrow \text{A1}.$$

In order to get the final *real-coded* transition rule set $\mathcal{R}$, we take the following procedures:

1. For each rule with **no** wild card in $\mathcal{R}_{Shi}$, **interpret** the rule as it is and **insert** it into $\mathcal{R}$.
2. For each rule with **two** wild cards in $\mathcal{R}_{Shi}$, **determine** a zone where the rule should be applied. See Fig. 2. Then, **expand** the rule into the one having no wild card and **insert** it into $\mathcal{R}$.
3. **Start** a simulation of synchronization processes on 2-D arrays with the rule set $\mathcal{R}$.
4. **Apply** a rule with **four** wild cards in $\mathcal{R}_{Shi}$, if found "an undefined transition rule" during the simulation. Then, **construct** a new rule satisfying the local constraints encountered and **insert** it into $\mathcal{R}$ and **continue** the simulation.

Based on our computer simulation, the following observation can be made.

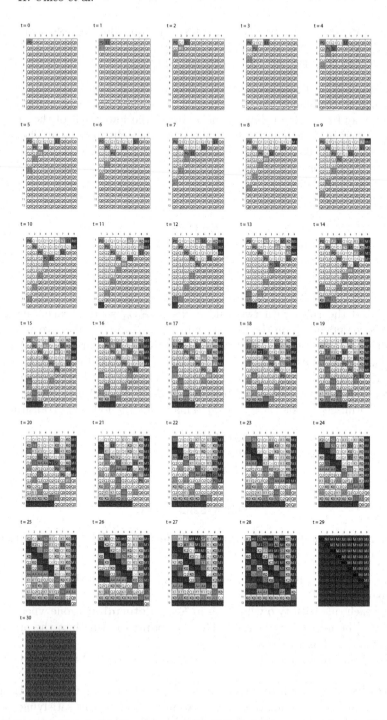

Fig. 3. Snapshots of the configurations of the Shinahr's 28-state synchronization algorithm on a rectangle array of size 12×9

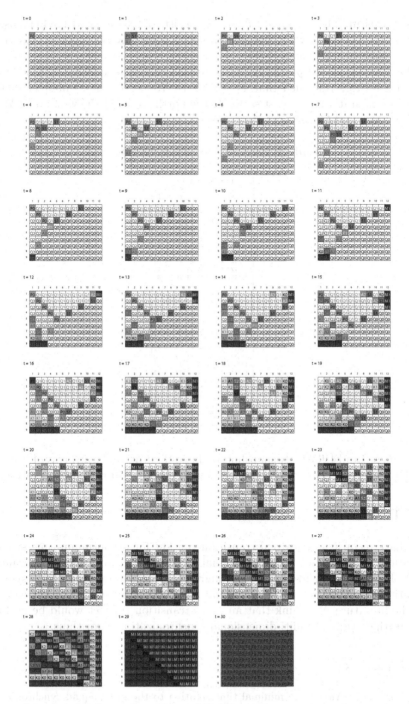

Fig. 4. Snapshots of the configurations of the Shinahr's 28-state synchronization algorithm on a rectangle array of size 9×12

Observation. The proposed rule set $\mathcal{R}$ consists of 28-state and 12849 transition rules, and has validity for the synchronization for any rectangle arrays of size $m \times n$ such that $2 \leq m, n \leq 500$.

We give below the partial list of transition rule set consisting of 12849 rules. In each 6-tuple rule such that Y1 Y2 Y3 Y4 Y5 → Y6, the symbol Y1 denotes the present state, Y2 the east state, Y3 the north state, Y4 the west state, Y5 the south state, and Y6 the next state, respectively.

Table 1. Reconstructed transition rule set $\mathcal{R}$

```
    1: A1 KO A2 A1 A1 → M1
    2: A1 KO A2 A1 KO → M1
    3: A1 KO A2 A1 XO → M1
    4: A1 KO A2 A2 A1 → M1
    5: A1 KO A2 A2 KO → M1
    6: A1 KO A2 A2 M1 → M1
    7: A1 KO A2 A2 Q1 → M1
    8: A1 KO A2 A2 XO → M1
    9: A1 KO A2 M1 KO → M1
   10: A1 KO A2 M1 M1 → M1
   11: A1 KO A2 M1 XO → M1
   12: A1 KO A2 Q1 KO → M1
   13: A1 KO A2 Q1 K1 → M1

                 .
                 .
                 .

12839: Y1 XO Q1 XO M2 → K2
12840: Y1 XO Q1 XO Q2 → K2
12841: Y2 A4 M1 KO Q2 → K2
12842: Y2 A4 Q1 Q1 Q2 → K2
12843: Y2 A4 S1 Q1 Q2 → K3
12844: Y2 S4 M1 KO Q2 → K2
12845: Y2 S4 Q1 Q1 Q2 → K2
12846: Y2 S4 S1 Q1 Q2 → K3
12847: Y2 XO M1 XO Q2 → K2
12848: Y2 XO Q1 XO Q2 → K2
12849: Y2 XO S1 XO Q2 → K3
```

Figures 3 and 4 illustrate snapshots of the configurations on an array of size 12×9 and 9×12, respectively, based on our new 28-state 12849-rule implementation.

4   Discussions

We have reconstructed a real transition rule set for an optimum-time synchronization algorithm for two-dimensional rectangle arrays proposed by Shinahr [11]. The algorithm itself has been known as the first optimum-time synchronization algorithm for two-dimensional rectangle arrays and quoted frequently in the literature. We think that an exact transition rule set would be useful in the further study of multi-dimensional synchronization algorithms.

References

1. Balzer, R.: An 8-state minimal time solution to the firing squad synchronization problem. Information and Control 10, 22–42 (1967)
2. Beyer, W.T.: Recognition of topological invariants by iterative arrays. Ph.D. Thesis, p. 144. MIT (1969)

3. Gerken, H.-D.: Über Synchronisations - Probleme bei Zellularautomaten. *Diplomarbeit*, Institut für Theoretische Informatik, Technische Universität Braunschweig, p. 50 (1987)
4. Grasselli, A.: Synchronization of cellular arrays: The firing squad problem in two dimensions. Information and Control 28, 113–124 (1975)
5. Minsky, M.: Computation: Finite and infinite machines, pp. 28–29. Prentice Hall, Englewood Cliffs (1967)
6. Moore, E.F.: The firing squad synchronization problem. In: Moore, E.F. (ed.) Sequential Machines, Selected Papers, pp. 213–214. Addison-Wesley, Reading (1964)
7. Moore, F.R., Langdon, G.G.: A generalized firing squad problem. Information and Control 12, 212–220 (1968)
8. Nguyen, H.B., Hamacher, V.C.: Pattern synchronization in two-dimensional cellular space. Information and Control 26, 12–23 (1974)
9. Romani, F.: On the fast synchronization of tree connected networks. Information Sciences 12, 229–244 (1977)
10. Schmid, H.: Synchronisations probleme für zelluläre Automaten mit mehreren Generälen. Diplomarbeit, Universität Karlsruhe (2003)
11. Shinahr, I.: Two- and three-dimensional firing squad synchronization problems. Information and Control 24, 163–180 (1974)
12. Szwerinski, H.: Time-optimum solution of the firing-squad-synchronization-problem for n-dimensional rectangles with the general at an arbitrary position. Theoretical Computer Science 19, 305–320 (1982)
13. Umeo, H.: A simple design of time-efficient firing squad synchronization algorithms with fault-tolerance. IEICE Trans. on Information and Systems E87-D(3), 733–739 (2004)
14. Umeo, H., Ishida, K., Tachibana, K., Kamikawa, N.: A construction of real-coded transition rule set for Shinahr's optimum-time synchronization algorithm on two-dimensional cellular arrays. draft, pp. 1–80 (2008)
15. Umeo, H.: Synchronization algorithms for two-dimensional cellular automata. Journal of Cellular Automata 4, 1–20 (2008)
16. Umeo, H.: Firing squad synchronization problem in cellular automata. In: Meyers, R.A. (ed.) Encyclopedia of Complexity and System Science, vol. 4, pp. 3537–3574. Springer, Heidelberg (2009)
17. Umeo, H., Maeda, M., Hisaoka, M., Teraoka, M.: A state-efficient mapping scheme for designing two-dimensional firing squad synchronization algorithms. Fundamenta Informaticae 74(4), 603–623 (2006)
18. Umeo, H., Uchino, H.: A new time-optimum synchronization algorithm for rectangle arrays. Fundamenta Informaticae 87(2), 155–164 (2008)
19. Waksman, A.: An optimum solution to the firing squad synchronization problem. Information and Control 9, 66–78 (1966)

A Two-Dimensional Optimum-Time Firing Squad Synchronization Algorithm and Its Implementation

Hiroshi Umeo[1], Jean-Baptiste Yunès[2], and Takuya Yamawaki[1]

[1] Univ. of Osaka Electro-Communication,
Neyagawa-shi, Hastu-cho, 18-8, Osaka, 572-8530, Japan
umeo@cyt.osakac.ac.jp
[2] LIAFA - Universite Paris 7 Denis Diderot,
175, rue du chevaleret, 75013 Paris, France

Abstract. The firing squad synchronization problem on cellular automata has been studied extensively for more than forty years, and a rich variety of synchronization algorithms have been proposed for not only one-dimensional arrays but two-dimensional arrays. In the present paper, we propose a new and simpler optimum-time synchronization algorithm that can synchronize any rectangle array of size $m \times n$ with a general at one corner in $m + n + \max(m, n) - 3$ steps. An implementation for the algorithm in terms of local transition rules is also given.

1 Introduction

We study a synchronization problem that gives a finite-state protocol for synchronizing cellular automata on large grids. The synchronization in cellular automata has been known as a firing squad synchronization problem (FSSP) since its development, in which it was originally proposed by J. Myhill in Moore [7] to synchronize all parts of self-reproducing cellular automata. The problem has been studied extensively for more than 40 years. In the present paper, we propose a new synchronization algorithm for rectangle cellular automata. The algorithm can synchronize any rectangle array of size $m \times n$ with a general at one corner in $m + n + \max(m, n) - 3$ steps. An implementation for the algorithm in terms of local transition rules is also given.

2 Firing Squad Synchronization Problem on Two-Dimensional Arrays

A finite two-dimensional (2-D) cellular array consists of $m \times n$ cells. Each cell is an identical (except the border cells) finite-state automaton. The array operates in lock-step mode in such a way that the next state of each cell (except border cells) is determined by both its own present state and the present states of its north, south, east and west neighbors. All cells (*soldiers*), except the north-west

F. Peper et al. (Eds.): IWNC 2009, PICT 2, pp. 342–351, 2010.

corner cell (*general*), are initially in the quiescent state at time $t = 0$ with the property that the next state of a quiescent cell with quiescent neighbors is the quiescent state again. At time $t = 0$, the north-west corner cell C_{11} is in the *fire-when-ready* state, which is the initiation signal for the array. The firing squad synchronization problem is to determine a description (state set and next-state function) for cells that ensures all cells enter the *fire* state at exactly the same time and for the first time. The tricky part of the problem is that the same kind of soldier having a fixed number of states must be synchronized, regardless of the size $m \times n$ of the array. The set of states and next state function must be independent of m and n.

The problem was first solved by J. McCarthy and M. Minsky who presented a $3n$-step algorithm for one-dimensional cellular array of length n. In 1962, the first optimum-time, i.e. $(2n - 2)$-step, synchronization algorithm was presented by Goto [4], with each cell having several thousands of states. Waksman [18] presented a 16-state optimum-time synchronization algorithm. Afterward, Balzer [1] and Gerken [3] developed an eight-state algorithm and a seven-state synchronization algorithm, respectively, thus decreasing the number of states required for the synchronization. Mazoyer [6] developed a six-state synchronization algorithm which, at present, is the algorithm having the fewest states for one-dimensional arrays. On the other hand, several synchronization algorithms on 2-D arrays have been proposed by Beyer [2], Shinahr [9], Szwerinski [10] and Umeo, Maeda, Hisaoka and Teraoka [16]. It has been shown by Beyer [2] and Shinahr [9] independently that there exists no two-dimensional cellular automaton that can synchronize any 2-D array of size $m \times n$ in less than $m + n + \max(m, n) - 3$ steps. In addition they first proposed an optimum-time synchronization algorithm that can synchronize any 2-D array of size $m \times n$ in optimum $m + n + \max(m, n) - 3$ steps. Shinahr [9] gave a 28-state implementation. Umeo, Hisaoka and Akiguchi [13] presented a 12-state synchronization algorithm operating in optimum-step, realizing a smallest solution to the rectangle synchronization problem at present.

3 Delayed Synchronization Scheme for One-Dimensional Arrays

First we introduce a *freezing-thawing* technique that yields a delayed synchronization algorithm for one-dimensional array. The readers can find that the technique will be used efficiently for the design of two-dimensional optimum-time synchronization algorithm.

Theorem 1. [Umeo [11]] Let t_0, t_1, t_2 and Δt be any integer such that $t_0 \geq 0$, $t_1 = t_0 + n - 1$, $t_1 \leq t_2$ and $\Delta t = t_2 - t_1$. We assume that a usual synchronization operation is started at time $t = t_0$ by generating a special signal which acts as a General at the left end of one-dimensional array of length n. We also assume that the right end cell of the array receives an other special signal from outside at time $t_1 = t_0 + n - 1$ and $t_2 = t_1 + \Delta t$, respectively. Then, there exists a one-dimensional cellular automaton that can synchronize the array of length n at time $t = t_0 + 2n - 2 + \Delta t$.

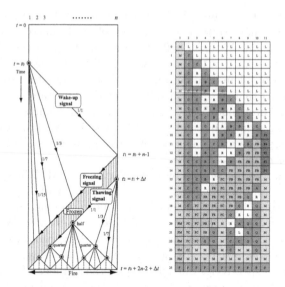

Fig. 1. Space-time diagram for delayed firing squad synchronization scheme based on the freezing-thawing technique (left) and delayed (for $\Delta t = 5$) configuration for optimum-time firing squad synchronization algorithm on $n = 11$ cells (right)

The array operates as follows:

1. Start an optimum-time firing squad synchronization algorithm at time $t = t_0$ at the left end of the array. A 1/1 speed, i.e., 1 cell per 1 step, signal is propagated towards the right direction to wake-up cells in quiescent state. We refer the signal as *wake-up signal*. A *freezing signal* is given from outside at time $t_1 = t_0 + n - 1$ at the right end of the array. The signal is propagated in the left direction at its maximum speed, that is, 1 cell per 1 step, and freezes the configuration progressively. Any cell that receives the freezing signal from its right neighbor has to stop its state-change and transmits the freezing signal to its left neighbor. The frozen cell keeps its state as long as no thawing signal will arrive.

2. A special signal supplied with outside at the right end at time $t_2 = t_1 + \Delta t$ is used as a *thawing signal* that thaws the frozen configuration progressively. The thawing signal forces the frozen cell to resume its state-change procedures immediately. See Fig. 1 (left). The signal is also transmitted toward the left end at speed 1/1.

The readers can see how those three special signals work. We can freeze the entire configuration during Δt steps and delay the synchronization on the array for Δt steps. Figure 1 (right) shows snapshots of delayed (for $\Delta t = 5$) configurations for optimum-time synchronization algorithm on eleven cells. In this example, note that the wake-up signal is supplied with the array at time $t = 0$. We refer the scheme as *freezing-thawing technique*. The technique was developed by Umeo [11] for designing several fault-tolerant firing squad synchronization algorithms for one-dimensional arrays. In the next section, the freezing-thawing technique will be employed efficiently in the design of optimum-time synchronization algorithms for two-dimensional cellular arrays.

4   New Optimum-Time Synchronization Algorithm

Several synchronization algorithms on 2-D arrays have been proposed by Beyer [2], Grasselli [5], Shinahr [9], Szwerinski [10], Umeo, Maeda, Hisaoka and Teraoka [16], and Umeo and Uchino [17]. As for the time optimality of the two-dimensional firing squad synchronization algorithms, Beyer [2] and Shinahr [9] have shown independently that there exists no cellular automaton that can synchronize any two-dimensional array of size $m \times n$ in less than $m+n+\max(m, n)-3$ steps, where the general is located at one corner of the array. In addition, Shinahr [9] presented a 28-state cellular automaton that can synchronize any two-dimensional array of size $m \times n$ at exactly $m + n + \max(m, n) - 3$ optimum-steps, where the general is located at one corner of the array.

   In this section we develop a new optimum-time firing squad synchronization algorithm $\mathcal{A}$ operating in $m+n+\max(m, n)-3$ steps for any array of size $m \times n$. The overview of the algorithm $\mathcal{A}$ is as follows:

1. A 2-D array of size $m \times n$ is regarded as $\min(m, n)$ rotated or mirrored L-shaped 1-D arrays, each consisting of a horizontal and a vertical segment.
2. The shorter segment is synchronized by the freezing-thawing technique with $\Delta t = \mid m - n \mid$ steps delay. The longer one is synchronized with the usual way without using the freezing-thawing technique.
3. All of the L-shaped arrays fall into a special firing (synchronization) state simultaneously and for the first time.

4.1   Segmentation of Rectangular Array of Size $m \times n$

Firstly, we consider the case where $m \leq n$. We assume that the initial General **G** is on the north-west corner denoted by • in Fig. 2. We regard a two-dimensional array of size $m \times n$ as consisting of m rotated ($90°$ in counterclockwise direction) L-shaped one-dimensional arrays. Each bending point of the L-shaped array is the farthest one from the general. Each L-shaped array is denoted by $L_i, 1 \leq i \leq m$, shown in Fig. 2. Each L_i is divided into two segments, that is, one horizontal and one vertical segment, each referred to as 1st and 2nd segments. The length of each segment of L_i is $n - m + i$ and i, respectively.

4.2   Starting Synchronization Process

At time $t = 0$ a two-dimensional array M has a *General* at C_{11} and any other cells of the array are in quiescent state. The *General* G (denoted by • in Fig. 2) generates three signals s_V, s_D and s_H, simultaneously, each propagating at $1/1$-speed in the vertical, diagonal and horizontal directions, respectively. See Fig. 3.

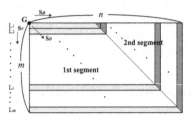

Fig. 2. A two-dimensional array of size $m \times n$ ($m \leq n$) is regarded as consisting of m rotated ($90°$ in counterclockwise direction) L-shaped 1-D arrays. Each L-shaped array is divided into two segments.

The s_V- and s_H-signals work for generating wake-up signals for the 1st and 2nd segments on each *L-shaped* array. The s_D-signal is used for printing a special marker "∎" for generating a thawing signal that thaws frozen configurations on shorter segment. Their operations are as follows:

- **Signal s_V:** The s_V-signal travels along the 1st column and reaches C_{m1} at time $t = m - 1$. Then, it returns there and begins to travel again at $1/2$-speed along the 1st column towards C_{11}. On the return's way, the signal initiates the synchronization process for the 1st segment of each L_i. Thus a new *General* G_{i1} for the synchronization of the 1st segment of each L_i is generated, together with its wake-up signal, at time $t = 3m - 2i - 1$ for $1 \leq i \leq m$.
- **Signal s_D:** The s_D-signal travels along a principal diagonal line by repeating a zigzag movement: going one cell to the right, then going down one cell. Each time it visits cell C_{ii} on the diagonal, it marks a special symbol "∎" to inform the wake-up signal on the segment of the position where a thawing signal is generated for the neighboring shorter segment. The symbol on C_{ii} is marked at time $t = 2i - 2$ for any $i, 1 \leq i \leq m$. Note that the wake-up signal of the 1st segment of L_m knows the right position by the arrival of the s_D-signal, where they meet at C_{mm} at the very time $t = 2m - 2$.
- **Signal s_H:** The s_H-signal travels along the 1st row at $1/1$-speed and reaches C_{1n} at time $t = n-1$. Then it reflects there and returns the same route at $1/2$-speed. Each time it visits a cell of the 1st row on its return way, it generates a *General* G_{i2} at time $t = 2m + n - 2i - 1$ to initiate a synchronization for the 2nd segment on each $L_i, 1 \leq i \leq m$.

The wake-up signals for the 1st and 2nd segments of L_i meet on $C_{i,n-m+i}$ at time $t = 2m+n-i-2$. The collision of the two signals acts as a delimiter between the 1st and 2nd segments. Note that the synchronization operations on the 1st segment are started at the left end of the segment. On the other hand, the synchronization on the 2nd segment is started at the right (upper) end of the segment. The wake-up signal generated by G_{i1} reaches a cell where the special mark is printed at time $t = 3m - i - 2$ and generates a thawing signal which travels at $1/2$-speed along the 1st segment to thaw the configuration on the 2nd segment.

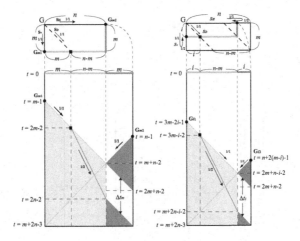

Fig. 3. Space-time diagram for synchronizing L_m (left) and L_i (right)

4.3 Synchronization of L_m

Now we consider the synchronization on L_m. Figure 3 (left) shows a time-space diagram for synchronizing L_m. As was mentioned in the previous subsection, the synchronization of the 1st and 2nd segments of L_m are started by the generals G_{m1} and G_{m2} at time $t = m - 1$ and $t = n - 1$, respectively. Each General generates a wake-up signal propagating at $1/1$-speed. The wake-up signal for the 1st and 2nd segments meets C_{mn} at time $t = m + n - 2$, where C_{mn} acts as an end of the both two segments. A freezing signal is generated simultaneously there for the 2nd segment at time $t = m + n - 2$. It propagates in upper (right in Fig. 3 (left)) direction at $1/1$-speed to freeze the synchronization operations on the 2nd segment. At time $t = 2m - 2$ the wake-up signal of the 1st segment reaches the symbol "■" and generates a thawing signal for the 2nd segment. The thawing signal starts to propagate from the cell at $1/2$-speed in the same direction. Those two signals reach at the left end of the 2nd segment with time difference $n - m$ which is equal to the delay for the 2nd segment. The synchronization for the 1st segment is started at time $t = m - 1$ and it can be synchronized at time $t = m + 2n - 3 = m + n + \max(m, n) - 3$ by the usual way. On the other hand, the synchronization for the 2nd segment is started at time $t = n - 1$ and its operations are delayed for $\Delta t = \Delta t_m = n - m$ steps. Now letting $t_0 = n - 1, \Delta t = n - m$ in Theorem 1, the 2nd segment of length m on L_m can be synchronized at time $t = t_0 + 2m - 2 + \Delta t = m + 2n - 3 = m + n + \max(m, n) - 3$. Thus, L_m can be synchronized at time $t = m + n + \max(m, n) - 3$.

4.4 Synchronization of L_i

Now we discuss the synchronization for $L_i, 1 \leq i \leq m$. Figure 3 (right) shows a space-time diagram for synchronizing L_i. The wake-up signals for the two

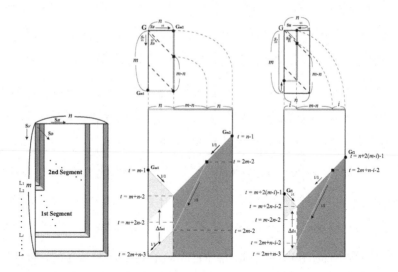

Fig. 4. Segmentation of a 2-D array of size $m \times n$ $(m > n)$ (right) and space-time diagram for synchronizing L_n (middle) and $L_i, 1 \le i \le n$ (left)

segments of L_i are generated at time $t = 3m - 2i - 1$ and $n + 2(m - i) - 1$, respectively. Generation of freezing and thawing signals is done in a similar way as employed in L_m. Synchronization operations on the 2nd segment are delayed for $\Delta t_i = n - m$ steps. The synchronization for the 1st segment of length $n - m + i$ is started at time $t = 3m - 2i - 1$ and it can be synchronized by a usual method at time $t = m + 2n - 3 = m + n + \max(m, n) - 3$. On the other hand, the synchronization for the 2nd segment is started at time $t = n + 2(m - i) - 1$ and its operations are delayed for $\Delta t = \Delta t_i = n - m$ steps. Now letting $t_0 = n + 2(m - i) - 1, \Delta t = n - m$ in Theorem 1, the 2nd segment of length i of L_i can be synchronized at time $t = t_0 + 2i - 2 + \Delta t = m + 2n - 3 = m + n + \max(m, n) - 3$. Thus, L_i can be synchronized at time $t = m + n + \max(m, n) - 3$.

4.5 Synchronization of Rectangle Longer Than Wide

In the case where $m > n$, a two-dimensional array of size $m \times n$ is regarded as consisting of n mirrored *L-shaped* arrays. See Fig. 4 (left). Segmentation and synchronization operations on each *L-shaped* array can be done almost in a similar way. It is noted that the thawing signal is generated on the 2nd segment to thaw frozen configurations on the 1st segment. Any rectangle of size $m \times n$ can be synchronized at time $t = 2m + n - 3 = m + n + \max(m, n) - 3$. Figure 4 (middle and right) shows the space-time diagram for the synchronization on L_n and $L_i, 1 \le i \le n$.

One notes that the algorithm needs no a priori knowledge on side length of a given rectangle, that is, whether wider than long or longer than wide. Now we can establish the next theorem.

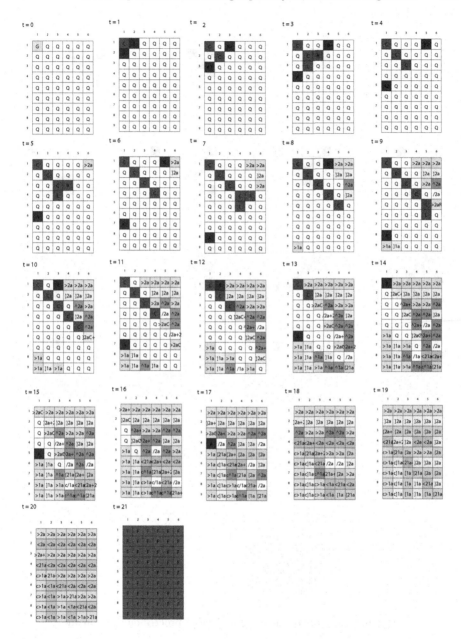

Fig. 5. Snapshots of the synchronization process on 9×6 array

Theorem 2. The synchronization algorithm $\mathcal{A}$ can synchronize any $m \times n$ rectangular array in optimum $m + n + \max(m, n) - 3$ steps.

Figures 5 and 6 show some snapshots of the synchronization process operating in optimum-steps on 9×6 and 6×9 arrays.

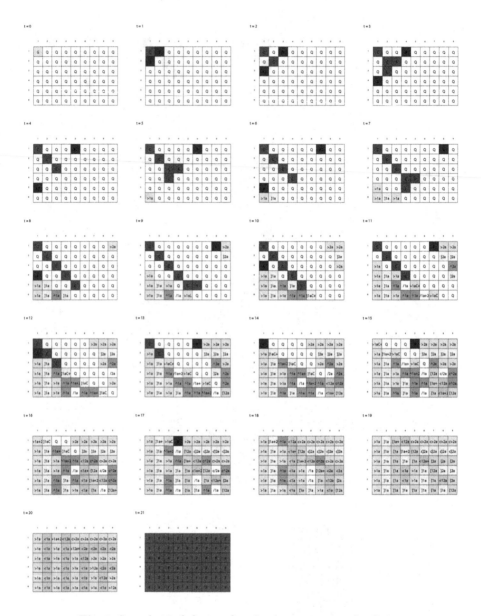

Fig. 6. Snapshots of the synchronization process on 6×9 array

5   Conclusions

We have proposed a new optimum-time synchronization algorithm for two-dimensional cellular arrays. It can synchronize any two-dimensional array of size $m \times n$ in optimum $m + n + \max(m, n) - 3$ steps.

References

1. Balzer, R.: An 8-state minimal time solution to the firing squad synchronization problem. Information and Control 10, 22–42 (1967)
2. Beyer, W.T.: Recognition of topological invariants by iterative arrays. Ph.D. Thesis, p. 144. MIT (1969)
3. Gerken, H.D.: Über Synchronisations - Probleme bei Zellularautomaten. *Diplomarbeit*, Institut für Theoretische Informatik, Technische Universität Braunschweig, p. 50 (1987)
4. Goto, E.: A minimal time solution of the firing squad problem. Dittoed course notes for Applied Mathematics 298, 52–59 (1962)
5. Grasselli, A.: Synchronization of cellular arrays: The firing squad problem in two dimensions. Information and Control 28, 113–124 (1975)
6. Mazoyer, J.: A six-state minimal time solution to the firing squad synchronization problem. Theoretical Computer Science 50, 183–238 (1987)
7. Moore, E.F.: The firing squad synchronization problem. In: Moore, E.F. (ed.) Sequential Machines, Selected Papers, pp. 213–214. Addison-Wesley, Reading (1964)
8. Schmid, H.: Synchronisationsprobleme für zelluläre Automaten mit mehreren Generälen. Diplomarbeit, Universität Karsruhe (2003)
9. Shinahr, I.: Two- and three-dimensional firing squad synchronization problems. Information and Control 24, 163–180 (1974)
10. Szwerinski, H.: Time-optimum solution of the firing-squad-synchronization-problem for n-dimensional rectangles with the general at an arbitrary position. Theoretical Computer Science 19, 305–320 (1982)
11. Umeo, H.: A simple design of time-efficient firing squad synchronization algorithms with fault-tolerance. IEICE Trans. on Information and Systems E87-D(3), 733–739 (2004)
12. Umeo, H.: Firing squad synchronization problem in cellular automata. In: Meyers, R.A. (ed.) Encyclopedia of Complexity and System Science, vol. 4, pp. 3537–3574. Springer, Heidelberg (2009)
13. Umeo, H., Hisaoka, M., Akiguchi, S.: Twelve-state optimum-time synchronization algorithm for two-dimensional rectangular cellular arrays. In: Calude, C.S., Dinneen, M.J., Păun, G., Jesús Pérez-Jímenez, M., Rozenberg, G. (eds.) UC 2005. LNCS, vol. 3699, pp. 214–223. Springer, Heidelberg (2005)
14. Umeo, H., Hisaoka, M., Sogabe, T.: A survey on optimum-time firing squad synchronization algorithms for one-dimensional cellular automata. Intern. J. of Unconventional Computing 1, 403–426 (2005)
15. Umeo, H., Hisaoka, M., Teraoka, M., Maeda, M.: Several new generalized linear- and optimum-time synchronization algorithms for two-dimensional rectangular arrays. In: Margenstern, M. (ed.) MCU 2004. LNCS, vol. 3354, pp. 223–232. Springer, Heidelberg (2005)
16. Umeo, H., Maeda, M., Hisaoka, M., Teraoka, M.: A state-efficient mapping scheme for designing two-dimensional firing squad synchronization algorithms. Fundamenta Informaticae 74, 603–623 (2006)
17. Umeo, H., Uchino, H.: A new time-optimum synchronization algorithm for rectangle arrays. Fundamenta Informaticae 87(2), 155–164 (2008)
18. Waksman, A.: An optimum solution to the firing squad synchronization problem. Information and Control 9, 66–78 (1966)

Quaternion Based
Thermal Condition Monitoring System

Wai Kit Wong, Chu Kiong Loo, Way Soong Lim, and Poi Ngee Tan

Faculty of Engineering and Technology, Multimedia University,
75450 Jln Ayer Keroh Lama, Malaysia
{wkwong,ckloo,wslim}@mmu.edu.my, poingee24@yahoo.com

Abstract. In this paper, we will propose a new and effective machine condition monitoring system using log-polar mapper, quaternion based thermal image correlator and max-product fuzzy neural network classifier. Two classification characteristics namely: peak to sidelobe ratio (PSR) and real to complex ratio of the discrete quaternion correlation output (p-value) are applied in the proposed machine condition monitoring system. Large PSR and p-value observe in a good match among correlation of the input thermal image with a particular reference image, while small PSR and p-value observe in a bad/not match among correlation of the input thermal image with a particular reference image. In simulation, we also discover that log-polar mapping actually help solving rotation and scaling invariant problems in quaternion based thermal image correlation. Beside that, log-polar mapping can have a two fold of data compression capability. Log-polar mapping can help smoother up the output correlation plane too, hence makes a better measurement way for PSR and p-values. Simulation results also show that the proposed system is an efficient machine condition monitoring system with accuracy more than 98%.

1   Introduction

Nowadays, most factories rely on machines to help boost up their production and process. Therefore an effective machine condition monitoring system plays an important role in those factories to ensure that their production and process are running smoothly all the time. Infrared thermal imaging found to be used in many vision based machinery monitoring system such as thise applications include temperature mapping of high temperature processes like brazing [1], metal smelting [2], fault electrical brakers, bad contacts and overloaded circuits inspection [3]. In this paper, a new and effective machine condition monitoring system using log-polar mapper, quaternion based thermal image correlator and max-product fuzzy neural network classifier is proposed. Two classification characteristics namely: peak to sidelobe ratio (PSR) and real to complex ratio of the discrete quaternion correlation output (p-value) are applied in the proposed machine condition monitoring system. In simulation, log-polar mapping help solving rotation and scaling invariant problems in quaternion based thermal image correlation. Log-polar mapping can have a two fold of data compression capability. Simulation results show that the proposed system is an efficient system with accuracy more than 98%. The paper is organized as follows: Section 2 briefly comments on the machines

F. Peper et al. (Eds.): IWNC 2009, PICT 2, pp. 352–362, 2010.
© Springer 2010

monitoring system, section 3 summarizes the log-polar mapping techniques. Then in section 4, we will describe the algorithm of the proposed quaternion based thermal image correlator. Section 5 describes the structure of the max-product fuzzy neural network classifier. Section 6 contains the experimental results. Finally in section 7, we summarize the work and envy some future work.

2 Machine Condition Monitoring System Model

The machine condition monitoring system model is shown in Fig. 1. The thermal camera used in this paper is a cost effective and fine resolution model: AXT 100 manufactured by Ann Arbor Sensor Systems [4]. AXT 100 has an embedded firmware called InternalWeb, which can be interface with Matlab or other image processing software in laptop or PC. Therefore, it is best outfitted for machine condition monitoring system.

Image partitioner is used for partitioning the input thermal image into S-partitioned sections, provided that the input thermal image consisted of S machines to be monitored. Each partitioned section consists of one machine to be monitored. An example of a thermal image with $S = 3$ partition sections is shown in Fig. 2.

Log-polar mapper applying log polar mapping techniques [5], it converts a Cartesian image into a retina-like (log-polar) image providing an important and useful property that scaling and rotating an object in a Cartesian plane corresponds to translate the object in the log-polar plane. Another advantage of log-polar image representation is that it has data compression manner. Log-polar mapping will be discussed in Section 3.

Quaternion based thermal image correlator is used to obtain correlation plane for each correlated input thermal image captured lively with reference images of all possible machines conditions stored in a database to calculate out some classification characteristics such as the p-value and PSR. Quaternion based thermal image correlation will be discussed in Section 4.

The max-product fuzzy neural network classifier is first applied to train for an accurate classification with the weight (w) obtained form training reference images of all possible machines conditions stored in the database. During application, the PSR and p-value output from quaternion based thermal image correlator are first fuzzified into Gaussian membership function. After that, product value is calculated based on multiplication of PSR in Gaussian membership value with p-value in Gaussian membership value. The product values are stored in an array and multiply with the weight (w). Max-composition is performing on the output based on two sets of fuzzy IF-THEN rules, defuzzification is performed to classify each machine's condition under monitoring. Detailed discussion on max-product fuzzy neural network classifier will be given in Section 5.

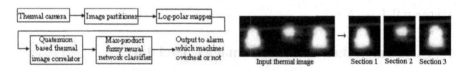

Fig. 1. Machine condition monitoring system model

Fig. 2. Example of a thermal image with $S = 3$ partition sections

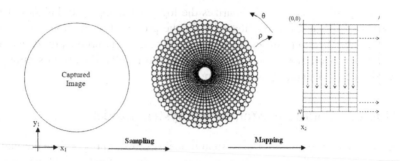

Fig. 3. A graphical view of log-polar mapping

3 Log-Polar Mapping

A graphical view illustrating the log-polar mapping is shown in Fig. 3 [6]. The log-polar mapping can be summarized as following. The centre of pixel for log-polar sampling is described by [5,7]:

$$\rho(x_1, y_1) = \log_\lambda \frac{R}{r_o} \tag{1}$$

$$\theta(x_1, y_1) = \frac{N_\theta}{2\pi} \arctan \frac{y_1}{x_1} \tag{2}$$

where R is the distance between the given point and the center of mapping $R = \sqrt{x_1^2 + y_1^2}$, r_o is a scaling factor which will define the size of the circle at $\rho(x_1, y_1) = 0$. λ is the base of the algorithm,

$$\lambda = \frac{1 + \sin \frac{\pi}{N_\theta}}{1 - \sin \frac{\pi}{N_\theta}} \tag{3}$$

N_θ is the total number of pixels per ring in log-polar geometry.

The number of rings in the fovea region is given by [5,7]:

$$N_{fov} = \frac{\lambda}{\lambda - 1} \tag{4}$$

To sample the Cartesian pixels (x_1, y_1) into log polar pixels (ρ, θ), at each center point calculated using (1) and (2), the corresponding log-polar pixel (ρ_n, θ_n) is cover a region of Cartesian pixels with radius:

$$r_n = \lambda r_{n-1} \tag{5}$$

where $n = 0, 1, \cdots, N - 1$.

4 Quaternion Based Thermal Image Correlator

The recently developed concept of quaternion correlation, which are based on quaternion algebra introduced by Hamilton in 1843 [8], have been found to be useful in color pattern recognition (e.g. human face recognition [9], color alphanumeric words recognition [10]). A quaternion array is a generalization of the complex number representation and it can be considered as a number with a real part and an imaginary part consisting three orthogonal components as follows [11]:

$$f = f_r + f_i \cdot i + f_j \cdot j + f_k \cdot k \qquad (6)$$

where f_r, f_i, f_j, f_k are real quaternion numbers and i, j, k, are imaginary operators.

In conventional correlation for pattern recognition, pattern should convert to gray level scale or treat the three color channels (R, G, B) separately to process and combine the results at last. But for quaternion correlation technique, it process all color channels jointly by using its quaternion array. R, G, B in color image can be represented using quaternion array by inserting the value of three color channels into the three imaginary parts of the quaternion numbers respectively [10]:

$$f(m, n) = f_R(m, n) \cdot i + f_G(m, n) \cdot j + f_B(m, n) \cdot k \qquad (7)$$

where $f_R(m, n)$, $f_G(m, n)$ and $f_B(m, n)$ represent the R, G, B parts of pattern and m, n are the 2-D pixels coordinates. Quaternion correlator is so far used in color human face recognition [9] and color alphanumeric words recognition [10]. It is not found in object recognition yet, especially in thermal object/image recognition. So, in this paper, we will present a new approach, whereby quaternion correlator will be used in thermal image recognition for machine condition monitoring system.

In this section, we will describe the algorithm of the quaternion based thermal image correlator:

1) Calculate energy of reference image $I(m, n)$ [10]:

$$E_I = \sum_{m=0}^{M-1} \sum_{n=0}^{N-1} |I(m, n)|^2 \qquad (8)$$

Then we normalized the reference image $I(m, n)$ and the input image $h_i(m, n)$ as:

$$I_a(m, n) = I(m, n)/\sqrt{E_I} \qquad (9)$$

$$H_a(m, n) = h_i(m, n)/\sqrt{E_I} \qquad (10)$$

2) Calculate the output of discrete quaternion correlation (DQCR):

$$g_a(m, n) = \sum_{\tau=0}^{M-1} \sum_{\eta=0}^{N-1} I_a(\tau, \eta) \cdot \overline{H_a(\tau - m, \eta - n)} \qquad (11)$$

where '—' means the quaternion conjugation operation and do the space reverse operation:

$$g(m, n) = g_a(-m, -n) \qquad (12)$$

3) Perform inverse discrete quaternion Fourier Transform (IDQFT) on (12) to obtain the correlation plane $P(m, n)$.

4) Search all the local peaks on the correlation plane and record the location of the local peaks as (m_s, n_s).

5) Then at all the location of local peaks (m_s, n_s) found in step 4, we calculate the real to complex value of the DQCR output:

$$p = \frac{|P_r(m_s, n_s)|}{|P_r(m_s, n_s)| + |P_i(m_s, n_s)| + |P_j(m_s, n_s)| + |P_k(m_s, n_s)|} \tag{13}$$

where $P_r(m_s, n_s)$ is the real part of $P(m_s, n_s)$. $P_i(m_s, n_s)$, $P_j(m_s, n_s)$ and $P_k(m_s, n_s)$ are the $i-$, $j-$ $k-$ parts of $P(m_s, n_s)$ respectively.

A strong peak can be observed in the correlation output if the input image comes from imposter class. A method of measuring the peak sharpness is the peak-to-sidelobe ratio (PSR) which is defined as below [9]:

$$\text{PSR} = \frac{peak - mean(sidelobe)}{\sigma(sidelobe)} \tag{14}$$

where *peak* is the value of the peak on the correlation output plane. *sidelobe* refers a fixed-sized surrounding area off the peak. *mean* is the average value of the sidelobe region. σ is the standard deviation of the sidelobe region. Large PSR values indicate the better match of the input image and the corresponding reference image.

5   Max-Product Fuzzy Neural Network Classifier

The max-product fuzzy neural network [12,13] classifier is training with 4 steps:

1) $\text{PSR}_{s(t_1,t_2)}$ and $p_{s(t_1,t_2)}$ output from the quaternion correlator are fuzzified through the activation functions (Gaussian membership function):

$$G_{\text{PSR}_{s(t_1,t_2)}} = \exp\left[\frac{-(\text{PSR}_{s(t_1,t_2)} - 1)^2}{\sigma^2}\right] \tag{15}$$

$$G_{p_{s(t_1,t_2)}} = \exp\left[\frac{-(p_{s(t_1,t_2)} - 1)^2}{\sigma^2}\right] \tag{16}$$

where $t_1, t_2 = 1, 2, \cdots, T$ represents the number of reference/input images, σ is the smoothing factor, that is the deviation of the Gaussian functions.

2) Calculate the product value for s-th machine section of the fuzzy neural network classifier at each correlated images:

$$G_{s(t_1,t_2)} = G_{\text{PSR}_{s(t_1,t_2)}} \times G_{p_{s(t_1,t_2)}} \tag{17}$$

3) Gather and store the product values in an array:

$$X_{straining} = \begin{bmatrix} G_{s(1,1)} & G_{s(1,2)} & \cdots & G_{s(1,T)} \\ G_{s(2,1)} & G_{s(2,2)} & \cdots & G_{s(2,T)} \\ \vdots & \vdots & \ddots & \vdots \\ G_{s(T,1)} & G_{s(T,2)} & \cdots & G_{s(T,T)} \end{bmatrix} \tag{18}$$

4) The output will set so that it will output 1 if it is authentic class and 0 if it is imposter class, and it is in an array $Y_{identify}$, whereby it is an identity matrix of dimension $T \times T$. To calculate the weight w for s-th machine section, the equation is:

$$w_s = X^{-1}_{straining} Y_{identify} \tag{19}$$

The max-product fuzzy neural network classification is with 7 steps:

1) $PSR_{s(i,t_2)}$ and $p_{s(i,t_2)}$ output from the quaternion correlator of the recognition stage are fuzzified through the activation functions (Gaussian membership function):

$$G_{PSR_{s(i,t_2)}} = \exp\left[\frac{-(PSR_{s(i,t_2)} - 1)^2}{\sigma^2}\right] \tag{20}$$

$$G_{p_{s(i,t_2)}} = \exp\left[\frac{-(p_{s(i,t_2)} - 1)^2}{\sigma^2}\right] \tag{21}$$

2) Calculate the product value for s-th machine section of the fuzzy neural network classifier at input image on the training images in database:

$$G_{s(i,t_2)} = G_{PSR_{s(i,t_2)}} \times G_{p_{s(i,t_2)}} \tag{22}$$

3) Gather and store the product values in an array:

$$X_{s\ classification} = [G_{s(i,1)} G_{s(i,2)} \cdots G_{s(i,T)}] \tag{23}$$

4) Obtain the classification outcomes for each machine condition in the s-th section by multiply (23) with the weight trained at (19):

$$Y_{s\ classification} = X_{s\ classification} \times w_s \tag{24}$$

5) Classify the input machine condition with the class of machine condition it belongs to by using max composition:

$$\text{Class}_s = \max\left\{Y_{s\ classification}\right\} \tag{25}$$

6) Determine which element in $Y_{s\ classification}$ matrix match with Class_s:

$$\psi = \text{the position number of element in } Y_{s\ classification} \text{ matrix which has}$$
$$\text{the equal value with Class}_s \tag{26}$$

ψ is corresponds to the assigned number of reference image in database.

7) Based on two sets of fuzzy IF-THEN rules, perform defuzzification:

R^1_s : IF ψ is match with the number stored in overheat class of s−th machine, THEN alarm : 'machine s overheat'. $\tag{27}$

R^2_s : IF ψ is match with the number stored in overheat class of s−th machine, THEN alarm : 'machine s function properly'. $\tag{28}$

Fig. 4. Image on the site capture by digital camera

Fig. 5. Image on the site capture by thermal camera for all machines are in overheats condition

Fig. 6. Log-polar mapping of Fig. 5 as divided into 3 partitioned machines' sections

6   Experimental Results

In this section, we briefly illustrate the experimental results. We use a database with thermal images collected at the Applied Mechanics Lab in Faculty of Engineering and Technology, Multimedia University to test the proposed machine condition monitoring system. An image captured by digital camera on the site is shown in Fig. 4. A thermal image is also captured by using AXT 100 on the same position based on all the functioning machines are in overheat condition is shown in Fig. 5, and the corresponding log-polar form of it in Fig. 6. In Fig. 4, machine A (leftmost one) and machine C (rightmost one) are vibro test machines with same model and same specifications, whereas machine B (center one) is a fatigue test machine. Three machines are considered to be overheat when their motors' temperature reach 90°C. The database has $T = 30$ reference images, each with dimension 180 horizontal pixels × 70 vertical pixels of varying possible machines' conditions.

For authentic case, the correlation plane should have sharp peaks and it should not exhibit such strong peaks for imposter case. These two cases will be investigated below:

Authentic case: Fig. 7a-7c show the samples correlation plane for input thermal image (for every machine is overheat) matching with one of the reference image of overheat class in the database, for section machine A, section machine B and section machine C respectively. Since the three pairs of images are in good match, we can observe their correlation plane are having smooth and sharp peaks.

Imposter case: Fig. 8a-8c show the samples correlation plane for input thermal image (for every machine is not overheating) matching with one of the reference image of overheat class in the database, for section machine A, section machine B and section machine C respectively. Since the three pairs of images are not in good match, we can observe that their correlation planes are having no sharp peak at all.

Table 1 shows the PSR and p-value for both authentic and imposter case as in Fig. 7 and Fig. 8 for section machine A, B and C. Note that the sharp correlation peak resulting in large normalized PSR and p-value in authentic case of section machine A, B and C,

Table 1. Normalized PSR and p-values for both authentic and imposter case

Authentic case	Normalized PSR	Normalized p-value	Imposter case	Normalized PSR	Normalized p-value
Section mac. A	0.9528	0.7971	Section mac. A	0.0338	0.0287
Section mac. B	1.0000	0.8987	Section mac. B	0.0421	0.0373
Section mac. C	0.9674	0.8317	Section mac. C	0.0371	0.0345

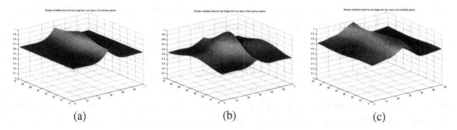

Fig. 7. Samples correlation plane for input thermal image (for every machine is overheat) matching with one of the reference image of overheat class for both all the machines in the database (authentic case) a.) section machine A, b.) section machine B, c.) section machine C

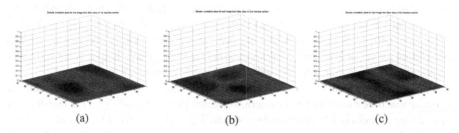

Fig. 8. Samples correlation plane for input thermal image (for every machine is not-overheat) matching with one of the reference image of overheat class for both all the machines in the database (imposter case) a.) section machine A, b.) section machine B, c.) section machine C

whereas small PSR and p-value exhibiting in the imposter case of section machine A, B and C.

Log- polar mapping is used in the proposed machine condition monitoring system to help solving rotation and scaling invariant. For each of the partitioned machine section (with total $x_1 = 81$ and total $y_1 = 100$), log-polar sampling is used to sample Cartesian input image into log-polar sampling image (with total $\rho = 60$ and total $\theta = 70$) and then mapping into log-polar mapping image with total $x_2 = 60$ and $y_2 = 70$. By applying log-polar mapping, we can have a data reduction ratio of 8300 : 4200 with almost 2 fold of data compression in a fine image resolution manner. Fig. 9 shows an example case of log-polar mapping for section machine A.

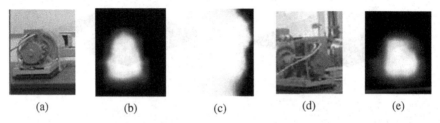

Fig. 9. An example case of rotational invariant a.) Cartesian input image capture by digital camera (1.6m), b.) the same image as in (a) capture using thermal camera during the machine is overheat, c.) log polar mapping image of (b), d.) rotational invariant of (a) capture by digital camera, e.) thermal image of (d) in Cartesian form, f.) log-polar form of (e).

Fig. 9b is captured using thermal camera according to the same position of the machine as in Fig. 9a during the machine is overheating. If we have a rotational invariant on the same machine as shown in Fig. 9d, then the corresponding thermal image in Cartesian and log polar form are shown in Fig. 9e and Fig. 9f respectively. The output correlation plane obtained among (i) Fig. 9b and Fig. 9b itself, (ii) Fig. 9b and Fig. 9e, (iii) Fig. 9c and Fig. 9c itself, (iv) Fig. 9c and Fig. 9f are shown in Fig. 10a, Fig. 10b, Fig. 10c and Fig 10d respectively. From Fig. 10a and Fig. 10b, the correlation planes are obtained among thermal images that captured directly without log-polar mapping. We can observe that the correlation planes are actually not smooth and there are more than one focusing sharp peaks. It is hard to trace out PSR and p-values from such correlation planes. However, in Fig. 10c and Fig. 10d, we can observe that after performing log-polar mapping on the captured images, the correlation planes are smooth and they are almost identical in shape. The calculated PSR and p-values from these two planes after normalized are also close to each other ($\mathrm{PSR}_{\mathrm{Fig.}10c} = 1.0000$, $\mathrm{PSR}_{\mathrm{Fig.}10d} = 0.9267$, $p_{\mathrm{Fig.}10c} = 0.8894$, $p_{\mathrm{Fig.}10d} = 0.8054$). We can claim that log-polar mapping not only providing a smoother correlation plane for PSR and p-values measurement, it actually help solving rotational invariant problem in quaternion based thermal image correlation too.

For scaling invariant, a test is taking on section machine A with different distance in between machine monitored and camera, as shown in Fig. 9a and Fig. 11a. The distance of machine-camera is 1.6 meters in Fig. 9a and 3.2 meters in Fig. 11a. The thermal images for them captured during overheat condition are shown in Fig. 9b and Fig. 11b

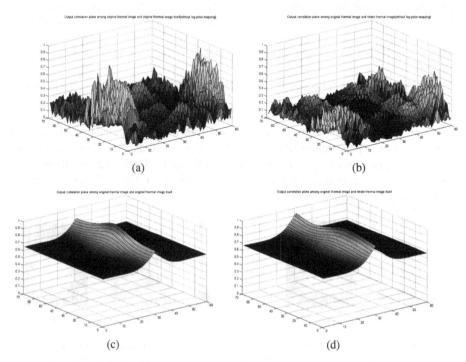

Fig. 10. Output correlation planes among: a.) Fig. 9b and Fig. 9b itself, b.) Fig. 9b and Fig. 9e, c.) Fig. 9c and Fig. 9c itself, d.) Fig. 9c and Fig. 9f

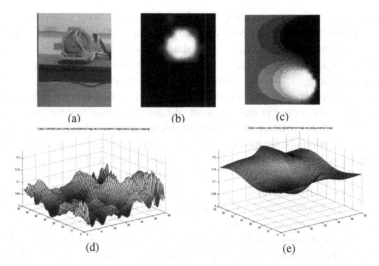

Fig. 11. An example case of scaling invariant a.) Cartesian input image capture by digital camera (3.2m), b.) the same image as in (a) capture using thermal camera during the machine is overheat, c.) log polar mapping image of (b). Output correlation planes among: d.) Fig. 9b and Fig. 11b, e.) Fig. 9c and Fig. 11c

respectively, and their log-polar images are shown in Fig. 9c and Fig. 11c respectively. The output correlation plane obtained among (i) Fig. 9b and Fig. 11b, and (ii) Fig. 9c and Fig. 11c are shown in Fig. 11d and Fig. 11e respectively. The output correlation plane in Fig. 11d is with rough surfaces and overall low normalized amplitudes as compare to Fig. 11e. By comparing output correlation planes on Fig. 11d and Fig. 10c, we can observe that the overall normalized amplitudes in correlation plane of Fig. 11e are reduced to almost $1/4$ of that in Fig. 10c. However, their plane shapes are almost identical. Sharp peaks still can detected around central regions, and their calculated PSR and p-values after normalized are also near to each other ($\mathrm{PSR}_{\mathrm{Fig.\,10c}} = 1.0000$, $\mathrm{PSR}_{\mathrm{Fig.\,10e}} = 0.8647$, $p_{\mathrm{Fig.\,10c}} = 0.8894$, $p_{\mathrm{Fig.\,10e}} = 0.7681$). Hence, we can comment that log-polar mapping can be used to solve scaling invariant problem in quaternion based thermal image correlation.

The machine monitoring system was evaluated with the 'operator perceived activity' whether any of the machines are overheat or not and compare with that classified by the (OPA) [14]. The operator will comments on the images captured by the thermal camera, proposed machine condition monitoring system. The system was evaluated for 10,000 samples images captured by the thermal camera for monitoring the three functioning machines as in Fig. 4. Among the 10,000 samples images, 9,851 were tracked perfectly (machines' conditions agreed by both observer and the machine condition monitoring system), i.e. an accuracy of 98.51%.

7   Conclusion

This paper presented a system capable of monitoring machine condition using log-polar mapping, quaternion based thermal image correlator and max-product fuzzy neural network classifier. Our results show that the proposed machine condition monitoring system is performed with very high accuracy, i.e. more than 98%. One of the advantage of using quaternion correlation rather than conventional correlation method is that quaternion correlation method deals with color images without converting those images into gray-scale images. Hence, important color information can be preserved. Max-product fuzzy neural network provides high level framework for approximate reasoning, hence it is best suit for use in classification. We also found out that the use of log-polar mapping help solving rotation and scaling problems in quaternion based thermal image correlation. Additionally, log-polar mapping has data compression capability. Therefore the use of log-polar mapping reduces the computation time and memory storage needs. Log-polar mapping also smoother up the output correlation plane hence makes a better measurement way for PSR and p-value. In future, we plan to implement this machine condition monitoring system in a wide area coverage using minimum hardware manner, whereby the machines surrounded in an omnidirectional (360°) view can be monitored by using a single thermal camera and a specific design mirror. Hence, a mathematical model need to be formulate out to design the geometry of such mirror and an unwarping processing method also need to be research for unwarping the omnidirectional image into panoramic forms for better machines' partitioned purpose. These topics will be addressed in future work.

References

1. Chaudhuri, P., Sentra, P., Yoele, S., Prakesh, A.: et al: Non-destructive evaluation of brazed joints between cooling tube and heat sink by IR thermography and its verification using FE analysis. NDT & E International 39(2), 88–95 (2006)
2. http://www.irtek-temp.com/index.php?option=com_content&task=view&id=39&Itemid=2
3. http://www.4infrared.com/
4. http://www.aas2.com/products/axt100/
5. Weiman, C.: Log-polar vision for mobile robot navigation. In: Proc. of Electronic Imaging Conferences, Boston, USA, November 1990, pp. 382–385 (1990)
6. Jurie, F.: A new log-polar mapping for space variant imaging: Application to face detection and tracking. Pattern Recognition 32(55), 865–875 (1999)
7. Berton, F.: A brief introduction to log-polar mapping. Technical report, LIRA-Lab, University of Genova (February 2006)
8. Hamilton, W.R.: Elements of Quaternions. Longmans/Green, London/U.K (1866)
9. Xie, C., Savvides, M., Kumar, B.V.K.V.: Quaternion correlation filters for face recognition in wavelet domain. In: Int. Conf. on Accoustic, Speech and Signal Processing, ICASSP 2005, pp. II85–II88 (2005)
10. Pei, S.C., Ding, J.J., Chang, J.: Color pattern recognition by quaternion correlation. In: Proc. of Int. Conf. on Image Processing, vol. 1, pp. 894–897 (2001)
11. Moxey, C.E., Sangwine, S.J., Ell, T.: Hypercomplex correlation techniques for vector images. IEEE Trans. on Signal Processing 51(7), 1941–1953 (2003)
12. Bourke, M.M., Fisher, D.G.: Solution algorithms for fuzzy relational equations with max-product composition. Fuzzy Sets Systems 94, 61–69 (1998)
13. Xiao, P., Yu, Y.: Efficient learning algorithm for fuzzy max-product associative memory networks. In: Proc. SPIE, vol. 3077, pp. 388–395 (1997)
14. Owen, J., Hunter, A., Fletcher, E.: A fast model-free morphology-based object tracking algorithm. In: British Machine Vision Conference, pp. 767–776 (2002)

Firing Correlation in Spiking Neurons with Watts–Strogatz Rewiring

Teruya Yamanishi[1] and Haruhiko Nishimura[2]

[1] Department of Management Information Science, Fukui University of Technology
Gakuen, Fukui 910–8505, Japan
[2] Graduate School of Applied Informatics, University of Hyogo
Chuo-Ku, Kobe 650–0044, Japan

Abstract. The brain is organized as neuron assemblies with hierarchies of complex network connectivity. In 1998, Watts and Strogatz conjectured that the structures of most complex networks in the real world have the so-called small-world properties of a small mean path between nodes and a high cluster value, regardless of whether they are artificial networks, such as the Internet, or natural networks, such as the brain. Here we explore the nature of a small-world network of neuron assemblies by simulating the network structural dependence of Izhikevich's spiking neuron model. The synchronized rhythmical firing is estimated in terms of rewiring probabilities, and the structural dependence of the firing correlation coefficient is discussed.

1 Introduction

The hippocampus generates brain waves in various frequency bands for different levels of activity. For example, we find mainly alpha wave ($8 \sim 13$ [Hz]) when a person closes his eyes or remains quiet. From the point of view of neuroscience, the brain waves originate from a synchronized rhythm event of neuron assemblies. That is, the spatial and temporal synchronization of firing by single neurons. At present, it is believed that the synchronized rhythm event is closely connected with actions or states of brain for animals, and plays a very important role in the successful execution of various functions of the brain[1]. Therefore, in order to understand the behavior of the brain from a neural network model, it is essential to consider not just the dynamics of a single neuron but also the characteristics of synaptic connection and the network structure of neuron assemblies.

In 1998, Watts and Strogatz attempted to explain the small-world phenomenon from network theory[2], and showed that small-world characteristics appear in a wide variety of networks (e.g., neural systems or the transmission network of *C. elegans*) irrespective of whether they are natural or artificial[3,4]. Recently, there have been attempts to understand the coupling structure of cerebral nerves and brain regions in terms of a small-world structure[5,6].

Various neuron models have been proposed[7]. In this work, we use the spiking model proposed by Izhikevich[8,9] that explains eight patterns of neuron

F. Peper et al. (Eds.): IWNC 2009, PICT 2, pp. 363–371, 2010.

spikes using four variables. We construct a small-world network using this spiking neuron model and perform simulation experiments to investigate the network structual dependence of firing correlations for neuron assemblies.

2   Izhikevich's Neuron Model and Network Structure

In this section, we describe Izhikevich's spiking neuron model and give an overview of the structure of our network.

2.1   Izhikevich's Spiking Neuron Model

In the neuron model proposed by Izhikevich, a firing pattern is represented by simultaneous linear ordinary differential equations for four variables[8]. Letting v [mV] represent the membrane potential of a neuron and I the synaptic current flowing into this neuron, we have

$$\frac{dv}{dt} = 0.04\ v^2 + 5\ v + 140 - u + I\ , \tag{1}$$

Here, u is the recovery variable related to ion permeation of the membrane of neuron, which is given by

$$\frac{du}{dt} = a\ (b\ v - u)\ . \tag{2}$$

The parameters a and b determine the time scale of u and the sensitivity of u to v, respectively. Moreover, if the value of v exceeds the prespecified firing threshold (V_{th}), v is returned to the value of the resting membrane potential, and an inactivity period is added to u. That is, for $v > V_{th}$,

$$\begin{aligned} v &\leftarrow \text{Resting\ \ membrane\ \ potential\ (RMP)}\ , \\ u &\leftarrow u + d\ . \end{aligned} \tag{3}$$

Figure 1 shows typical firing patterns of an excitatory neuron and an inhibitory neuron in this model.

2.2   Network Structure and Neuron Ensembles

Depending on the network structure, we can classify graphs as follows.

– Complete graph

  A network in which each node is linked to all other nodes.

– Random graph[10]

  A network in which two randomly selected nodes are linked with a certain binding rate.

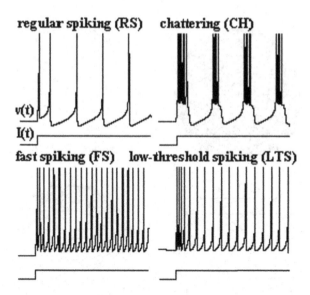

Fig. 1. Firing patterns of neurons in the Izhikevich model of Ref. [8]

– Watts—Strogatz (WS) model (small-world model)[2]

A network in which the average distance (the average number of nodes) between nodes is small, and the clustering (proportion of nodes adjacent to a random node that are also adjacent to each other) is high.

– Scale free model[11]

A network in which, in addition to small average distance and high clustering, the number of links from a node follows a power law.

Examples of these graphs with 20 nodes are shown in Figure 2.

The firing state of neuron assemblies in a complete graph based on Izhikevich's spiking neuron model is shown in Figure 3. In this simulation 800 excitatory neurons and 200 inhibitory neurons are placed randomly, and each neuron is connected to all the others. Coupling strengths s_{ij} (i and j are the indices of the receiving and firing neurons, respectively) are assigned a uniformly random value, which is between 0 and 0.5 for firing by an excitatory neuron, and -1 and 0 for an inhibitory neuron. Then, equations (1)–(3) can be rewritten as follows. For the ith neuron we have

$$\frac{dv_i}{dt} = 0.04 \, v_i^2 + 5 \, v_i + 140 - u_i + \sum_{j=\text{Firing}} (I_i + s_{ij}) \,, \tag{4}$$

$$\frac{du_i}{dt} = a_i \, (b_i \, v_i - u_i) \,, \tag{5}$$

$$\begin{aligned} v_i &\leftarrow \text{Resting membrane potential (RMP)} \,, \\ u_i &\leftarrow u_i + d_i \,. \end{aligned} \tag{6}$$

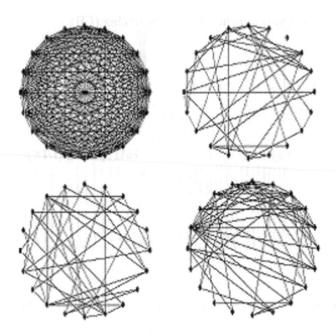

Fig. 2. Graphs of typical networks. In the top row are complete (left) and random (right) graphs; in the bottom row are Watts-Strogatz (left) and scale free (right) models.

The values of other parameters included in equations (5) and (6) are shown in Table 1. The symbol $\sim$ in the table shows that a random value between the lower and upper limits is taken. For example, for the time scale a of an inhibitory neuron, a random value between 0.02 and 0.1 would be taken. In Figure 3, an alpha rhythm ($\sim$ 10Hz) appears very early, and a gamma rhythm ($\sim$ 40Hz) of even shorter cycle appears approximately 500ms later. This agrees with the results of Izhikevich[8].

Now, let us consider the coupling structure of cerebral nerves and brain regions in terms of network structures. If neurons in the brain were connected by a complete network, the capacity of the brain would be so large that it could not placed in our head. Also, the value of the clustering coefficient of the brain is larger than that of a random graph. The connection rate of brain is approximately 5.7×10^{-7} to 7.1×10^{-7} based on 14 billion neurons in human brain cortex and $8,000$ to $10,000$ synapses per neuron. Therefore, the brain has a large value of the clustering coefficient and sparse connections. These properties agree with those of a small-world network or a scale free one.

So, we investigate the firing properties for neuron assemblies in a WS model (small-world network) which is constructed using Izhikevich's spiking neuron model.

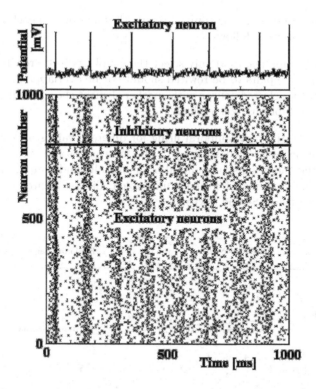

Fig. 3. Simulations of a complete network of 1000 randomly coupled Izhikevich spiking neurons. Top: typical spike train of an excitatory neuron. Bottom: spike raster plot with 800 excitatory neurons and 200 inhibitory ones.

Table 1. Parameter values of the Izhikevich neuron model used in our simulations

Parameters	Excitatory neuron	Inhibitory neuron
Time scale a	0.02	$0.02 \sim 0.10$
Sensitivity b	0.20	$0.20 \sim 0.25$
RMP	$-65 \sim -50$	-65
Inactivity period d	$2 \sim 8$	2

3 Simulation Experiment with a WS Network Structure

The same simulation experiments were performed as for the case of a complete graph described in the previous section. A pulse neuron assembly was formed by placing 800 excitatory neurons and 200 inhibitory neurons in one dimension. Coupling strengths were randomly assigned values between 0 and 12 in the case of coupling with an excitatory neuron and between -20 and 0 for an inhibitory neuron. Since we are considering a WS model (small-world network), we assume, initially, that each neuron is coupled to 10 other neurons on each side. Then we rewire each neuron as follows.

1. From all couplings, randomly select the couplings to be rewired with probability p. The pairs to be rewired should not be overlapping.
2. Connect the selected neuron with another element chosen randomly, avoiding loops and duplicate connections.

In Figure 4, spatial and temporal firing for different values of p is shown. It is clear that neurons fire locally for $p = 0$. However, as the value of p becomes larger, we find the firing synchronized both spatially and temporally. This could be due to the coupling of initially uncoupled, distant neurons by a shortcut introduced by the WS-type rewiring.

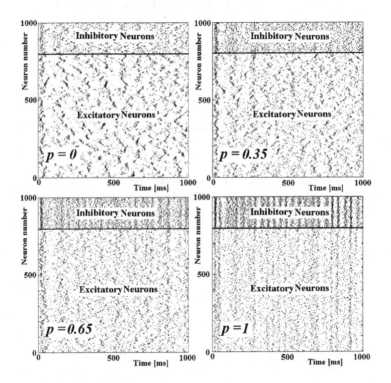

Fig. 4. Spike raster plots of a Watts–Strogatz network using an Izhikevich neuron model for different values of the rewiring probability p

Next, we quantitatively analyze the firing synchronization of neuron assemblies in the WS network. The firing correlation coefficient $C_i(\delta, d)$ is the correlation between any firing neuron i and one at distance d with delay δ. $C_i(\delta, d)$ is given by[12,13]

$$C_i(\delta, d) = \langle C_i(\delta) \rangle_d , \tag{7}$$

$$C_i(\delta) = \frac{1}{\sigma_i \, \sigma_{i+d}} \langle \, \Delta J_i(t) \, \Delta J_{i+d}(t+\delta) \, \rangle_t \, , \tag{8}$$

$$\sigma_i = \sqrt{\langle \, (\Delta J_i(t)) \, \rangle_t} \, , \tag{9}$$

$$\Delta J_i(t) = J_i(t) - \langle J_i(t) \rangle_t \, , \tag{10}$$

where $J_i(t)$ is the immediate firing rate of neuron i at the time t, which is given by

$$J_i(t) = \frac{1}{\omega} \sum_k \Theta(t - t_k) \, , \quad \Theta(t) = \begin{cases} 1 & \text{for } 0 \le t < \omega \, , \\ 0 & \text{otherwise} \, . \end{cases} \tag{11}$$

The notation $\langle \cdot \rangle_d$ and $\langle \cdot \rangle_t$ indicates averages at distance d or time t, respectively. The dependence of the firing correlation coefficient $C_i(\delta, d)$ on the rewiring probability p and the distance d is shown in Figures 5 and 6 for $\delta = 0$. For small d $C_i(\delta, d)$ decreases with increasing p, and increases with increasing p for large d. As in Figure 4, we find that firing in neuron assemblies evolves from local firing to spatial synchronization of the entire system as p increases.

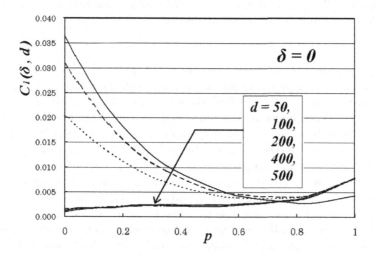

Fig. 5. The firing correlation coefficient $C_i(\delta, d)$ as a function of the rewiring probability p at $\delta = 0$. The solid, broken, and dotted lines are for distances $d = 1, 5$, and 10. Correlation coefficients for distances $d = 50\text{-}500$ are also shown, but they are almost identical.

4   Summary

We examined the network structure dependence of neuron assemblies based on Izhikevich's spiking neuron model. In a simulation experiment using a complete graph, synchronous firing in alpha and gamma rhythms could be observed by selecting appropriate parameters. However, this graph is not a suitable model for the structure of the neural circuit of an actual brain. So, we reconstructed the

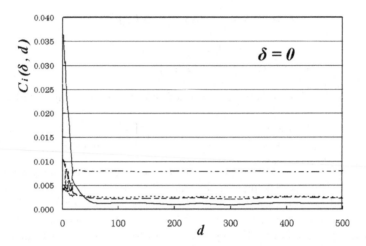

Fig. 6. The firing correlation coefficient $C_i(\delta, d)$ as a function of the distance d at $\delta = 0$. The solid, broken, dotted, and dotted–broken lines indicate are for rewiring probabilities $p = 0, 0.35, 0.65$, and 1, respectively.

network according to the Watts-Strogatz model, which possesses the characteristics of a natural complex network, and examined the dependence of synchronous firing on network structure. When there was no rewiring, firing was local and continuity of synchronization was not seen. However, as the rewiring probability was increased, firing synchronized spatially as well as temporally and reached gamma rhythm (~ 30 Hz). This could be because the synchronization propagated over the entire system through shortcuts created by coupling distant neurons.

Recently, genes have been reported to cause abnormal behaviors of neurons, such as mistranslocation, mislocation, and miswiring, during brain development [14,15]. These behaviors give rise to many brain disorders, and these may be described by dynamically rewiring neural network models. We are now making progress in investigations of raster plots and firing correlations not only for topological network connections but also for differing locations of excitatory and inhibitory neurons. It is also necessary to examine the dependency of the synchronization on the ranges of the coupling strength for those neurons in detail. In addition, we aim to gain understanding of the complex networks occurring in neural circuits through simulation experiments of a scale free model.

Acknowledgements

The work of the author, T.Y., was partially supported by research grants from Fukui University of Technology, and also the work of the author, H.N., was supported by a Grant-in-Aid for Scientific Research (C-19500606) from the Japan Society for the Promotion of Science.

References

1. Sakamoto, K., et al.: Discharge Synchrony during the Transition of Behavioral Goal Representations Encoded by Discharge Rates of Prefrontal Neurons. Cerebral Cortex 18, 2036–2045 (2008)
2. Watts, D.J., Strogatz, S.H.: Collective dynamics of 'Small-world' networks. Nature 393, 440–442 (1998)
3. Boccaletti, S., Latora, V., Chavez, M., Hwang, D.-U.: Complex networks: Structure and dynamics. Physics Reports 424, 175–308 (2006)
4. Arenas, A., Díaz-Guilera, A., Kurths, J., Moreno, Y., Zhou, C.: Synchronization in complex networks. Physics Reports 469, 93–153 (2008)
5. Bassett, D.S., Bullmore, E.: Small-World Brain Networks. The Neuroscientist 12, 512–523 (2006)
6. Zemanová, L., Zhou, C., Kurths, J.: Structural and functional clusters of complex brain networks. Physica D 224, 202–212 (2006)
7. Brette, R., et al.: Simulation of networks of spiking neurons: A review of tools and strategies. J. Comput. Neurosci. 23, 349–398 (2007)
8. Izhikevich, E.M.: Simple Model of Spiking Neurons. IEEE Trans. Neural Networks 14, 1569–1572 (2003)
9. Izhikevich, E.M.: Which Model to Use for Cortical Spiking Neurons? IEEE Trans. Neural Networks 15, 1063–1070 (2004)
10. Erdös, P., Renyi, A.: On random graphs. Publ. Math (Debrecen) 6, 290–297 (1959)
11. Barabási, A.-L., Albert, R.: Emergence of scaling in random networks. Science 286, 349–352 (1999)
12. Kanamaru, T., Aihara, K.: The effects of Watts-Strogatz's rewiring to the synchronous firings in a pulse neural network with gap junctions. Technical Report of IEICE, vol. 107, pp. 175–180 (2008)
13. Kanamaru, T., Aihara, K.: Stochastic Synchrony of Chaos in a Pulse-Coupled Neural Network with Both Chemical and Electrical Synapses Among Inhibitory Neurons. Neural Computation 20, 1951–1972 (2008)
14. Kamiya, A., et al.: A schizophrenia-associated mutation of $DISC_1$ perturbs cerebral cortex development. Nature Cell Biology 7, 1167–1178 (2005)
15. Miswiring the Brain: Breakthrough of the year. Science 310, 1880–1885 (2005)

Methods for Shortening Waiting Time in Walking-Distance Introduced Queueing Systems

Daichi Yanagisawa[1,2], Yushi Suma[1], Akiyasu Tomoeda[3,4], Ayako Kimura[5], Kazumichi Ohtsuka[4], and Katsuhiro Nishinari[4,6]

[1] Department of Aeronautics and Astronautics, School of Engineering,
The University of Tokyo, 7-3-1, Hongo, Bunkyo-ku, Tokyo, 113-8656, Japan
[2] Research Fellow of the Japan Society for the Promotion of Science
[3] Meiji Institute for Advanced Study of Mathematical Sciences, Meiji University
[4] Research Center for Advanced Science and Technology, The University of Tokyo
[5] Mitsubishi Research Institute, Inc.
[6] PRESTO, Japan Science and Technology Agency
tt087068@mail.ecc.u-tokyo.ac.jp

Abstract. We have investigated the *walking-distance introduced queueing theory* and verified that the mean waiting time in a parallel-type queueing system, i.e., queues for each service windows, becomes smaller than that in a fork-type queueing system, which collects people into a single queue, when sufficiently many people are waiting in queues. In a fork-type queueing system, a person at the head of the queue, which is usually set at the end of the system, starts to move when one of the service windows become vacant. Since this walking time from the head of the queue to the windows increases the waiting time, we propose to set the head of the queue at the center of the system and keep one person waiting at each service window when it is occupied by other person. The validity of the methods is examined by the theoretical analysis, simulations, and experiments.

Keywords: Queueing theory, pedestrian dynamics, cellular automaton.

1 Introduction

Queueing theory has been considerably studied since Erlang started designing telephone exchanging system in 1909 [1], and many important theories have been developed [2]. Nowadays, it is one of the fundamental theories for designing queueing systems, and also applied to traffic systems [3,4]. However, the queueing theory does not include the effect of walking distance, thus, it is not appropriate to apply it to designing large queueing systems for people such as an immigration inspection floor in an international airport since walking time from the head of the queue to the service windows is too large to ignore.

Therefore, we have introduced such effect of delay to the queueing theory and investigated the *walking-distance introduced queueing theory*. We compare the

F. Peper et al. (Eds.): IWNC 2009, PICT 2, pp. 372–379, 2010.

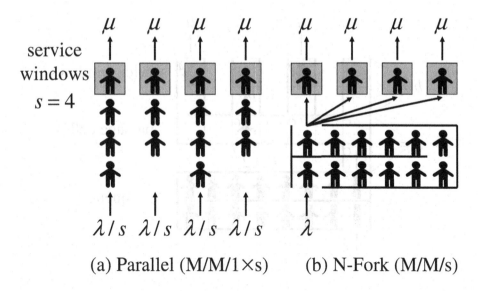

service windows $s = 4$

(a) Parallel (M/M/1×s)   (b) N-Fork (M/M/s)

Fig. 1. Schematic views of queueing systems in the case $s = 4$. (a) Parallel $(M/M/1\times 4)$. (b) N-Fork $(M/M/4)$. An ideal fork-type queueing system. A person at the head of the queue moves to the service window instantaneously when one of them becomes vacant.

mean waiting time in a parallel-type queueing system and a fork-type queueing system by using it. Furthermore, two new methods are also proposed to shorten the mean waiting time of people. We set the head of the queue at the center of the system and keep one person waiting at each service window when it is occupied by other person in a fork-type queueing system to decrease the effect of walking distance. In addition to the theoretical analysis and simulations, experiments of the queueing system for people have been performed. Therefore, we have succeeded to verify the characteristic of the walking-distance introduced queueing system not only theoretically but also experimentally.

This paper is organized as follows. In the next section, the walking-distance introduced queueing theory is explained and the two methods are introduced in Sec. 3. The results of the simulations and experiments are shown in Sec. 4 and 5, respectively. Sec. 6 is devoted to summary and future works.

2   Walking-Distance Introduced Queueing Theory

2.1   Distance Introduced Fork-Type Queueing System: D-Fork

In a parallel-type queueing system (Parallel) (Fig. 1 (a)), people wait just behind the former person, so that there is no delay in walking. While, in a fork-type queueing system, people take some time to walk from the head of the queue to the service windows. However, the walking time is not taken into account in a fork-type queueing system in the normal queueing theory (N-Fork) (Fig. 1 (b)), and people move from the head of the queue to the service windows instantaneously.

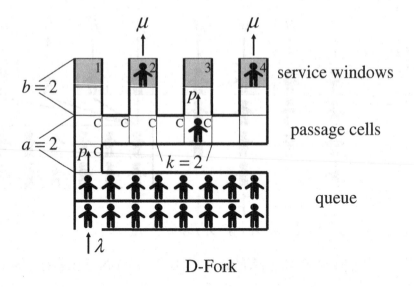

Fig. 2. Schematic view of D-Fork ($s = 4$)

Therefore, we consider D-Fork as in Fig. 2 by representing the walking distance using cellular automaton. The gray cells are window cells, and the numbers described in there are window numbers. The white cells are passage cells. Note that the letter "C" described in the passage cells represents the *common* passage cells. For example, both persons who are going to the window 3 and 4 pass the common passage cells. People sometimes cannot go forward in the common passage cells since there is a possibility that other people stand in front of them. The place that people are waiting, which is not divided into cells, is a queue. $s \in \mathbf{N}$, $\lambda \in [0, \infty)$, and $\mu \in [0, \infty)$ represent the number of service windows, the arrival rate, and the service rate, respectively. a and b represent the length of the passage, and k is the interval length between two service windows. The distance from the head of the queue to the service window $n \in [1, s]$ is described as $d_n = a + b + k(n-1)$. Fig. 2 represents the case $s = 4$, $a = 2$, $b = 2$, and $k = 2$. Service windows have two states: vacant and occupied. When a person at the head of the queue decides to move to the vacant service window n, it changes into occupied state. The person proceed to the service window by one cell with the rate $p \in [0, \infty)$ as the asymmetric simple exclusion process. A service starts when the person arrives at the service window, and after it finishes the state of the service window changes into vacant state.

2.2   Update Rules

The simulation of walking-distance introduced queueing systems consists of the following five steps per unit time step.

1. If there is at least one vacant service window and one person in the queue, and the first cell of the passage is vacant, then the person decide to proceed

to a vacant service window which is the nearest to the head of the queue, and the state of the service window become occupied.

2. Add one person to the queue with the probability $\lambda \Delta t$, where Δt is the length of the unit time step.
3. Proceed each person in the passage cells to his/her service windows with the probability $p\Delta t$ if there is not other person at their proceeding cell.
4. Remove people at the service windows and change their states into vacant state with the probability $\mu \Delta t$.
5. If 1. takes into practice, proceed the person at the head of the queue to the first passage cell with the probability $p\Delta t$.

2.3 Stationary Equations

We define the sum of the walking time and the service time at service window n as a throughput time τ_n and its reciprocal as a throughput rate μ_n. Here, we calculate the mean throughput rate $\hat{\mu}_n$ when n service windows are occupied, and obtain stationary equations of D-Fork. We suppose that all passage cells are vacant by mean field approximation. Then, the mean value of the throughput time $E(\tau_n)$ is described as follows.

$$E(\tau_n) = \frac{1}{\mu} + \frac{a + b + k(n-1)}{p}. \tag{1}$$

The throughput rate μ_n is obtained as

$$\mu_n = \frac{1}{E(\tau_n)} = \frac{\mu}{1 + \alpha + 2\beta(n-1)} \qquad \left(\alpha = \frac{\mu(a+b)}{p},\ \beta = \frac{k\mu}{2p} \right). \tag{2}$$

In the case $2\beta(n-1)/(1+\alpha) \ll 1$, we calculate the mean throughput rate $\hat{\mu}_n$ as

$$\hat{\mu}_n = \frac{1}{n} \sum_{l=1}^{n} \mu_l \approx \frac{\mu}{1 + \alpha + \beta(n-1)}. \tag{3}$$

By using (3) the stationary equations are described as follows:

$$\lambda P_0 = \hat{\mu}_1 P_1$$
$$\lambda P_{n-1} + (n+1)\hat{\mu}_{n+1} P_{n+1} = (\lambda + n\hat{\mu}_n) P_n \qquad (1 \le n \le s-1) \tag{4}$$
$$\lambda P_{n-1} + s\hat{\mu}_s P_{n+1} = (\lambda + s\hat{\mu}_s) P_n \qquad (n \ge s).$$

We obtain the mean waiting time W_q by solving (4) analytically. In the case $\alpha = \beta = 0$, we have the stationary equations of $M/M/s$ [2] from (4), thus α and β represent the effect of walking time.

In our simulation the distribution of the throughput time is gamma distribution. We approximate it as exponential distribution in this calculation, however, when β is small the results from the exponential distribution approximated well to those from gamma distribution.

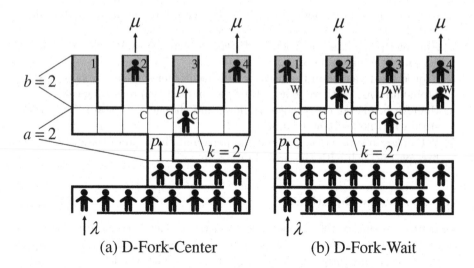

Fig. 3. Schematic views of walking-distance introduced queueing systems ($s = 4$). (a) D-Fork-Center. The head of the queue is at the center of the system. (b) D-Fork-Wait. People can wait at the cells, which are described as "W".

3   Methods for Shortening Mean Waiting Time in D-Fork

3.1   Set the Head of the Queue at the Center: D-Fork-Center

The head of the queue is usually set at the end of the system since we can efficiently use the space for the queue. However, people have to walk a long distance to the farthest window as in Fig. 2. Thus, we propose to set the head of the queue at the center (D-Fork-Center) as in Fig. 3 (a). Then, the mean throughput rate is described as follows:

$$\hat{\mu}_n \approx \frac{\mu}{1 + \alpha + \frac{\beta}{2}\left(n + \frac{1 + (-1)^{n+1}}{2n}\right)} \tag{5}$$

Comparing (3) and (5), we see that the coefficient of n in (5) is half of that in (3). Therefore, the effect of the walking distance approximately becomes half if we set the head of the queue at the center. The mean waiting time is approximately calculated by replacing β of D-Fork with $\beta/2$ in Sec. 4. Note, that the (5) represents the $\hat{\mu}_n$ in the case that both s and k are even number. The mathematical formulation of the other cases are described in Ref. [5].

3.2   Keep One Person Waiting at the Window: D-Fork-Wait

The walking distance in D-Fork is essentially problematic, since it delays the start of services, i.e., people have to walk the passage before they start to receive the service. Thus, we propose to keep one person waiting at the service window.

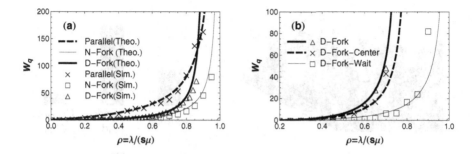

Fig. 4. The mean waiting time W_q. (a) Comparison among Parallel, N-Fork, and D-Fork in the case $s = 3$, $\alpha = 0$, $\beta = 0.05$, $\mu = 0.05$, $\Delta t = 1$. (b) Comparison among D-Fork, D-Fork-Center, and D-Fork-Wait in the case $s = 4$, $\alpha = 0.15$, $\beta = 0.05$, $\mu = 0.05$, $\Delta t = 0.1$.

We call a queueing system which this method is applied to as D-Fork-Wait (Fig. 3 (b)). Since people are waiting just next to the service windows, they can receive service instantaneously when their former people leave there. The delay in walking is almost removed by this method, i.e., the effect of walking distance does not need to be considered. Therefore, the mean waiting time is approximately calculated by the expression for N-Fork.

4   Theoretical Analysis and Simulation

Figure 4 (a) show the mean waiting time W_q against the utilization $\rho(= \lambda/(s\mu))$, which represents the degree of congestion, in Parallel, N-Fork, and D-Fork. The results of analysis agree with those of the simulation very well. We see that W_q in N-Fork is smaller than that in Parallel and D-Fork in $0 \leq \rho < 1$. The N-Fork is the most efficient of the three; however, it is an ideal system and does not exist in reality. By focusing on the curves of Parallel and D-Fork, we can clearly observe the crossing of them. This means that when the utilization ρ is small, i.e., there are not sufficiently many people in the system; we should form D-Fork to decrease the waiting time. On the contrary, when the utilization ρ is large, i.e., there are many people in the system, we should form Parallel. The reversal phenomenon of W_q is obtained for the first time by introducing the effect of walking distance. Note, that the proper type of the queueing system is changed according to not only ρ but also β, which is studied in Ref. [6].

Figure 4 (b) show the mean waiting time W_q in D-Fork, D-Fork-Center, and D-Fork-Wait. We see that W_q in D-Fork-Center is always smaller than that in D-Fork in $0 \leq \rho < 1$. This result verifies that we can decrease W_q by setting the head of the queue at the center. We also find that W_q in D-Fork-Wait is the smallest of the three since it almost completely removes the effect of walking time by keep one person waiting at the window. The well correspondence between the results of the theoretical analysis and simulation in $\rho \leq 0.85$ also verifies our

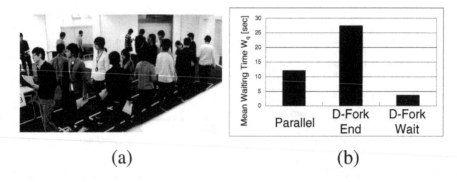

Fig. 5. (a) Schematic view of the experiment. (b) The experimental mean waiting time for Parallel, D-Fork, and D-Fork-Wait.

assumption that D-Fork-Wait becomes close to N-Fork. However, in the large-ρ region, the result of the simulation becomes larger than that of the theoretical analysis. In D-Fork-Wait, there is a possibility that $2s$ people are in the window cells or passage cells, while s people is the maximum in the other cases. Thus, when ρ is large, people in the common passage cells cannot often proceed since other people are in front of them. This jam in the common passage cells, which is not considered in the theoretical calculation, increase the waiting time in D-Fork-Wait when ρ is large.

5   Experiments

We have performed the experiments to examine the two results in the former section. 1. There is a case that W_q in Parallel becomes smaller than that in D-Fork. 2. We can decrease W_q by keeping one person waiting at the window. We made the queueing system as in Fig. 1 (a), 2, and 3 (b), whose parameters are $s = 4$, $a = 1$ [m], $b = 0.5$ [m], $k = 3$ [m], $\lambda = 188/600$ [persons/sec], and $\mu = 1/8$ [persons/sec]. Figure 5 (a) is the snap shot of the experiment. Participants of the experiments enter the system and line up in the queue when the staff says to do so. They proceed to the windows and receive service. After that they wait at the starting position until the staff let him/her enter the system again. We put 188 people in 600 [sec] in one experiment. Note, that the arrival was random while the service was deterministic for simplicity. According to the Pollaczek-Khintchine formula [2], W_q becomes small when the service is deterministic. Since this effect acts on the all kinds of the queueing systems in the same way, the results are not critically influenced by deterministic service. Therefore, we can examine the result of the theoretical analysis and simulations by these experiments.

Figure 5 (b) shows the result of the experiments. We see that W_q in Parallel is smaller than that in D-Fork. This result verifies our theoretical analysis and simulation by using the walking-distance introduced queueing theory. The reversal of W_q between Parallel and D-Fork is observed experimentally for the first time in this paper. We also find that W_q becomes dramatically small in D-Fork-Wait. This new result indicates that the method "Keep one person waiting" is an effective way to shorten the waiting time empirically.

6   Conclusion

We have introduced the effect of walking distance from the head of the queue to the service windows and shown that the performance of a parallel-type queueing system is better than that of a fork-type queueing system when there are sufficiently many people in the system. The effectiveness of the two new methods is also studied. The mean waiting time becomes small when we set the head of the queue at the center or keep one person waiting at each service window in a fork-type queueing system since the effect of walking distance decreases. We would like to emphasize that our study is based on the theoretical analysis, simulations, and experiments. Therefore, the results in this paper are reliable enough to apply to the queueing systems in the real world. It is an important future work to study the effect of the costs raised by the two methods. When we set the head of the queue at the center, the space for waiting people decreases, and at least one clerk is needed at the head of the queue to keep one person waiting at each window.

Acknowledgement

We thank Dai Nippon Printing Co. in Japan for the assistance of the experiment, which is described in Sec. 5. This work is financially supported by Japan Society for the Promotion of Science and Japan Science and Technology Agency.

References

1. Erlang, A.K.: The theory of probabilities and telephone conversations. Nyt. Tidsskr. Mat. Ser. B 20, 33–39 (1909)
2. Bolch, G., Greiner, S., de Meer, H., Trivedi, K.: Queueing Networks and Markov Chains. A Wiley-Interscience Publication, U.S.A. (1998)
3. Mukherjee, G., Manna, S.S.: Phase transition in a directed traffic flow network. Phys. Rev. E 71, 066108 (2005)
4. Helbing, D., Treiber, M., Kesting, A.: Understanding interarrival and interdeparture time statistics from interactions in queueing systems. Physica A 363, 62–72 (2006)
5. Yanagisawa, D., Tomoeda, A., Kimura, A., Nishinari, K.: Analysis on queueing systems by walking-distance introduced queueing theory. Journal of JSIAM 18(4), 507–534 (2008) (in Japanese)
6. Yanagisawa, D., Tomoeda, A., Kimura, A., Nishinari, K.: Walking-distance introduced queueing theory. In: Umeo, H., Morishita, S., Nishinari, K., Komatsuzaki, T., Bandini, S. (eds.) ACRI 2008. LNCS, vol. 5191, pp. 455–462. Springer, Heidelberg (2008)

Effect of Mutation to Distribution of Optimum Solution in Genetic Algorithm

Yu-an Zhang[1], QingLian Ma[2], Makoto Sakamoto[2], and Hiroshi Furutani[2]

[1] Interdisciplinary Graduate School of Agriculture and Engineering, University of Miyazaki,
Miyazaki City, 889-2192 Japan
[2] Faculty of Engineering, University of Miyazaki, Miyazaki City, 889-2192 Japan
ng3807u@student.miyazaki-u.ac.jp,
{sakamoto,furutani}@cs.miyazaki-u.ac.jp

Abstract. Mutation plays an important role in the computing of Genetic Algorithms (GAs). In this paper, we use the success probability as a measure of the performance of GAs, and apply a method for calculating the success probability by means of Markov chain theory. We define the success probability as there is at least one optimum solution in a population. In this analysis, we assume that the population is in linkage equilibrium, and obtain the distribution of the first order schema. We calculate the number of copies of the optimum solution in the population by using the distribution of the first order schema. As an application of the method, we study the GA on the multiplicative landscape, and demonstrate the process to calculate the success probability for this example. Many researchers may consider that the success probability decreases exponentially as a function of the string length L. However, if mutation is included in the GA, it is shown that the success probability decreases almost linearly as L increases.

1 Introduction

It is always a problem for us to select parameters such as crossover rate, mutation rate and population size when genetic algorithms (GAs) are applied to practical tasks. Unfortunately, there are little theories to select these parameters, and we have to decide parameters by intuition or experience. In this paper, we define the success probability S as there is at least one optimum solution in a population, and use it to analyze the performance of GA. We have obtained a theoretical method for calculating success probability S [1]. In this method, we assume that the population is in linkage equilibrium, and calculate the distribution of the first order schemata by using the Markov chain model. Then we obtain the number of optimum solution in a population by using the Markov chain method.

In population genetics, many researchers apply the Markov chain theory which was invented by Wright and Fisher [2]. In the field of GA, the Markov-chain theory was studied by Nix and Vose [3]. In addition, Davis and Principe independently proposed Markov chain model of GA [4]. Furthermore, another model of GA with a viewpoint of mixture system is also proposed [5]. However, in the Markov chain model of Vose, the dimension of the transition matrix increases exponentially with increasing bit string L and population size N. Therefore, some kind of approximations is necessary to apply these methods in GA.

F. Peper et al. (Eds.): IWNC 2009, PICT 2, pp. 380–387, 2010.
© Springer 2010

In the analysis of GA on the flat fitness function, Asoh and Mühlenbein assumed the population is in linkage equilibrium, and obtained the evolution probability of GA by using the first order schema [6]. We applied their stochastic approaches to the GA on multiplicative landscape, and examined the evolution of first order schemata. We used the theoretical method to calculate the success probability S [1], and studied the effects of mutation on GA. This paper reports some results of our analysis in the GA on multiplicative landscape.

2 Mathematical Model

2.1 Representations

We use the fitness proportionate selection. The population evolves in discrete and non-overlapping generations. N denotes the population size. Individuals are represented by binary strings of the fixed length L. Then the number of genotypes is $n = 2^L$, and the ith genotype is identified with the binary representation of integer $i = < i(L), \ldots, i(1) >$ ($0 \leq i \leq n - 1$). The average fitness of the population is $\bar{f}(t) \equiv \sum_{i=0}^{n-1} f_i x_i(t)$, where f_i is the fitness of ith genotype, and $x_i(t)$ is its relative frequency at generation t. We define the fitness of multiplicative landscape as $f_i = \prod_{k=1}^{L} (1 + i(k) s)$, where $i(k)$ is the kth bit in the binary representation of i. The parameter s represents selection strength.

2.2 Linkage Equilibrium

In genetics, when each gene evolves independently, we state that a population is in linkage equilibrium. On the other hand, if there are any correlations among genes at different loci, a population is in linkage disequilibrium. In this analogy for GA, when the population is in linkage equilibrium, we can treat each bit independently. Thus the theoretical analysis becomes very simple.

We introduce a deterministic equation for selection [7]. If the population is in linkage equilibrium, the distribution of individuals depends only on the frequencies of the first order schemata [6]. The relative frequency $x_i(t)$ depends only on the relative frequency of the first order schema of each bit. Therefore, the relative frequency $x_i(t)$ is represented by the form of the multiplicative form $x_i(t) = \prod_{k=1}^{L} h_{i(k)}(t)$, where $h_{i(k)}$ denotes the relative frequency of first-order schema of defining bit $i(k)$. Deterministic evolution equation of the first order schema is

$$h_{1(k)}(t + 1) = \frac{(1 + s) h_{1(k)}(t)}{1 + s h_{1(k)}(t)}, \tag{1}$$

where $h_{1(k)}$ stands for $h_{i(k)}$ with $i(k) = 1$. If mutation is included, the deterministic evolution equation is given by

$$h'_{1(k)}(t + 1) = (1 - 2p_m)h_{1(k)}(t + 1) + p_m, \tag{2}$$

where p_m denotes mutation rate.

3 Stochastic Models

3.1 Wright-Fisher Model

The deterministic equation can approximate the evolution of a population only when the population size N is large enough. From this reason, Wright and Fisher put forward their evolution equation by considering the limited individuals [2]. The Wright-Fisher model treats chromosomes having one locus and two alleles, corresponding to the GA of $L = 1$ with genotypes $k \in \{0, 1\}$ [8]. The number of the first genotype 1 takes the values of $N_1 = \{0, 1, \ldots, N\}$, and that of the genotype 0 is given by $N_0 = N - N_1$.

We analyze selection processes by taking into account the effect of random sampling, and consider the process of choosing offspring randomly from the population in proportion with their fitness values. If there are $N_1 = i$ copies of the genotype 1 at the current generation t, the probability $P(j|i)$ of N_1 taking the value of j at the next generation $t + 1$ is given by the binomial distribution

$$P(j|i) = \binom{N}{j} \left(\frac{i}{N}\right)^j \left(1 - \frac{i}{N}\right)^{N-j}. \tag{3}$$

Let $q_i(t)$ be the probability that the population is in $N_1 = i$ at generation t. Then with the normalization condition $\sum_{i=0}^{N} q_i(t) = 1$, the evolution process is described by $q_j(t + 1) = \sum_{i=0}^{N} P(j|i) q_i(t)$. We consider the population with the fitness values,

$$f_k = \begin{cases} 1 & (k = 0) \\ 1 + s & (k = 1) \end{cases}$$

We assume $s \geq 0$ and consider the maximization of the fitness, thus $k = 1$ is the favorable allele. The transition probability is given by

$$P_{i,j} = P(j|i) = \binom{N}{j} b^j (1 - b)^{N-j}, \quad b = \frac{(1 + s)i}{(1 + s)i + N - i} \tag{4}$$

where i is the number of genotype 1 at the current generation t, and j is the number of 1 at the next generation $t + 1$. Let $\mu_i(t)$ be the probability that the population is in $N_1 = i$ at generation t. Then evolution process is described by

$$\mu_j(t + 1) = \sum_{i=0}^{N} \mu_i(t) P_{i,j}. \tag{5}$$

The normalization condition is $\sum_{i=0}^{N} \mu_i(t) = 1$. The extinction state, $i = 0$, is the state that genotype A is lost from population. On the contrary, the fixation state, $i = N$, is that genotype A occupies the population. Both fixation and extinction states are absorbing states in the sense of Markov chain. μ_0 and μ_N are the probabilities of absorption states.

3.2 With Mutation

By the deterministic evolution equation with mutation, the transition probability $P_{i,j}$ is given by

$$P_{i,j} = P(j|i) = \binom{N}{j} (b')^j (1 - b')^{N-j}. \quad b' = (1 - 2p_m)b + p_m. \tag{6}$$

If $p_m \neq 0$, the Markov chain is ergodic. In the ergodic Markov chain, the population converges into the stationary distribution $\pi = \lim_{t\to\infty} \mu(t)$. All $\pi_i > 0$, and the stationary distribution satisfies the following equation

$$\pi_j = \sum_{i=0}^{N} \pi_i P_{i,j}. \tag{7}$$

The researchers derived the diffusion approximation from Wright-Fisher model, and obtained ultimate fixation probability which is expressed as

$$u(p) = \frac{1 - \exp(-2Nsp)}{1 - \exp(-2Ns)}, \tag{8}$$

where p is the relative frequency of genotype A at $t = 0$. In the same way we have ultimate extinction probability, $v(p) = 1 - u(p)$.

4   Calculation of Success Probability

In the case without mutation, the first order schema converges into the fixation state or the extinction state. In order to make optimum solution exist in stationary state, it is necessary that all L units of the first order schema converge into the fixation state. Therefore, success probability is expressed by the fixation probability $u(p)$ and string length L [1,9],

$$S = u(p)^L \tag{9}$$

However, this method cannot calculate the success probability for the GA with mutation, which does not have any absorbing state. Therefore, we use the following method in the case with mutation [1].

We call the $\ell(\leq L)$ bit string of $< 1,1,\ldots,1,1 >$ as the partial optimum solution. The random number X_ℓ represents the number of partial optimum solution of length ℓ in the population. The probability that there are j units of optimum solution is $S_j^{(\ell)} = \Pr\{X_\ell = j\}$. We can get the success probability S and failure probability $F = 1 - S$

$$S = \sum_{j=1}^{N} S_j^{(L)}, \quad F = S_0^{(L)}. \tag{10}$$

To obtain S and F, it is necessary to have $S_j^{(\ell)}$, and it can be calculated by iteration. The initial condition is $S_j^{(1)} = \pi_j^{(1)}, \quad (0 \leq j \leq N)$. The transition probability from j units of partial optimum solution of length $\ell-1$ to j units of optimum solution of length ℓ is $Q_{i,j}^{(\ell)} = \Pr\{X_\ell = j|X_{\ell-1} = i\}$. The transition probability $Q_{i,j}^{(\ell)}$ can be calculated by using the distribution of the first order schema in ℓth bit $\pi^{(\ell)}$. If $j > i$, we have $Q_{i,j}^{(\ell)} = 0$. In the case of $j \leq i$, we rearrange the solutions of the length $\ell - 1$ as

$$\underbrace{\square \ \square \ldots \square \ \square}_{i} \qquad \underbrace{\sqsubset \ \sqsubset \ldots \sqsubset}_{N-i}.$$

Here, the symbol $\square$ denotes the partial optimum solution, and $\sqsubset$ denotes the non-optimum solutions. Let m be the number of bit one at the ℓth bit, and its probability

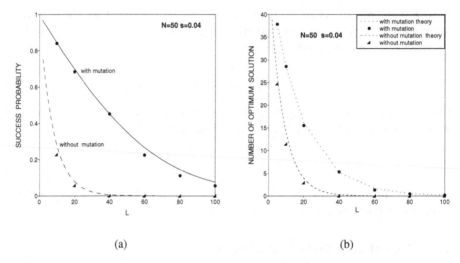

(a)   (b)

Fig. 1. (a)Dependence of success probability S on string length L. (b)Average number of optimum solution with $N = 50$ and $s = 0.04$. The horizontal axis represents string length L.

be $\pi_m^{(\ell)}$. To obtain $S_j^{(\ell)}$, it is necessary to arrange the j units of bit 1 and $i - j$ units of bit 0 in the region of the left i units at the ℓth bit. The number m for this configuration must be $j \leq m \leq N - (i - j)$. The contribution of this term to the total transition probability is

$$\binom{N}{m}^{-1}\binom{i}{j}\binom{N-i}{m-j}\pi_m^{(\ell)}.$$

Then summing all terms from possible values of m, we have

$$Q_{i,j}^{(\ell)} = \sum_{m=j}^{N-i+j}\binom{N}{m}^{-1}\binom{i}{j}\binom{N-i}{m-j}\pi_m^{(\ell)}. \tag{11}$$

Here $\binom{0}{0} = 1$. The probability $S_j^{(\ell)}$ of having j units of ℓ- bit partial optimum solution can be calculated by

$$S_j^{(\ell)} = \sum_{i=0}^{N}S_i^{(\ell-1)}Q_{i,j}^{(\ell)}. \tag{12}$$

5   Results

We made numerical calculations on the multiplicative landscape with fitness proportionate selection. The strength of selection is $s = 0.04$, and population size is $N = 50$. Crossover is done with uniform crossover of crossover rate $p_c = 1$. We assumed that the population is in the linkage equilibrium. The initial value of the first order schema is $p = 0.5$. The calculations are performed repeatedly by changing the seed of random

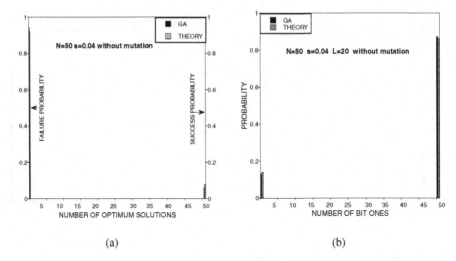

Fig. 2. (a) Distribution of the number of optimum solution. (b) Distribution of the first order schemata. Without mutation.

numbers in each run, and results are averaged over 1000 runs. The values of stationary distribution are calculated at generation $t = 2000$.

Figure 1(a) shows dependence of the success probability S on string length L in stationary distribution. In numerical experiments, theoretical values are compared with $p_m = 0$ and $p_m = 0.001$. In case without mutation, theoretical values are calculated by using equation (9). In case with mutation, theoretical values are calculated by equation (12). The theoretical data agree with experimental data very well. By applying mutation of $p_m = 0.001$, we can obtain larger S compared with $p_m = 0$ showing the positive effect of mutation. Figure 1(b) shows the average number of optimum solution with $N = 50$ and $s = 0.04$. This figure shows that the number of optimum solution reduces rapidly with string length L.

As can be seen from the Figure 1(a), S reduces rapidly without mutation according to the increasing L. The reduction becomes slower when $p_m > 0$. The reason of this change can be obtained from the relationship between L and the average number of optimum solution ($\bar{n}$) which showed in Figure 1(b). In the case with mutation, although $\bar{n}$ reduces rapidly according to the increasing L, S keep the same value so long as $\bar{n} \geq 1$, since S is defined as there is at least one optimum solution in population [1].

Figure 2(a) is the distribution of number of optimum solution $< 1, 1, \ldots, 1 >$ with $p_m = 0$. It has already indicated the population converges into the state of all optimum solution or no optimum solution if mutation is absent. Therefore, number of optimum solution at $j = N$ is related to the success probability by $N \times S_N^{(L)}$. The right vertical axis of the figure shows success probability $S_N^{(L)}$, and the left axis of the figure shows failure probability $S_0^{(L)}$. When $L = 20$, the success probability is very small.

Figure 2(b) use the same parameters as Figure 2(a). This figure shows experimental values of the distribution of the first order schema in case without mutation and string

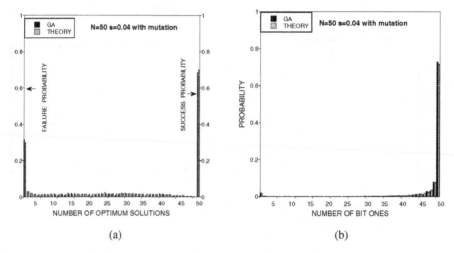

Fig. 3. (a) Distribution of the number of optimum solution. (b) Distribution of the first order schemata. With mutation.

length $L = 20$. Horizontal axis represents the number of first order schema having bit 1. The first order schema converges into fixation state $N_1 = N$ that all member have bit 1 or extinction state $N_1 = 0$.

To examine the effect of mutation, GA data of including mutation are given in Figure 3(a) and Figure 3(b). Figure 3(a) shows the distribution of the number of optimum solution in the stationary distribution. The right vertical axis shows the success probability $S = \sum_{j=1}^{N} S_j^{(L)}$. Compared with Figure 2(a), non-zero data are distributed widely, and the value of $S_{50}^{(L)}$ becomes almost zero. On the other side, the failure probability $F = S_0^{(L)}$ becomes small. As showed in Figure 3(b), this is related to the fact that mutation reduces π_0 which corresponds to the ultimate extinction probability.

Figure 3(b) shows the distribution of first order schema, π_N corresponding to the ultimate fixation probability $u(0.5)$, and π_0 corresponding to the ultimate extinction probability $v(0.5)$. Though π_N is smaller than 0.5, non-zero values are distributed widely within the range of $40 \leq i < 50$. In addition, the ultimate extinction probability π_0 becomes very small 0.012.

6   Summary and Discussion

We have performed the analysis for the success probability of GA on the multiplicative landscape. As a method of calculating the number of optimum solution in a population, we used the new method [1]. If crossover works well enough, the population is in the linkage equilibrium. We hope this is an effective method for predicting computational performance of GAs. Markov chain model and diffusion approximation play an important role to solve the distribution of first order schema, and get good results in this problem. In future, we would like to study the Markov chain model for more complicated fitness functions.

In this analysis, we found that the behavior of S as a function of L is against our expectation. It decreases so slowly as increasing L. Since the number of genotypes is given by 2^L in the multiplicative landscape, we have assumed that the success probability S also decreases like a^L, $a < 1$. This interesting behavior is explained by considering the distribution of the number of the optimum solution in the population. The success probability requires one optimum solution, and the GA with mutation produces very small number of optimum solution in the case of large L. In the case without mutation, the number of optimum solution is N if the GA succeeds in reaching optimum. However, this is achieved at the cost of small S.

References

1. Furutani, H., Zhang, Y., Sakamoto, M.: Study of the Distribution of Optimum Solution in Genetic Algorithm by Markov Chains. Transactions on Mathematical Modeling and its Applications (TOM) (to be published)
2. Ewens, J.W.J.: Mathematical Population Genetics. I. Theoretical Introduction, 2nd edn. Springer, New York (2004)
3. Nix, A.E., Vose, M.D.: Modelling Genetic Algorithm with Markov Chains. Annals of Mathematical and Artificial Intelligence 5, 79–88 (1992)
4. Davis, T.E., Principe, J.C.: A Markov Chain Framework for the Simple Genetic Algorithm. Evolutionary Computation 1, 269–288 (1993)
5. Imai, J., Shioya, H., Kurihara, M.: Modeling of Genetic Algorithms Based on the Viewpoint of Mixture Systems. Transactions on Mathematical Modeling and its Applications (TOM) 44(SIG 07), 51–60 (2003)
6. Asoh, H., Mühlenbein, H.: On the Mean Convergence Time of Evolutionary Algorithms without Selection and Mutation. In: Davidor, Y., Männer, R., Schwefel, H.-P. (eds.) PPSN 1994. LNCS, vol. 866, pp. 88–97. Springer, Heidelberg (1994)
7. Furutani, H.: Schema Analysis of Average Fitness in Multiplicative Landscape. In: Cantú-Paz, E., Foster, J.A., Deb, K., Davis, L., Roy, R., O'Reilly, U.-M., Beyer, H.-G., Kendall, G., Wilson, S.W., Harman, M., Wegener, J., Dasgupta, D., Potter, M.A., Schultz, A., Dowsland, K.A., Jonoska, N., Miller, J., Standish, R.K. (eds.) GECCO 2003. LNCS, vol. 2723, pp. 934–947. Springer, Heidelberg (2003)
8. Crow, J.F., Kimura, M.: An Introduction to Population Genetics Theory. Harper and Row, New York (1970)
9. Zhang, Y., Sakamoto, M., Furutani, H.: Effects of Population Size on Performance of Genetic Algorithms and Roles of Crossover and Mutation. In: The Thirteenth International Symposium on Artificial Life and Robotics (AROB 13th 2008), pp. 861–864 (2008)
10. Zhang, Y., Sakamoto, M., Furutani, H.: Probability of Obtaining Optimum Solutions in Genetic Algorithm and Roles of Mutation. IPSJ SIG Technical Reports, 2008-MPS-68, vol. 2008(17), pp. 161–164 (2008)

Author Index